Functional Polymer Composites: Synthesis, Characterization and Application

Functional Polymer Composites: Synthesis, Characterization and Application

Editors

Tomasz Makowski
Sivanjineyulu Veluri

Basel • Beijing • Wuhan • Barcelona • Belgrade • Novi Sad • Cluj • Manchester

Editors
Tomasz Makowski
Department of Polymeric
Nano-Materials
Centre of Molecular and
Macromolecular Studies
Lodz
Poland

Sivanjineyulu Veluri
Department of Polymeric
Nano-Materials
Centre of Molecular and
Macromolecular Studies
Lodz
Poland

Editorial Office
MDPI
St. Alban-Anlage 66
4052 Basel, Switzerland

This is a reprint of articles from the Special Issue published online in the open access journal *Polymers* (ISSN 2073-4360) (available at: www.mdpi.com/journal/polymers/special_issues/Functional_Polymer_Composites_Synthesis_Characterization_Application).

For citation purposes, cite each article independently as indicated on the article page online and as indicated below:

Lastname, A.A.; Lastname, B.B. Article Title. *Journal Name* **Year**, *Volume Number*, Page Range.

ISBN 978-3-7258-0136-7 (Hbk)
ISBN 978-3-7258-0135-0 (PDF)
doi.org/10.3390/books978-3-7258-0135-0

© 2024 by the authors. Articles in this book are Open Access and distributed under the Creative Commons Attribution (CC BY) license. The book as a whole is distributed by MDPI under the terms and conditions of the Creative Commons Attribution-NonCommercial-NoDerivs (CC BY-NC-ND) license.

Contents

Han-Bi Lee, Ah-Jeong Choi, Young-Kwan Kim and Min-Wook Lee
Composite Membrane Based on Melamine Sponge and Boehmite Manufactured by Simple and Economical Dip-Coating Method for Fluoride Ion Removal
Reprinted from: *Polymers* **2023**, *15*, 2916, doi:10.3390/polym15132916 1

Dalsu Choi, Cheol Ho Lee, Han Bi Lee, Min Wook Lee and Seong Mu Jo
Electropositive Membrane Prepared via a Simple Dipping Process: Exploiting Electrostatic Attraction Using Electrospun SiO_2/PVDF Membranes with Electronegative SiO_2 Shell
Reprinted from: *Polymers* **2023**, *15*, 2270, doi:10.3390/polym15102270 17

Irina Morosanu, Florin Bucatariu, Daniela Fighir, Carmen Paduraru, Marcela Mihai and Carmen Teodosiu
Optimization of Lead and Diclofenac Removal from Aqueous Media Using a Composite Sorbent of Silica Core and Polyelectrolyte Coacervate Shell
Reprinted from: *Polymers* **2023**, *15*, 1948, doi:10.3390/polym15081948 27

Muhammad Ameerul Atrash Mohsin, Lorenzo Iannucci and Emile S. Greenhalgh
Delamination of Novel Carbon Fibre-Based Non-Crimp Fabric-Reinforced Thermoplastic Composites in Mode I: Experimental and Fractographic Analysis
Reprinted from: *Polymers* **2023**, *15*, 1611, doi:10.3390/polym15071611 44

Henri Perrin, Masoud Bodaghi, Vincent Berthé and Régis Vaudemont
On the Addition of Multifunctional Methacrylate Monomers to an Acrylic-Based Infusible Resin for the Weldability of Acrylic-Based Glass Fibre Composites
Reprinted from: *Polymers* **2023**, *15*, 1250, doi:10.3390/polym15051250 58

Aljawharah M. Alangari, Layla A. Al Juhaiman and Waffa K. Mekhamer
Enhanced Coating Protection of C-Steel Using Polystyrene Clay Nanocomposite Impregnated with Inhibitors
Reprinted from: *Polymers* **2023**, *15*, 372, doi:10.3390/polym15020372 73

Guanhua Lu, Akop Yepremyen, Khaled Tamim, Yang Chen and Michael A. Brook
Ascorbic Acid-Modified Silicones: Crosslinking and Antioxidant Delivery
Reprinted from: *Polymers* **2022**, *14*, 5040, doi:10.3390/polym14225040 94

Kristina V. Mkrtchyan, Vladislava A. Pigareva, Elena A. Zezina, Oksana A. Kuznetsova, Anastasia A. Semenova and Yuliya K. Yushina et al.
Preparation of Biocidal Nanocomposites in X-ray Irradiated Interpolyelectrolyte Complexes of Polyacrylic Acid and Polyethylenimine with Ag-Ions
Reprinted from: *Polymers* **2022**, *14*, 4417, doi:10.3390/polym14204417 105

Ryoma Tokonami, Katsuhito Aoki, Teruya Goto and Tatsuhiro Takahashi
Surface Modification of Carbon Fiber for Enhancing the Mechanical Strength of Composites
Reprinted from: *Polymers* **2022**, *14*, 3999, doi:10.3390/polym14193999 118

Kohei Takahashi, Kazuki Nagura, Masumi Takamura, Teruya Goto and Tatsuhiro Takahashi
Development of Electrically Conductive Thermosetting Resin Composites through Optimizing the Thermal Doping of Polyaniline and Radical Polymerization Temperature
Reprinted from: *Polymers* **2022**, *14*, 3876, doi:10.3390/polym14183876 136

Vicente Genovés, María Dolores Fariñas, Roberto Pérez-Aparicio, Leticia Saiz-Rodríguez, Juan López Valentín and Tomás Gómez Álvarez-Arenas
Micronized Recycle Rubber Particles Modified Multifunctional Polymer Composites: Application to Ultrasonic Materials Engineering
Reprinted from: *Polymers* **2022**, *14*, 3614, doi:10.3390/polym14173614 . . . . **147**

Sungryul Yun, Seongcheol Mun, Seung Koo Park, Inwook Hwang and Meejeong Choi
A Thermo-Mechanically Robust Compliant Electrode Based on Surface Modification of Twisted and Coiled Nylon-6 Fiber for Artificial Muscle with Highly Durable Contractile Stroke
Reprinted from: *Polymers* **2022**, *14*, 3601, doi:10.3390/polym14173601 . . . . **166**

Article

Composite Membrane Based on Melamine Sponge and Boehmite Manufactured by Simple and Economical Dip-Coating Method for Fluoride Ion Removal

Han-Bi Lee [1,†], Ah-Jeong Choi [2,†], Young-Kwan Kim [2,*] and Min-Wook Lee [1,*]

1. Institute of Advanced Composite Materials, Korea Institute of Science and Technology, Jeonbuk 55324, Republic of Korea; 092414@kist.re.kr
2. Department of Chemistry, Seoul Campus, Dongguk University, 30 Pildong-ro, Seoul 04620, Republic of Korea; caj9038@gmail.com
* Correspondence: kimyk@dongguk.edu (Y.-K.K.); mwlee0713@kist.re.kr (M.-W.L.)
† These authors contributed equally to this work.

Citation: Lee, H.-B.; Choi, A.-J.; Kim, Y.-K.; Lee, M.-W. Composite Membrane Based on Melamine Sponge and Boehmite Manufactured by Simple and Economical Dip-Coating Method for Fluoride Ion Removal. *Polymers* 2023, 15, 2916. https://doi.org/10.3390/polym15132916

Academic Editors: Tomasz Makowski and Sivanjineyulu Veluri

Received: 13 May 2023
Revised: 19 June 2023
Accepted: 26 June 2023
Published: 30 June 2023

Copyright: © 2023 by the authors. Licensee MDPI, Basel, Switzerland. This article is an open access article distributed under the terms and conditions of the Creative Commons Attribution (CC BY) license (https://creativecommons.org/licenses/by/4.0/).

Abstract: The wastewater generated from the semiconductor production process contains a wide range and a large number of harmful substances at high concentrations. Excessive exposure to fluoride can lead to life-threatening effects such as skin necrosis and respiratory damage. Accordingly, a guideline value of fluoride ions in drinking water was 1.5 mg L^{-1} recommended by the World Health Organization (WHO). Polyvinylidene fluoride (PVDF) has the characteristics of excellent chemical and thermal stability. Boehmite (AlOOH) is a mineral and has been widely used as an adsorbent due to its high surface area and strong adsorption capacity for fluoride ions. It can be densely coated on negatively charged surfaces through electrostatic interaction due to its positively charged surface. In this study, a composite membrane was fabricated by a simple and economical dip coating of a commercial melamine sponge (MS) with PVDF and boehmite to remove fluoride ions from semiconductor wastewater. The prepared MS-PVDF-Boehmite composite membrane showed a high removal efficiency for fluoride ions in both incubation and filtration. By the incubation process, the removal efficiency of fluoride ions was 55% within 10 min and reached 80% after 24 h. In the case of filtration, the removal efficiency was 95.5% by 4 cycles of filtering with a flow rate of 70 mL h^{-1}. In addition, the removal mechanism of fluoride ions on MS-PVDF-Boehmite was also explored by using Langmuir and Freundlich isotherms and kinetic analysis. (R2-1) From the physical, chemical, thermal, morphological, and mechanical analyses of present materials, this study provides an MS-PVDF-Boehmite composite filter material that is suitable for fluoride removal applications due to its simple fabrication process, cost-effectiveness, and high performance.

Keywords: melamine sponge; boehmite; adsorption; composite membrane

1. Introduction

Wastewater from the semiconductor manufacturing process contains a wide range of harmful substances at high concentrations. Among them, fluoride ions are particularly important because they are released at high concentrations not only from semiconductor manufacturing but also from coal-fired power plants, which contaminate groundwater. The concentration of fluoride ions discharged from these industrial activities can reach several hundred to several thousand mg L^{-1}. When it presents in drinking water at low concentrations, fluoride ions can prevent tooth decay, but if they present at high concentrations higher than 1.5 mg L^{-1}, they can cause fluorosis in bones and teeth, making them a hazardous substance. In addition, if more than 5 mg of fluoride ions per 1 kg of body weight are consumed excessively, gastrointestinal disorders, nausea, vomiting, and in severe cases, death can occur [1]. Accordingly, the Centers for Disease Control and Prevention (CDC) in the United States recommends fluoride ion concentrations in drinking water of 0.7–12 mg L^{-1}, while the World Health Organization (WHO) recommends a guideline value of 1.5 mg L^{-1} [1–4].

There are various methods available for removing fluoride ions from wastewater. Representative methods include chemical precipitation, ion exchange resin, membrane filtration, adsorption, and precipitation. The chemical precipitation method is a method of precipitating CaF$_2$, an insoluble salt, by neutralizing calcium ions and fluoride ions by adding lime water [5,6]. However, this method is only effective for treating high concentrations of fluoride ions and cannot remove them to levels below 10 mg L^{-1}. The ion exchange resin method removes dissolved fluoride ions by an ion exchange process [7], but it has the disadvantage of being unable to remove highly concentrated fluoride ions and having a high treatment cost. Membrane filtration has a high removal efficiency for fluoride ions [8,9], but it also has disadvantages such as high maintenance cost, fouling around the membrane, and a complex treatment process. In contrast, the removal of fluoride ions by adsorption has the advantages of requiring less energy and cost than other removal technologies and having a simple treatment process [10–12]. The adsorbents commonly used for fluoride ions include aluminum-based adsorbents, calcium-based adsorbents, hydroxides, boehmite, graphite, activated carbon, and others [13–18]. **(R2-2)** If it is difficult to use this adsorbent independently, and it may contribute to adsorption by immobilizing it on a membrane [19,20]. Materials for the membrane include polyacrylonitrile (PAN), polyvinylidene fluoride (PVDF), and polytetrafluoroethylene (PTFE) [21–23]. As for related studies, a separator that adsorbs copper ions by grafting PAMAM on the surface of a PVDF membrane has been reported [24], and a MOF membrane for removing Cd and Zn prepared by electrospinning of Zr-based MOF-808 and hydrophilic PAN has been reported [25]. These materials are suitable for aqueous applications and exhibit physical and chemical stability. However, we studied boehmite and sponge-based composite membranes, including higher adsorption capacity and superior mechanical properties, in addition to the advantages of previously reported composite membranes.

Boehmite is synthesized by a hydrothermal method under high temperature and pressure by putting a solid reactant and solvent into an autoclave [26]. Boehmite has the advantages of being environmentally friendly and cost-effective, and it is widely used as an adsorbent due to its high surface area of about 448 m^2 g^{-1} [27]. Additionally, it carries a positive charge, making it suitable for electrostatic adsorption of negatively charged compounds [28,29]. Melamine sponge (MS) is a commercially available material and is widely harnessed as a support for various adsorbents due to its inherent characteristics, including high porosity, high absorbency, wide surface area, and low density [30–32]. Additionally, depending on the coating materials, it can selectively exhibit hydrophilic and hydrophobic properties, which is beneficial to applications in water/oil separation and absorption [33–36]. In this study, a composite membrane of MS and boehmite was fabricated through a simple and sequential dip coating of MS in a solution of polyvinylidene fluoride (PVDF) and an aqueous suspension of boehmites. The resulting composite membrane of MS-PVDF-Boehmite was thoroughly characterized with analytical tools to reveal its structure and directly applied to the removal of fluoride ions under various conditions. The MS-PVDF-Boehmite composite membrane (8 cm^3) showed a high removal rate of 60% from a solution containing 1–80 mg L^{-1} fluoride ions, and its adsorption characteristics were also investigated with Langmuir and Freundlich isotherms and kinetic analysis. Based on its high performance, the MS-PVDF-Boehmite composite membrane was inserted into a syringe and applied as a cartridge to remove fluoride ions from flowing wastewater. After 4 cycles of the purification process, the concentration of fluoride ions fell to below the WHO standard (1.5 mg L^{-1}), and the removal efficiency was 95.5%. **(R2-2)** In this work, a composite membrane for removal of fluoride ions was developed by a simple dip-coating process. The results indicated that the MS-PVDF-Boehmite composite membrane is also an effective tool to purify wastewater contaminated with fluoride ions.

2. Experimental Section

2.1. Materials

(R1-1) Aluminum isopropoxide (AIP, ≥98%), N,N-Dimethylformamide (DMF, 99.8%), and acetic acid (≥99.7%) were purchased from Sigma-Aldrich (St. Louis, MO, USA). Ethyl alcohol (EtOH, 94.5%) was purchased from Samchun Pure Chemicals (Pyeongtaek, Republic of Korea). Polyvinylidene fluoride (PVDF, Kynar-761) was purchased from Arkema (Singapore). Hydrochloric acid (HCl, 37%) and sodium hydroxide (NaOH, 97%) were purchased from Daejung Reagent Chemicals (Siheung, Republic of Korea). A melamine sponge (MS) was purchased from BASF (Ludwigshafen, Germany).

2.2. Synthesis of Boehmite (γ–AlOOH)

Boehmite (γ–AlOOH) was prepared by a conventional sol–gel reaction of AIP. For the synthesis, 68 g of AIP was added to 300 mL of de-ionized (DI) water at 75 °C, and the aqueous solution of AIP was heated at 95 °C with stirring until the total volume of the solution became 200 mL through evaporation. Next, 3.1 g of acetic acid was added to the AIP solution in a drop-by-drop manner and stirred for 10 min. Finally, a hydrothermal reaction was carried out by using an autoclave at 150 °C for 6 h. During this hydrothermal reaction, AIP was transformed into boehmite crystals.

2.3. Manufacturing Process

Boehmite-based composite membranes were fabricated with two different processes (Figure 1). First, the melamine sponge (MS) was washed with flowing DI water and EtOH. After washing, the MS was immersed in the boehmite solution for 1 h. The resulting MS-Boehmite was washed 3 times with DI water and dried in an oven at 50 °C. MS-boehmite was prepared through a one-step dip-coating process (Figure 1a). PVDF was put into a DMF solution and stirred at a constant speed for 4 h at 50 °C to prepare a 5 wt% PVDF solution. The MS was immersed in the PVDF solution at room temperature for 2 h. Next, the sample was taken out, and the excess solution was shaken off and dried in an oven overnight. The resulting MS-PVDF was washed 3 times with DI water and dried in an oven at 50 °C. After drying, MS-PVDF was immersed in boehmite solution for 1 h. The resulting MS-PVDF-Boehmite was washed 3 times with DI water and dried in an oven at 50 °C. MS-PVDF-Boehmite was prepared by electrostatic interaction by sequential coating with negatively charged PVDF and positively charged boehmite to improve the adhesion and bonding strength between MS and boehmite (Figure 1b). (R2-3) Boehmite was synthesized through a hydrolysis reaction between DI water and AIP precursor followed by hydrothermal treatment (Figure 1c). The synthesized boehmite showed a typical white color, and it was well dispersed in DI water, forming a translucent suspension (Figure 1d,e).

2.4. Characterization

To evaluate the adsorption performance of a composite membrane for fluoride removal, fluoride wastewater (initial concentration: 5000 mg L^{-1}) was diluted to prepare solutions of fluoride ions at 1, 5, 10, 20, 40, and 80 mg L^{-1}. The adsorption test was conducted in two ways. First, MS-PVDF-Boehmite samples were cut into $2 \times 2 \times 2$ cm^3, placed in 20 mL fluoride ion solutions at various concentrations and stirred at 50 °C for 24 h. After that, 1 mL of the fluoride ion solutions was collected to measure the changes of fluoride ion concentrations. Second, MS-PVDF-Boehmite samples were cut into $\pi \times 0.6^2 \times 4$ cm^3 and inserted into a 5 mL syringe. Then, 5 mL of 20 mg L^{-1} fluoride ion solution was injected at flow rates of 30, 70, and 110 mL h^{-1} to examine the adsorption efficiency under different flow rates. This experiment was repeated in cycles until the concentration of fluoride ions became lower than that of the WHO standard (1.5 mg L^{-1}).

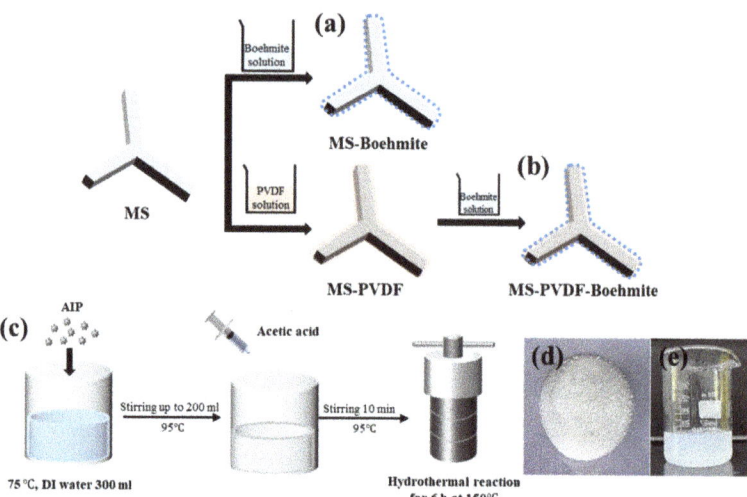

Figure 1. Schematic diagram of preparation of (**a**) MS-Boehmite and (**b**) MS-PVDF-Boehmite composite membranes. (**c**) A schematic illustration of boehmite synthesis. Photographs of (**d**) the dried and (**e**) suspended boehmites in water.

The physical, chemical, thermal, morphological, and mechanical characteristics of the prepared MS, MS-Boehmite, and MS-PVDF-Boehmite were systematically investigated. An X-ray diffraction (XRD) pattern (Rigaku, Tokyo, Japan) of boehmite was obtained with a Rigaku X-ray diffractometer equipped with a Cu Kα source. Their thermal properties were explored using thermogravimetric analysis (TGA Q50, TA Instruments, USA). In the thermal tests, samples were placed in a ceramic pan at a constant heating rate of 10 °C min^{-1} within 40–800 °C under a nitrogen–air atmosphere at a flow rate of 90 mL min^{-1}. The surface and interface of the specimen were observed using an optical microscope (VHX-900F, Keyence Corporation, Osaka, Japan) and a scanning electron microscope (SEM, Nova NanoSEM 450, FEI) at 15 kV. The functional group analysis of the samples was performed using FT-IR (Sinco, Seoul, Korea). The compression tests were conducted at a crosshead speed of 10 mm min^{-1} with a dimension of 2 × 2 × 2 cm^3. The concentration of fluoride ions was measured using a fluoride colorimeter (HI-739, HANNA Instruments, Woonsocket, RI, USA).

The adsorption capacity was calculated by Equation (1) from the measured concentration of fluoride ions remaining in the solution:

$$q_e = \frac{(C_0 - C_e)V}{W} \quad (1)$$

q_e: Equilibrium adsorption amount adsorbed per unit g of adsorbent (mg g^{-1})
C_0: Initial concentration of fluoride ion (mg L^{-1})
C_e: Equilibrium concentration of fluoride ion in solution after adsorption (mg L^{-1})
V: Volume of solution (L)
W: Adsorbent Dosage (g)

The heavy metal removal efficiency Re (%) was obtained by Equation (2).

$$Re(\%) = \frac{(C_0 - C_e)}{C_0} \times 100 \quad (2)$$

3. Results and Discussions

3.1. Characterization of Synthesized Boehmite (γ-AlOOH)

SEM images showed the morphological characteristics of rod-like boehmites with a length of few hundred nanometers and a diameter from 20 to 50 nm (Figure 2a). To confirm the successful synthesis of boehmite, an XRD pattern of the synthesized sample was obtained (Figure 2b). The XRD pattern exhibited characteristic diffraction peaks at $2\theta = 13.75°, 28.25°, 38.35°$, and $49.20°$, which correspond to the (020), (120), (031), and (200) planes of boehmite, and those peaks verified that boehmite was successfully synthesized under our synthetic condition. Then, the zeta potential value of boehmite at various pH conditions was measured to examine its surface charges, and it showed a positive zeta potential ranging from 20 to 40 mV and a pH range from 3 to 8 (Figure 2c). This result implied that boehmite can maintain its positive charge for the electrostatic adsorption of negatively charged contaminants such as fluoride ions in diverse environments. **(R1-2)** The negative charge of boehmite was caused by the increase in the number of OH^- groups with the increase in pH and the decrease in zeta potential value [37].

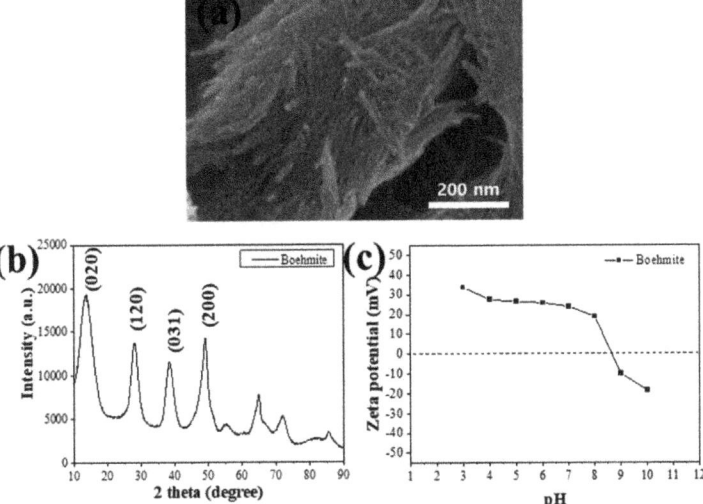

Figure 2. (a) SEM image and (b) XRD pattern of the synthesized boehmite. (c) Zeta potential values of boehmites suspended in water with varying pH conditions.

3.2. Morphology of MS, MS-Boehmite, and MS-PVDF-Boehmite

Figure 3 showed SEM and EDX images of MS, MS-Boehmite, and MS-PVDF-Boehmite composite membranes. MS has a smooth surface morphology and an interconnected 3D network framework (Figure 3a). After coating with boehmite, MS-Boehmite exhibited boehmites coated on its surface (Figure 3b), and Al was detected by EDX mapping (3.40%) (Figure 3b inset). Compared to MS-Boehmite, the surface coverage of boehmites on MS-PVDF-Boehmite was considerably enhanced, and as a result, its surface was rougher than MS and MS-Boehmite and was composed of large boehmite crystals (Figure 3c). In addition, Al content of MS-PVDF-Boehmite (4.59%) was also higher than that of MS-Boehmite (3.40%) (Figure 3c inset). This indicates that PVDF played an important role as an adhesive layer for the electrostatic adsorption of positively charged boehmites on the surface of MS due to its negative charges.

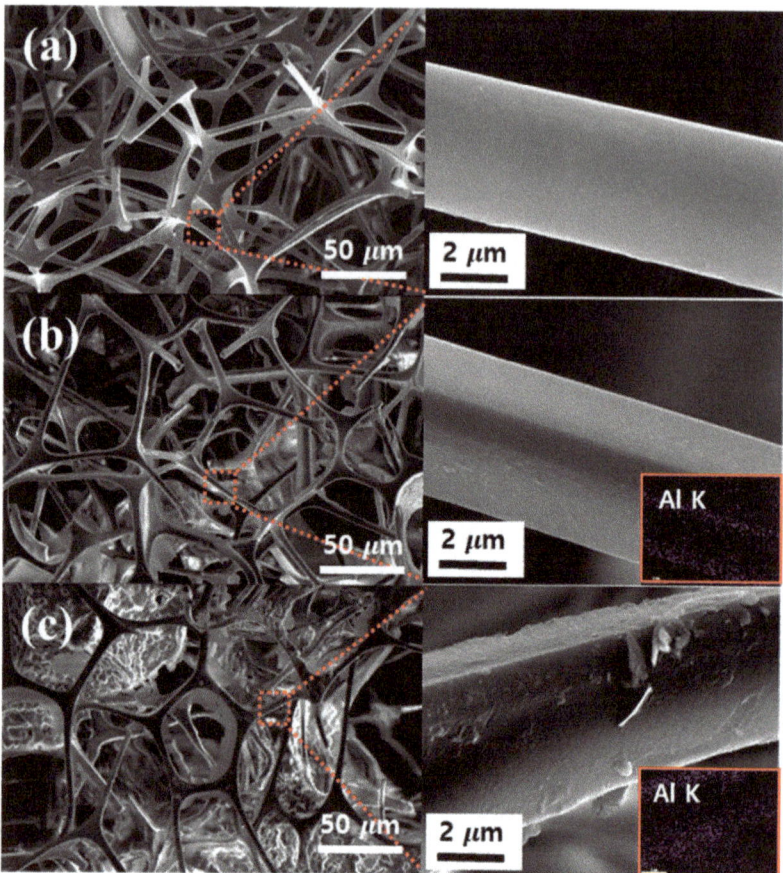

Figure 3. SEM and EDX images of (**a**) MS, (**b**) MS-Boehmite, and (**c**) MS-PVDF-Boehmite.

3.3. Characterization of MS, MS-Boehmite, and MS-PVDF-Boehmite

Figure 4a shows the FT-IR spectra of MS, MS-Boehmite, and MS-PVDF-Boehmite. Compared with MS [38], new peaks appeared at around 3068 and 1060 cm^{-1} from MS-Boehmite and MS-PVDF-Boehmite (Figure 4a), and those peaks correspond to the O-H vibration of AlOOH and to the Al-O-Al symmetric bending vibration, respectively. A strong peak of C-H stretching vibration was observed at 1398 cm^{-1} from MS-PVDF-Boehmite, and there were also peaks located at 1280 and 1011 cm^{-1} corresponding to C-F bond vibrations (Figure 4a). In addition, the peaks at 1494 and 1328 cm^{-1} from C=N and C-N bonds of MS were weakened with sequential coating with PVDF and boehmite (Figure 4a). The FT-IR analysis confirmed that PVDF and boehmite were successfully coated on the surface of MS.

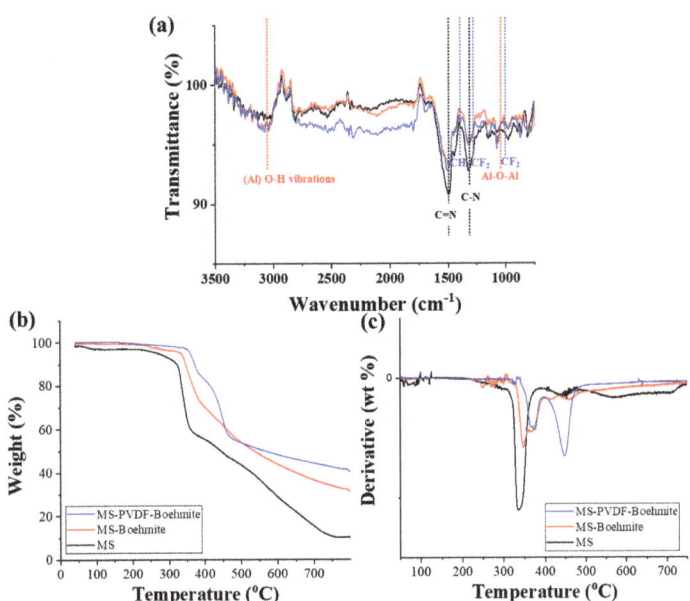

Figure 4. Characterization of MS, MS-Boehmite, and MS-PVDF-Boehmite: (**a**) FT-IR, (**b**) TGA, and (**c**) DTA.

TGA analysis was performed to investigate the thermal stability of the prepared composite membranes. Figure 4b,c show the TGA and DTA curves of MS, MS-Boehmite, and MS-PVDF-Boehmite in a nitrogen atmosphere. The TGA curve of MS showed a rapid weight loss in the temperature range of 330 to 400 °C, which occurs when the HN-CH$_2$-NH bond is broken [39]. The weight loss at higher temperatures is due to thermal decomposition of the triazine ring. MS-Boehmite maintains thermal stability up to 335 °C, which is 10 °C higher than that of MS. It implied that the thermal stability of MS was enhanced by coating with boehmite. Interestingly, MS-PVDF-Boehmite retained its thermal stability up to 353 °C, which is approximately 20 °C higher than that of MS-Boehmite. It can be inferred that due to the important role of negatively charged PVDF as a binder, a greater amount of positively charged boehmite was coated on MS-PVDF than MS, and thus its thermal stability increased.

3.4. Mechanical Properties of MS, MS-Boehmite, and MS-PVDF-Boehmite

The mechanical properties of MS, MS-Boehmite, and MS-PVDF-Boehmite were explored with compression tests to reveal the coating effect of boehmite with different methods (Figure 5). A compression test of 10 cycles was conducted with a 70% strain and a strain rate of 10 mm min^{-1} (Figure 5a–c). MS and MS-Boehmite showed a similar compressive stress of 22.7 and 32.8 kPa, respectively, while MS-PVDF-Boehmite exhibited a relatively high compressive stress of 65.0 kPa, which is nearly threefold higher than MS and twofold higher than MS-Boehmite. The highly enhanced compressive stress implied that boehmite reinforced the mechanical properties of MS, and this effect is augmented with the PVDF adhesive layer leading to a high surface coverage of boehmite. The compression stress of MS-PVDF-Boehmite was partially diminished after 10 repeated compression tests, but there was no significant damage, confirming their high durability for the practical application. It is also worthy to note that the decreased compression stress of MS-PVDF-Boehmite was still much higher than that of MS and MS-Boehmite.

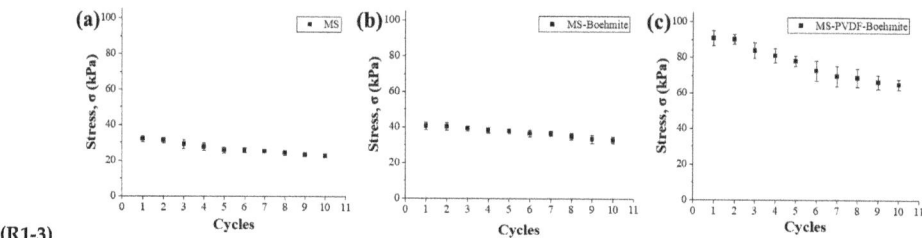

(R1-3)

Figure 5. Stress according to the number of repetitions of compression of (**a**) MS, (**b**) MS-Boehmite, and (**c**) MS-PVDF-Boehmite with a constant strain rate of 10 mm min^{-1}. The insets show the compression test images of the MS, MS-Boehmite, and MS-PVDF-Boehmite samples.

3.5. Isothermal Adsorption Test

The adsorption performance of MS, MS-Boehmite, and MS-PDVF-Boehmite ($2 \times 2 \times 2$ cm^3 in their dimension) was explored by incubating them in standard solutions having initial concentrations of 1, 5, 10, 20, 40, and 80 mg L^{-1} with constant stirring at 230 rpm. After 24 h of incubation, the adsorption membranes (MS, MS-Boehmite, and MS-PVDF-Boehmite) were retrieved and 1 mL of the residual solutions was collected to measure the concentration of fluoride ions. As shown in Figure 6a, MS exhibited a removal efficiency of over 67% at 1 mg L^{-1} and showed a removal efficiency of over 37–40% at other concentrations, indicating fluoride ions can be adsorbed on MS without coating of boehmite. In the case of MS-Boehmite, it showed 100% removal efficiency at 1 mg L^{-1}. This high removal efficiency was derived from the high surface area and strong affinity of boehmites for fluoride ions. However, it was confirmed that the removal efficiency decreased with an increasing concentration of fluoride ions. In the case of MS-PVDF-Boehmite, fluoride ions at concentrations of 1–10 mg L^{-1} were completely removed within 24 h, and the removal efficiency was about 78% or higher even at 20 mg L^{-1}. The enhanced removal efficiency clearly indicated that the removal of fluoride ions was derived from the electrostatic adsorption of fluoride ions on the surface of boehmites, and thus the removal efficiency significantly increased with the loading amount of boehmites. In addition, when the pH value was less than 5.0, hydroxyl groups on the surface of boehmites were prone to be protonated for the formation of $-OH_2^+$ in acidic solutions. Therefore, the surface of boehmites became further positively charged and facilitated the electrostatic adsorption of fluoride ions [29,40].

The adsorption capacity (mg g^{-1}) of MS, MS-Boehmite, and MS-PVDF-Boehmite for fluoride ions was examined with different initial concentration (Figure 6b). The adsorption capacity was determined to be 2.98, 2.60, and 1.06 mg g^{-1} for MS, MS-Boehmite, and MS-PVDF-Boehmite at 20 mg L^{-1} of fluoride ions, respectively. Interestingly, although MS-PVDF-Boehmite has the highest removal efficiency of fluoride ions among the tested samples, its adsorption capacity was significantly lower than that of MS and MS-Boehmite. This low adsorption capacity was attributed to its increased weight compared to MS and MS-Boehmite because MS, MS-Boehmite, and MS-PVDF-Boehmite were prepared in an equal dimension ($2 \times 2 \times 2$ cm^3), and thus MS-PVDF-Boehmite has the highest weight among the tested samples due to the PVDF adhesive layer and high loading amount of boehmites (the weight of MS-PVDF-Boehmite was fivefold higher than that of MS). To quantitatively compare the adsorption capacity, MS, MS-Boehmite, and MS-PVDF-Boehmite were cut with an equal weight (0.07 g) and different dimensions such as $2.0 \times 2.0 \times 2.0$, $1.7 \times 1.7 \times 1.7$, and $1.2 \times 1.2 \times 1.2$ cm^3, respectively. The removal efficiency and adsorption capacity of MS-PVDF-Boehmite were 27.1% and 1.5 mg g^{-1} and these values were still lower than those of MS (43.8% and 2.4 mg g^{-1}) and MS-Boehmite (34.8% and 1.9 mg g^{-1}) **(R2-4)** (Figure 6e). Considering a nearly fivefold smaller volume of MS-PVDF-Boehmite than MS, its adsorption performance is sufficient for the practical application of removing fluoride ions.

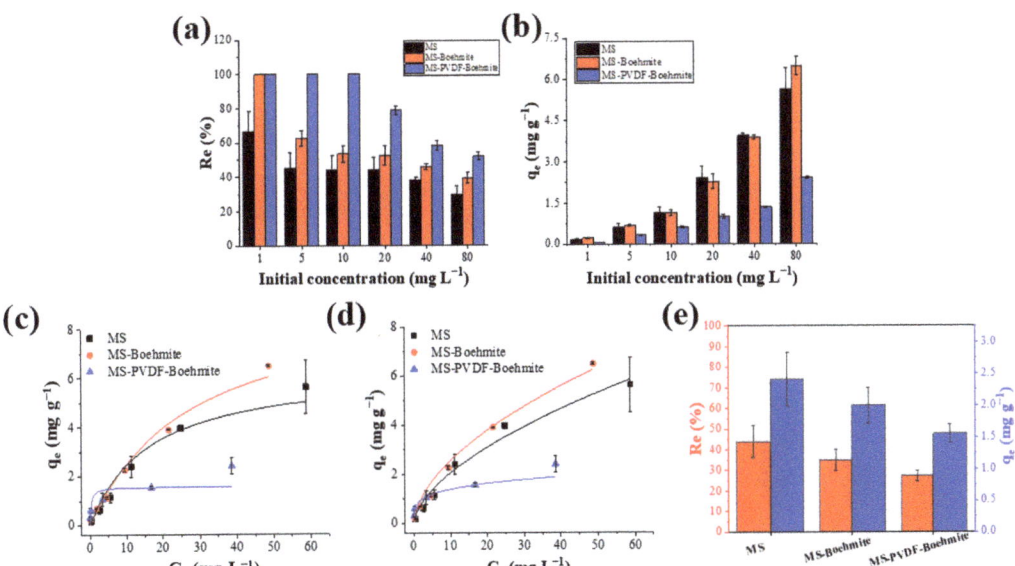

Figure 6. Effect of initial fluoride ion concentration on (**a**) adsorption efficiency and (**b**) adsorption capacity. (**c**) Langmuir and (**d**) Freundlich isotherm models to investigate the adsorption process of fluoride ions on MS, MS-Boehmite, and MS-PVDF-Boehmite. (R2-4) (**e**) The removal efficiency and adsorption capacity of MS, MS-Boehmite, and MS-PVDF-Boehmite for fluoride ions by incubation process.

Then, the adsorption mechanism of fluoride onto MS, MS-Boehmite, and MS-PVDF-Boehmite was investigated with Langmuir and Freundlich isotherm models. These models are extensively harnessed to study a solid–liquid interface system at adsorption equilibrium. To determine the suitability of each isotherm model, three error functions such as coefficient of determination (R^2), sum of absolute error (SAE), and chi-square (χ^2) were calculated from each isotherm model, respectively (Table 1).

Table 1. Error functions for estimation of nonlinear regression models.

Error Function	Equation
Coefficient of determination (R^2)	$\dfrac{\sum_{i=1}^{n}\left[(q_{e,meas}-\overline{q_{e,cal}})\right]_i^2}{\sum_{i=1}^{n}\left[(q_{e,meas}-\overline{q_{e,cal}})^2+(q_{e,meas}-q_{e,cal})^2\right]_i}$
Nonlinear chi-square (χ^2)	$\sum_{i=1}^{n}\left[\dfrac{(q_{e,meas}-q_{e,cal})^2}{q_{e,meas}}\right]_i$
Sum of absolute errors (SAE)	$\sum_{i=1}^{n}\left|q_{e,meas}-q_{e,cal}\right|_i$

The Langmuir isotherm equation indicates that the adsorption is mainly conducted by the bonding force between the surface of adsorbents and aqueous adsorbates. Therefore, the Langmuir model assumes that adsorbate forms a monomolecular layer onto the adsorbents without lateral interactions, and no further adsorption occurs when monolayer adsorption is completed [41]. The nonlinear form of the Langmuir isotherm model can be expressed as Equation (3):

$$q_e = \frac{q_{max}K_L C_e}{1+K_L C_e} \quad (3)$$

Here, the K_L is Langmuir constant, which is a crucial parameter that can determine the adsorption rate (L mg^{-1}), and q_{max} is the maximum adsorption capacity (mg g^{-1}) for fluoride ions, representing the theoretical maximum monomolecular layer adsorption capacity of the used adsorbents.

The Freundlich adsorption isotherm is a semi-experimental model derived from the Langmuir isotherm. It implies multilayered adsorption with uneven distribution of adsorption energy on the surface of adsorbents. It assumes that adsorbates are initially adsorbed on the stronger adsorption site of adsorbents, and the adsorption heat decreases gradually with increasing coverage of active sites of adsorbents. The nonlinear Freundlich isotherm equation is expressed as Equation (4):

$$q_e = K_F C_e^{\frac{1}{n}} \tag{4}$$

Here, K_F is the Freundlich constant related to adsorption capacity of the adsorbent (L mg^{-1}), and n is a measure of adsorption intensity, which can vary with the surface heterogeneity and affinity of adsorbents. A higher K_F value indicates a better relative adsorption capacity [42]. The experimental adsorption data were fitted using Langmuir and Freundlich models, as shown in Figure 6c,d. The calculated adsorption isotherm parameters and error functions from the two models are summarized in Table 2.

Table 2. Parameters calculated from the Langmuir and Freundlich isotherm models.

Case	Isotherm Model	The Calculated Parameters		Error Functions		
	Langmuir	q_{max}	K_L	R^2	χ^2	SAE
MS		6.36	0.064	0.943	0.760	2.688
MS-Boehmite		9.47	0.036	0.920	0.253	1.345
MS-PVDF-Boehmite		1.58	3.655	0.927	0.674	1.605
	Freundlich	K_F	$1/n$	R^2	χ^2	SAE
MS		0.528	0.590	0.978	0.454	2.089
MS-Boehmite		0.704	0.562	0.992	0.150	1.325
MS-PVDF-Boehmite		0.837	0.223	0.993	0.140	0.733

The q_{max} values of MS, MS-Boehmite, and MS-PVDF-Boehmite were obtained as 6.36, 9.47, and 1.58 mg L^{-1} by using the Langmuir isotherm, respectively. This result is consistent with the experimental adsorption results, which showed that MS-PVDF-Boehmite possessed the lowest adsorption capacity due to the increased density of the sponge samples. Using the Freundlich isotherm, the K_F values of MS, MS-Boehmite, and MS-PVDF-Boehmite were calculated to be 0.528, 0.704, and 0.837 L mg^{-1}, respectively. Those results suggest that the adsorption capacity of fluoride ions was higher on the surface of MS-PVDF-Boehmite than that of MS and MS-Boehmite due to the large loading amount of boehmites which have a strong affinity toward fluoride ions. The Freundlich isotherm also gives an important factor of $1/n$ as an indicator of adsorption preference. When the $1/n$ value ranges from 0 to 1, the adsorption process is favorable, and a smaller value suggests a more heterogeneous surface of the adsorbent and nonlinear isotherm [43]. On the other hand, if this value is greater than 1, the adsorption process becomes unfavorable. The $1/n$ values of MS, MS-Boehmite, and MS-PVDF-Boehmite were calculated to be 0.590, 0.562, and 0.223, respectively, which are all less than 1, implying that the adsorption process of fluoride ions on their surface was favorable. Those results concurred well with the experimental results that MS-PVDF-Boehmite presented a much higher fluoride removal efficiency than MS and MS-Boehmite. Then, the error functions were compared to ensure the reliability of the isotherm modeling results. In all cases of MS, MS-Boehmite, and MS-PVDF-Boehmite, R^2 values were close to 1, and χ^2 and SAE values were also relatively low in the Freundlich isotherm model compared to the Langmuir model. This result indicates that the Freundlich isotherm is more appropriate for describing the adsorption process, and thus the multilayer adsorption is dominant for fluoride ions onto MS, MS-Boehmite, and MS-PVDF-Boehmite.

3.6. Adsorption Kinetics

The effect of adsorption time on removal efficiency of fluoride ions was investigated by conducting adsorption experiments at 20 mg L^{-1} of fluoride ions with varying adsorption times from 10 to 1440 min (Figure 7a). Within 30 min of adsorption, MS-PVDF-Boehmite exhibited a rapid adsorption process compared to MS and MS-Boehmite. A total of 62% of fluoride ions were removed by MS-PVDF-Boehmite, whereas only 2% and 25% of fluoride ions were removed by MS and MS-Boehmite (Figure 7a). At the equilibrium state (after 1440 min of adsorption time), MS, MS-Boehmite, and MS-PVDF-Boehmite exhibited removal efficiencies of 44%, 53%, and 79%, respectively. The presence of positively charged boehmites on the surface of MS-PVDF-Boehmite facilitated the adsorption of fluoride ions through electrostatic interactions, leading to formation of a strong bonding between them. As a result, a higher boehmite content on the surface of MS-PVDF-Boehmite leads to a faster adsorption process and higher adsorption capacity at equilibrium than MS and MS-Boehmite.

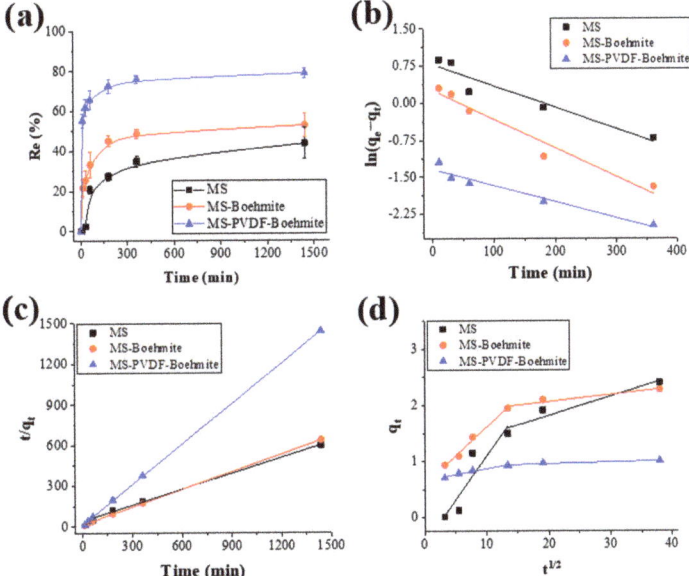

Figure 7. (**a**) Effect of contact time for fluoride adsorption onto adsorbents, (**b**) pseudo-first-order model, (**c**) pseudo-second-order model, and (**d**) Weber–Morris intraparticle diffusion model for adsorption kinetic study of fluoride ions.

The pseudo-first-order and pseudo-second-order kinetic models were employed to investigate the adsorption process and determine the kinetic parameters based on the experimental adsorption data at different contact times. The pseudo-first-order kinetic model is typically applied to reversible reactions where an equilibrium is established between the liquid and solid phases, while the pseudo-second-order model assumes that the rate-determining step involves chemisorption with valence forces through electron sharing or exchange between the adsorbent and adsorbate [44,45]. Kinetic curves and parameters from the experimental adsorption data are shown in Figure 7b,c and Table 3, respectively. The linearized forms of the pseudo-first-order and pseudo-second-order kinetic equations are given by Equations (5) and (6), respectively:

$$ln(q_e - q_t) = ln\ q_e - k_1 t \tag{5}$$

$$\frac{t}{q_t} = \frac{1}{k_2 q_e^2} + \frac{1}{q_e} t \tag{6}$$

Here, q_t represents the adsorption capacity of fluoride at contact time (mg g^{-1}), while q_e represents the adsorption capacity at the equilibrium state. The rate constants for the pseudo-first-order and pseudo-second-order models are denoted as k_1 (min^{-1}) and k_2 (g mg^{-1}·min^{-1}), respectively, while t (min) indicates the contact time.

Table 3. Parameters calculated from the pseudo-first-order, pseudo-second-order, and intraparticle diffusion kinetic models.

Case	$q_{e,exp}$	Pseudo-First-Order Model				
		k_1	$q_{e,cal}$	R^2	SAE	χ^2
MS	2.406	0.004	2.188	0.917	1.353	1.381
MS-Boehmite	2.276	0.006	1.283	0.956	3.695	2.322
MS-PVDF-Boehmite	1.001	0.003	0.260	0.946	2.484	1.682

	$q_{e,exp}$	Pseudo-second-order Model				
		k_2	$q_{e,cal}$	R^2	SAE	χ^2
MS	2.406	0.002	2.578	0.999	0.420	0.057
MS-Boehmite	2.276	0.015	2.322	0.999	0.582	0.158
MS-PVDF-Boehmite	1.001	0.057	1.007	0.999	0.334	0.079

	Intraparticle diffusion Model						
	k_{id1}	k_{id2}	R_1^2	SAE$_1$	χ_1^2	R_2^2	SAE$_2$	χ_2^2
MS	0.154	0.034	0.848	0.810	0.741	0.935	0.258	0.002
MS-Boehmite	0.102	0.013	0.990	0.136	0.005	0.936	0.093	0.002
MS-PVDF-Boehmite	0.021	0.003	0.960	0.063	0.001	0.891	0.028	0.001

The values of error functions in Table 3 indicate that the pseudo-second-order model provided a better fit for MS, MS-Boehmite, and MS-PDVF-Boehmite based on the high R^2 values (0.999) and markedly lower values of SAE and χ^2 compared to the pseudo-first-order model. Furthermore, the equilibrium adsorption capacity ($q_{e,cal}$) from the pseudo-second-order model was determined to be 2.578, 2.322, and 1.007 mg g^{-1} for MS, MS-Boehmite, and MS-PVDF-Boehmite, respectively, which were closely matched with the experimentally obtained adsorption capacity ($q_{e,exp}$). Those results also implied that the adsorption process of fluoride ions on MS, MS-Boehmite, and MS-PVDF-Boehmite was well described with the pseudo-second-order kinetic model rather than the pseudo-first-order kinetic model. The kinetic analysis further confirmed that fluoride ions were dominantly removed through the strong electrostatic interactions between negatively charged fluoride ions and positively charged boehmites.

To further investigate the rate-determining step during the adsorption process, the experimental adsorption data were plotted by the Weber–Morris intraparticle diffusion model, and the intradiffusion curves and parameters are shown in Figure 7d and Table 3, respectively. The equation of this diffusion model is expressed as follows in Equation (7):

$$q_t = k_{id} t^{1/2} + C \tag{7}$$

The intraparticle diffusion model incorporates parameters such as k_{id}, C, and q_t, which represent the intraparticle diffusion rate constant (mg g^{-1}·min$^{-1/2}$), the thickness of the boundary layer (mg g^{-1}), and the adsorption capacity at a given contact time (mg g^{-1}). According to the Weber–Morris model, if the plot of the adsorption data follows a straight line, it suggests the intraparticle diffusion process is rate-controlling. Conversely, if the plot passes through the origin, it implies that the intraparticle diffusion is the rate-determining step [46]. The intradiffusion curves were roughly divided by two straight lines with different slopes and none of lines passed through the origin of graph (Figure 7d). This result implied that the intraparticle diffusion was not solely the rate-determining step, and there was influence of boundary layer diffusion. Considering the slopes of two straight lines, the main rate-determining step was intraparticle diffusion because its slope is smaller than that of boundary layer diffusion.

3.7. Adsorption Performance According to Flow Rate

For the practical application of the MS-PVDF-Boehmite composite membrane, its adsorption performance needs to be evaluated with flowing wastewater containing fluoride ions with different flow rates. Figure 8a showed the experimental setup of the adsorption test with flowing wastewater. MS-PVDF-Boehmite was cut to fit into a syringe ($\pi \times 0.6^2 \times 4$) and used as a cartridge to remove fluoride ions from flowing wastewater. After putting 5 mL of a 20 mg L^{-1} solution of fluoride ions into a syringe, the flow rate was controlled with a syringe pump at 30, 70, and 110 mL h^{-1}, and the concentration of fluoride ions in the treated water was measured. This filtration process was repeated for several cycles until the concentration of fluoride ions was lower than the WHO standard (1.5 mg L^{-1}).

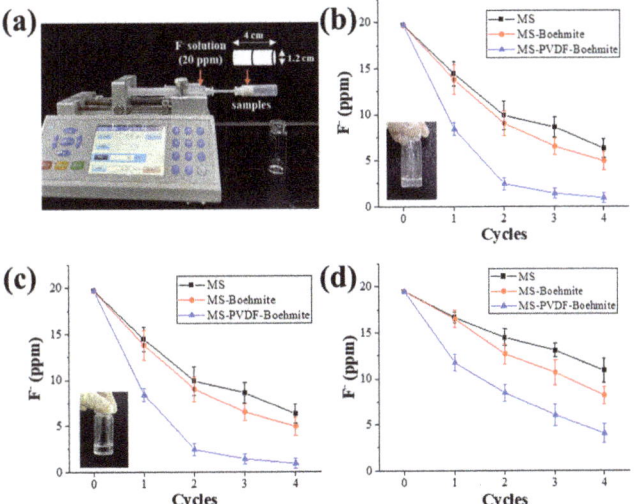

Figure 8. Comparison of mg L^{-1} concentration before and after filtering test of composite membrane for removing fluoride in different flows: (**a**) experimental setup, (**b**) 30 mL h^{-1}, (**c**) 70 mL h^{-1}, and (**d**) 110 mL h^{-1}.

At a flow rate of 30 mL h^{-1}, the filtered solution through MS-PVDF-Boehmite showed 0.9 mg L^{-1} of fluoride ions after four cycles of filtration, while those through MS and MS-Boehmite showed 6.2 and 4.9 mg L^{-1} (Figure 8b). However, the filtered solution was slightly opaque (an inset of Figure 8b), and it implied that boehmites were partially detached into the filtered water during the repeated filtration processes owing to a prolonged contact with wastewater with a low flow rate. When the flow rate increased to 70 mL h^{-1}, the removal efficiency of fluoride ions was not changed regardless of the composite membranes, but the filtered solutions through them became clear. This result signified that the increase in flow rate prevented a detachment of boehmites from MS-PVDF-Boehmite during the filtration process without deterioration of its adsorption performance. However, with a further increase in flow rate to 110 mL h^{-1}, the removal efficiency of MS, MS-Boehmite, and MS-PVDF-Boehmite for fluoride ions declined sharply to 10.8, 8.2, and 4.0 mg L^{-1}, respectively, because the contact time of wastewater with the composite membranes decreased.

4. Conclusions

A composite membrane for the removal of fluoride ions was developed by a simple dip-coating process. **(R1-4)** The characterization results suggested that the thermal and mechanical properties of MS were enhanced with a loading of boehmites, and the loading amount of boehmite increased greatly with a PVDF adhesive layer. Then, the prepared composite membranes were applied to the removal of fluoride ions by two different processes such as incubation and filtration. This study found that the MS-PVDF-Boehmite showed the highest performance to remove fluoride ions through both processes. At low concentrations below 10 mg L^{-1}, fluoride ions were completely removed with MS-PVDF-Boehmite within 1 h of incubation. At a high concentration of 20 mg L^{-1}, its removal efficiency was 78.6% and it was maintained to 51.8% even at 80 mg L^{-1} after 24 h of incubation. The experimental results were applied to Langmuir and Freundlich adsorption isotherms as well as kinetic analysis to study the adsorption characteristics of the prepared composite membranes. **(R2-5)** The Freundlich adsorption isotherm and pseudo-second-order kinetic model were found to be the best-fitted models for MS-PVDF-Boehmite. Furthermore, the Weber–Morris intraparticle diffusion model indicated that the diffusion rate was not solely affected by intraparticle diffusion, and it was also influenced by boundary layer diffusion. The modeling studies revealed that the adsorption of fluoride ions on MS, MS-Boehmite, and MS-PVDF-Boehmite occurred through chemical interaction with valance forces between positively charged boehmite and negatively charged boehmite. Finally, the composite membranes were inserted into a syringe as a cartridge, and adsorption performance was evaluated at varying flow rates. After four cycles of filtration at a flow rate of 70 mL h^{-1}, the concentration of fluoride ions fell to 0.9 mg L^{-1} with MS-PVDF-Boehmite, which is below the WHO standard (1.5 mg L^{-1}). We believe that MS-PVDF-Boehmite can be a simple, efficient, and practical tool for the removal of fluoride ions from wastewater owing to its simple fabrication process, cost-effectiveness, and high performance.

Author Contributions: Conceptualization, M.-W.L.; Methodology, H.-B.L., A.-J.C. and M.-W.L.; Formal analysis, H.-B.L., A.-J.C., Y.-K.K. and M.-W.L.; Data curation, H.-B.L. and Y.-K.K.; Writing—original draft, H.-B.L., A.-J.C., Y.-K.K. and M.-W.L.; Supervision, Y.-K.K. and M.-W.L.; Funding acquisition, Y.-K.K. and M.-W.L. All authors have read and agreed to the published version of the manuscript.

Funding: This research was financially supported by the National Research Foundation of Korea (NRF) funded by the Ministry of Science and ICT (NRF-2020M3H4A3106354). This research was supported by the Basic Science Research Program through the National Research Foundation of Korea (NRF) funded by the Ministry of Education (2022R1A6A1A03053343). This work was also supported by the Technology Innovation Program (RS-2022-00155769) funded by the Ministry of Trade, Industry & Energy (MOTIE, Korea).

Institutional Review Board Statement: Not applicable.

Data Availability Statement: Not applicable.

Conflicts of Interest: The authors declare no conflict of interest.

References

1. Bhatnagar, A.; Kumar, E.; Sillanpää, M. Fluoride removal from water by adsorption—A review. *Chem. Eng. J.* **2011**, *171*, 811–840. [CrossRef]
2. Raza, M.; Farooqi, A.; Niazi, N.K.; Ahmad, A. Geochemical control on spatial variability of fluoride concentrations in groundwater from rural areas of Gujrat in Punjab, Pakistan. *Environ. Earth Sci.* **2016**, *75*, 1364. [CrossRef]
3. Damtie, M.M.; Woo, Y.C.; Kim, B.; Hailemariam, R.H.; Park, K.-D.; Shon, H.K.; Park, C.; Choi, J.-S. Removal of fluoride in membrane-based water and wastewater treatment technologies: Performance review. *J. Environ. Manag.* **2019**, *251*, 109524. [CrossRef] [PubMed]
4. World Health Organization. *Guidelines for Drinking-Water Quality*; World Health Organization: Geneva, Switzerland, 2011; Volume 216, pp. 303–304.
5. Huang, H.; Liu, J.; Zhang, P.; Zhang, D.; Gao, F. Investigation on the simultaneous removal of fluoride, ammonia nitrogen and phosphate from semiconductor wastewater using chemical precipitation. *Chem. Eng. J.* **2017**, *307*, 696–706. [CrossRef]

6. Chen, D.; Zhao, M.; Tao, X.; Ma, J.; Liu, A.; Wang, M. Exploration and Optimisation of High-Salt Wastewater Defluorination Process. *Water* **2022**, *14*, 3974. [CrossRef]
7. Wei, F.; Cao, C.; Huang, P.; Song, W. A new ion exchange adsorption mechanism between carbonate groups and fluoride ions of basic aluminum carbonate nanospheres. *RSC Adv.* **2015**, *5*, 13256–13260. [CrossRef]
8. Hu, K.; Dickson, J.M. Nanofiltration membrane performance on fluoride removal from water. *J. Membr. Sci.* **2006**, *279*, 529–538. [CrossRef]
9. Zhang, J.; Chen, N.; Su, P.; Li, M.; Feng, C. Fluoride removal from aqueous solution by zirconium-chitosan/graphene oxide membrane. *React. Funct. Polym.* **2017**, *114*, 127–135. [CrossRef]
10. Dhillon, A.; Soni, S.K.; Kumar, D. Enhanced fluoride removal performance by Ce–Zn binary metal oxide: Adsorption characteristics and mechanism. *J. Fluor. Chem.* **2017**, *199*, 67–76. [CrossRef]
11. Wu, Y.; Pang, H.; Liu, Y.; Wang, X.; Yu, S.; Fu, D.; Chen, J.; Wang, X. Environmental remediation of heavy metal ions by novel-nanomaterials: A review. *Environ. Pollut.* **2019**, *246*, 608–620. [CrossRef]
12. Camacho, L.M.; Torres, A.; Saha, D.; Deng, S. Adsorption equilibrium and kinetics of fluoride on sol–gel-derived activated alumina adsorbents. *J. Colloid Interface Sci.* **2010**, *349*, 307–313. [CrossRef]
13. Ayoob, S.; Gupta, A.; Bhakat, P.; Bhat, V.T. Investigations on the kinetics and mechanisms of sorptive removal of fluoride from water using alumina cement granules. *Chem. Eng. J.* **2008**, *140*, 6–14. [CrossRef]
14. Tao, W.; Zhong, H.; Pan, X.; Wang, P.; Wang, H.; Huang, L. Removal of fluoride from wastewater solution using Ce-AlOOH with oxalic acid as modification. *J. Hazard. Mater.* **2020**, *384*, 121373. [CrossRef]
15. Farrah, H.; Slavek, J.; Pickering, W. Fluoride interactions with hydrous aluminum oxides and alumina. *Soil Res.* **1987**, *25*, 55–69. [CrossRef]
16. Turner, B.D.; Binning, P.; Stipp, S. Fluoride removal by calcite: Evidence for fluorite precipitation and surface adsorption. *Environ. Sci. Technol.* **2005**, *39*, 9561–9568. [CrossRef] [PubMed]
17. Jiao, Z.; Chen, Z.; Yang, M.; Zhang, Y.; Li, G. Adsorption of fluoride ion by inorganic cerium based adsorbent. *High Tech Bull. Engl. Ed.* **2004**, *10*, 83–86.
18. Bhargava, D.; Killedar, D. Fluoride adsorption on fishbone charcoal through a moving media adsorber. *Water Res.* **1992**, *26*, 781–788. [CrossRef]
19. Efome, J.E.; Rana, D.; Matsuura, T.; Lan, C.Q. Experiment and modeling for flux and permeate concentration of heavy metal ion in adsorptive membrane filtration using a metal-organic framework incorporated nanofibrous membrane. *Chem. Eng. J.* **2018**, *352*, 737–744. [CrossRef]
20. Koushkbaghi, S.; Zakialamdari, A.; Pishnamazi, M.; Ramandi, H.F.; Aliabadi, M.; Irani, M. Aminated-Fe$_3$O$_4$ nanoparticles filled chitosan/PVA/PES dual layers nanofibrous membrane for the removal of Cr (VI) and Pb (II) ions from aqueous solutions in adsorption and membrane processes. *Chem. Eng. J.* **2018**, *337*, 169–182. [CrossRef]
21. Koushkbaghi, S.; Jafari, P.; Rabiei, J.; Irani, M.; Aliabadi, M. Fabrication of PET/PAN/GO/Fe$_3$O$_4$ nanofibrous membrane for the removal of Pb (II) and Cr (VI) ions. *Chem. Eng. J.* **2016**, *301*, 42–50. [CrossRef]
22. Efome, J.E.; Rana, D.; Matsuura, T.; Lan, C.Q. Enhanced performance of PVDF nanocomposite membrane by nanofiber coating: A membrane for sustainable desalination through MD. *Water Res.* **2016**, *89*, 39–49. [CrossRef]
23. Huang, Q.-L.; Huang, Y.; Xiao, C.-F.; You, Y.-W.; Zhang, C.-X. Electrospun ultrafine fibrous PTFE-supported ZnO porous membrane with self-cleaning function for vacuum membrane distillation. *J. Membr. Sci.* **2017**, *534*, 73–82. [CrossRef]
24. Sun, H.; Ji, Z.; He, Y.; Wang, L.; Zhan, J.; Chen, L.; Zhao, Y. Preparation of PAMAM modified PVDF membrane and its adsorption performance for copper ions. *Environ. Res.* **2022**, *204*, 111943. [CrossRef]
25. Efome, J.E.; Rana, D.; Matsuura, T.; Lan, C.Q. Insight studies on metal-organic framework nanofibrous membrane adsorption and activation for heavy metal ions removal from aqueous solution. *ACS Appl. Mater. Interfaces* **2018**, *10*, 18619–18629. [CrossRef]
26. Mohammadi, M.; Khodamorady, M.; Tahmasbi, B.; Bahrami, K.; Ghorbani-Choghamarani, A. Boehmite nanoparticles as versatile support for organic–inorganic hybrid materials: Synthesis, functionalization, and applications in eco-friendly catalysis. *J. Ind. Eng. Chem.* **2021**, *97*, 1–78. [CrossRef]
27. Lueangchaichaweng, W.; Singh, B.; Mandelli, D.; Carvalho, W.A.; Fiorilli, S.; Pescarmona, P.P. High surface area, nanostructured boehmite and alumina catalysts: Synthesis and application in the sustainable epoxidation of alkenes. *Appl. Catal. A Gen.* **2019**, *571*, 180–187. [CrossRef]
28. Kumar, A.; Ghosh, U.K. Polyvinylidene fluoride/boehmite nanocomposite membrane for effective removal of arsenate ion from water. *J. Water Process Eng.* **2022**, *47*, 102652. [CrossRef]
29. Leyva-Ramos, R.; Medellín-Castillo, N.A.; Jacobo-Azuara, A.; Mendoza-Barrón, J.; Landín-Rodríguez, L.E.; Martínez-Rosales, J.M.; Aragón-Piña, A. Fluoride removal from water solution by adsorption on activated alumina prepared from pseudo-boehmite. *J. Environ. Eng. Manag.* **2008**, *18*, 301–309.
30. Ding, Y.; Xu, W.; Yu, Y.; Hou, H.; Zhu, Z. One-step preparation of highly hydrophobic and oleophilic melamine sponges via metal-ion-induced wettability transition. *ACS Appl. Mater. Interfaces* **2018**, *10*, 6652–6660. [CrossRef]
31. Feng, Y.; Yao, J. Design of melamine sponge-based three-dimensional porous materials toward applications. *Ind. Eng. Chem. Res.* **2018**, *57*, 7322–7330. [CrossRef]
32. Kim, S.; Lim, C.; Kwak, C.H.; Kim, D.; Ha, S.; Lee, Y.-S. Hydrophobic melamine sponge prepared by direct fluorination for efficient separation of emulsions. *J. Ind. Eng. Chem.* **2023**, *118*, 259–267. [CrossRef]

33. Chen, X.; Weibel, J.A.; Garimella, S.V. Continuous oil–water separation using polydimethylsiloxane-functionalized melamine sponge. *Ind. Eng. Chem. Res.* **2016**, *55*, 3596–3602. [CrossRef]
34. Li, J.; Tenjimbayashi, M.; Zacharia, N.S.; Shiratori, S. One-step dipping fabrication of Fe_3O_4/PVDF-HFP composite 3D porous sponge for magnetically controllable oil–water separation. *ACS Sustain. Chem. Eng.* **2018**, *6*, 10706–10713. [CrossRef]
35. Lei, Z.; Deng, Y.; Wang, C. Ambient-temperature fabrication of melamine-based sponges coated with hydrophobic lignin shells by surface dip adsorbing for oil/water separation. *Rsc Adv.* **2016**, *6*, 106928–106934. [CrossRef]
36. Zhu, Q.; Chu, Y.; Wang, Z.; Chen, N.; Lin, L.; Liu, F.; Pan, Q. Robust superhydrophobic polyurethane sponge as a highly reusable oil-absorption material. *J. Mater. Chem. A* **2013**, *1*, 5386–5393. [CrossRef]
37. Pandey, M.; Sharma, K.; Islam, S.S. Wide Range RH Detection with Digital Readout: Niche Superiority in Terms of Its Exceptional Performance and Inexpensive Technology. *Adv. Mater. Phys. Chem.* **2019**, *9*, 11–24. [CrossRef]
38. Hoffman, D.M. Infrared properties of three plastic bonded explosive binders. *Int. J. Polym. Anal. Charact.* **2017**, *22*, 545–556. [CrossRef]
39. Liu, W.; Jiang, H.; Ru, Y.; Zhang, X.; Qiao, J. Conductive graphene–melamine sponge prepared via microwave irradiation. *ACS Appl. Mater. Interfaces* **2018**, *10*, 24776–24783. [CrossRef]
40. Wang, Z.; Gu, X.; Zhang, Y.; Zhang, X.; Ngo, H.H.; Liu, Y.; Jiang, W.; Tan, X.; Wang, X.; Zhang, J. Activated nano-Al_2O_3 loaded on polyurethane foam as a potential carrier for fluorine removal. *J. Water Process Eng.* **2021**, *44*, 102444. [CrossRef]
41. Bonilla-Petriciolet, A.; Mendoza-Castillo, D.I.; Reynel-Ávila, H.E. *Adsorption Processes for Water Treatment and Purification*; Springer: Berlin/Heidelberg, Germany, 2017; Volume 256.
42. Xu, J.; Zhang, B.; Lu, Y.; Wang, L.; Tao, W.; Teng, X.; Ning, W.; Zhang, Z. Adsorption desulfurization performance of PdO/SiO_2@graphene oxide hybrid aerogel: Influence of graphene oxide. *J. Hazard. Mater.* **2022**, *421*, 126680. [CrossRef]
43. Saadi, R.; Saadi, Z.; Fazaeli, R.; Fard, N.E. Monolayer and multilayer adsorption isotherm models for sorption from aqueous media. *Korean J. Chem. Eng.* **2015**, *32*, 787–799. [CrossRef]
44. Jiménez-Becerril, J.; Solache-Ríos, M.; García-Sosa, I. Fluoride removal from aqueous solutions by boehmite. *Water Air Soil Pollut.* **2012**, *223*, 1073–1078. [CrossRef]
45. Gholitabar, S.; Tahermansouri, H. Kinetic and multi-parameter isotherm studies of picric acid removal from aqueous solutions by carboxylated multi-walled carbon nanotubes in the presence and absence of ultrasound. *Carbon Lett.* **2017**, *22*, 14–24.
46. Lee, J.S.; Lee, H.B.; Oh, Y.; Choi, A.-J.; Seo, T.H.; Kim, Y.-K.; Lee, M.W. Used coffee/PCL composite filter for Cu (II) removal from wastewater. *J. Water Process Eng.* **2022**, *50*, 103253. [CrossRef]

Disclaimer/Publisher's Note: The statements, opinions and data contained in all publications are solely those of the individual author(s) and contributor(s) and not of MDPI and/or the editor(s). MDPI and/or the editor(s) disclaim responsibility for any injury to people or property resulting from any ideas, methods, instructions or products referred to in the content.

Article

Electropositive Membrane Prepared via a Simple Dipping Process: Exploiting Electrostatic Attraction Using Electrospun SiO₂/PVDF Membranes with Electronegative SiO₂ Shell

Dalsu Choi [1,†], Cheol Ho Lee [2,†], Han Bi Lee [3], Min Wook Lee [3,*] and Seong Mu Jo [3,*]

1. Chemical Engineering Department, Myongji University, Yongin-si 17058, Gyeonggi-do, Republic of Korea; dalsuchoi@mju.ac.kr
2. Center for Underground Physics, Institute for Basic Science, Daejeon 34126, Republic of Korea; lch2301@ibs.re.kr
3. Composite Materials Applications Research Center, Korea Institute of Science and Technology, Wanju-gun 55324, Jeollabuk-do, Republic of Korea; 092414@kist.re.kr
* Correspondence: mwlee0713@kist.re.kr (M.W.L.); smjo@kist.re.kr (S.M.J.); Tel.: +82-63-219-8183 (M.W.L.); +82-63-219-8150 (S.M.J.)
† These authors contributed equally to this work.

Citation: Choi, D.; Lee, C.H.; Lee, H.B.; Lee, M.W.; Jo, S.M. Electropositive Membrane Prepared via a Simple Dipping Process: Exploiting Electrostatic Attraction Using Electrospun SiO₂/PVDF Membranes with Electronegative SiO₂ Shell. *Polymers* **2023**, *15*, 2270. https://doi.org/10.3390/polym15102270

Academic Editors: Tomasz Makowski and Sivanjineyulu Veluri

Received: 13 April 2023
Revised: 4 May 2023
Accepted: 6 May 2023
Published: 11 May 2023

Copyright: © 2023 by the authors. Licensee MDPI, Basel, Switzerland. This article is an open access article distributed under the terms and conditions of the Creative Commons Attribution (CC BY) license (https://creativecommons.org/licenses/by/4.0/).

Abstract: This research aimed to develop a simple and cost-effective method for fabricating electropositive membranes for highly efficient water filtration. Electropositive membranes are novel functional membranes with electropositive properties and can filter electronegative viruses and bacteria using electrostatic attraction. Because electropositive membranes do not rely on physical filtration, they exhibit high flux characteristics compared with conventional membranes. This study presents a simple dipping process for fabricating boehmite/SiO₂/PVDF electropositive membranes by modifying an electrospun SiO₂/PVDF host membrane using electropositive boehmite nanoparticles (NPs). The surface modification enhanced the filtration performance of the membrane, as revealed by electronegatively charged polystyrene (PS) NPs as a bacteria model. The boehmite/SiO₂/PVDF electropositive membrane, with an average pore size of 0.30 μm, could successfully filter out 0.20 μm PS particles. The rejection rate was comparable to that of Millipore GSWP, a commercial filter with a pore size of 0.22 μm, which can filter out 0.20 μm particles via physical sieving. In addition, the water flux of the boehmite/SiO₂/PVDF electropositive membrane was twice that of Millipore GSWP, demonstrating the potential of the electropositive membrane in water purification and disinfection.

Keywords: electrospinning; electrostatic attraction; dipping; electropositive membrane; filter

1. Introduction

Currently, owing to its energy efficiency, pressure-driven membrane filtration is the most widely used technique for water purification [1,2]. Among various membrane-filtration approaches, physical sieving is the most common technique [1,3]. Although they are easily fabricated, filters based on physical sieving have a "flux–pore-size" issue when used to filter ultrafine particles, which require ultrafine pore sizes [4]. The reduction of flux is inevitably accompanied by pore size reduction. Furthermore, considerable pressure must drive ultrafine pore membranes, which reduces filtration efficiency [5,6]. Currently, various techniques, such as mechanical and chemical methods, are being investigated for filtration. Water purification using bulk mechanical filters is the most common technique in water treatment. Sand, hydro-anthracite, burned rocks, and crushed expanded clay are also used for filtration [7]. Additionally, various types of adsorption membranes have been used to remove heavy metals and organic dyes from wastewater [8]. Among them, metal–organic frameworks (MOFs) have been used in many fields owing to their high surface area and tunable pore volume and chemical properties, and MOF mixed membranes containing nanoparticles (NPs) have been studied [9]. Recently, membrane filtration based on mechanisms other than physical sieving has attracted considerable attention [5,10–14].

Electropositive filtration is a technique that uses a filtration mechanism other than sieving. An electropositive filter is fabricated by depositing highly electropositive components on the outer surface of the host membrane [15]. Polyelectrolytes [16], zirconia [17], copper oxide [18], and hematite [19] have been recently investigated as electropositive coatings. However, such electropositive coatings usually suffer from poor adhesion to the membrane, low zeta potential, and toxicity. Aluminum oxide, gibbsite, and activated alumina are the most widely used electropositive components for electropositive membranes owing to their low toxicity, low cost, and high mechanical and thermal stability [11,20–22]. By incorporating electropositive components, electronegative substances in a feed, such as bacteria and viruses, which commonly show a negative charge in water media, can be effectively filtered via electrostatic attraction [20,23,24]. Therefore, ultrafine electronegative particles can be filtered through pores larger than the particles without sacrificing the flux of the feed. However, the attachment of boehmite on membrane surfaces requires complicated processes, such as the direct hydrothermal synthesis of boehmite on the host membrane [14,21,25]. To date, boehmite deposition during the fabrication of electropositive membranes is achieved by direct hydrothermal synthesis on the host membrane [9]. Therefore, polymeric hosts cannot be used, as they cannot withstand harsh hydrothermal conditions. A simple dipping process for boehmite deposition would enable the usage of various host membranes. In addition, dipping processes allow continuous fabrication, making them more cost-effective than batch-type hydrothermal processes. PVDF-g-PNE and PVDF-g-PAA membranes contain electropositive materials; thus, bacteria and viruses, which normally exhibit negative charges, can be effectively filtered via electrostatic attraction. Polymer coatings firmly bond to PVDF membranes through adhesive force (coordination, hydrogen bonding, electrostatic interaction, and hydrophobic interaction) [26]. SiO_2 NPs have high electronegativity (2.82) and attract electropositive lithium ions from electrolytes. On this basis, a SiO_2/PVDF composite membrane was developed as a battery separator by varying the SiO_2-to-PVDF mass ratio [27]. Graft copolymerization of methacrylic acid (MAA) monomers with plasma-activated PVDF membranes was performed to introduce carboxyl groups into the membrane. Subsequently, the surface of the NPs was made hydrophilic using a positively charged ligand, and the NPs were coated on the membrane surface through electrostatic and covalent bonding [28].

Here, a simple technique for attaching boehmite to host membranes is proposed. The performance of the resultant boehmite/SiO_2/PVDF electropositive membrane was evaluated and compared with that of commercially available filters. The obtained boehmite/SiO_2/PVDF electropositive membrane exhibited better performance than the commercial filters. The water flux of the boehmite/SiO_2/PVDF electropositive membrane was almost twice that of Millipore GSWP, a conventional filter with a pore size similar to that of the fabricated membrane, and the particle rejection rates of both membranes were comparable.

2. Materials and Methods

2.1. Materials

Tetraethyl orthosilicate (TEOS), aluminum isopropoxide (AIP), and acetic acid were purchased from Sigma-Aldrich (St. Louis, MO, USA). N,N-dimethylformamide (DMF, 99.8%), ethyl alcohol (EtOH, 94.5%), hydrochloric acid (HCl, 37%), and sodium hydroxide (NaOH) were purchased from Daejung Reagent Chemicals, and poly(vinylidene fluoride) (PVDF, Kynar-761) was purchased from Alkema. All chemicals were used as received without further purification. Uniform polystyrene (PS) latex beads (0.20 μm) used for particle rejection tests were obtained from Magsphere Inc. Millipore GSWP with a pore size of 0.22 μm was used as a commercial filter to compare with the boehmite/SiO_2/PVDF electropositive membrane.

2.2. Synthesis of Boehmite (γ–AlOOH)

Boehmite (γ–AlOOH) was synthesized via the sol–gel reaction of aluminum isopropoxide (AIP). First, 68 g of AIP was added to the 300 mL of deionized water at 75 °C. The solution was heated to 95 °C with mechanical stirring, and the water was evaporated until the volume of the solution was reduced to 200 mL. Next, 3.1 g of acetic acid was added dropwise, and the solution was stirred for 10 min. Finally, hydrothermal synthesis was performed at 150 °C for 6 h using an autoclave.

2.3. Preparation of SiO_2/PVDF Solution

The silica precursor was prepared by the sol–gel reaction of TEOS. H_2O and HCl were added to the TEOS/EtOH solution in a TEOS-EtOH-H_2O-HCl molar ratio of 0.54:1.08:1.08:0.005. The prepared mixture was heated at 95 °C for 2 h to accelerate the condensation polymerization, after which it was cooled to room temperature, and 100 g of DMF was added before irreversible gelation occurred. The prepared silica sol was mixed with PVDF in a silica sol–PVDF weight ratio of 3:7. Then, DMF was added to the prepared silica sol/PVDF/DMF solution with 19.44 weight percent of silica sol/PVDF. Finally, by stirring the silica sol/PVDF/DMF solution at 100 °C for 1 h, a SiO_2/PVDF blend solution was obtained.

2.4. Preparation of Electrospun SiO_2/PVDF Membrane

SiO_2/PVDF membranes were fabricated by electrospinning the SiO_2/PVDF solution. The SiO_2/PVDF solution was fed at a rate of 5 µL/min through a 24 G needle, and the distance from the needle tip to the substrate was set to 15 cm. Then, 13 kV voltage was applied at a relative humidity of 18 ± 2% and a temperature of 30 ± 1 °C. Through a 4 h collection of the electrospun fibers, a 30-µm-thick membrane was obtained.

2.5. Preparation Boehmite/SiO_2/PVDF Membrane

Boehmite was attached to an electrospun SiO_2/PVDF membrane by submerging the membrane in an 11.5 wt% boehmite solution for 10 min. After the dipping process, the boehmite/SiO_2/PVDF composite membrane was washed by flowing 500 mL of deionized water through the membrane using a filter funnel. Finally, the washed membrane was dried in a convection oven at 60 °C for 6 h. The amount of attached boehmite was determined by comparing the weight of the boehmite-attached SiO_2/PVDF membrane with that of the bare SiO_2/PVDF membrane using a microbalance.

2.6. Membrane Performance Tests

The pore size distribution and average pore size of the electrospun SiO_2/PVDF and boehmite/SiO_2/PVDF membranes were measured using a capillary flow porometer (CFP-1500AE, Porous Materials Inc., Ithaca, NY, USA) following the ASTM-F316-03 standard. The membranes were soaked with a Galwick™ wetting liquid with a surface tension of 15.9 dyn/cm, and air was passed through the membrane. For the water flux test, deionized water was fed through a dead-end cell (AMICON stirred cell 8010) by applying pressure to the reservoir using nitrogen gas. The pure water flux was calculated by measuring the weight of deionized water passed through the membrane inside the dead-end cell using a microbalance (XS6002S, Mettler-Toledo, Columbus, OH, USA). Next, a particle rejection test was performed using a setup similar to that of the water flux test. First, the membrane for the rejection test was assembled inside the dead-end cell (AMICON stirred cell 8010). Then, 100 ppm of 0.20 µm PS particles dispersed in deionized water was fed into the dead-end cell at 2.5 psi for 30 min. Then, the permeate was collected at intervals, and the collected permeates were analyzed by ultraviolet–visible (UV-Vis) spectroscopy (V-670, Jasco, Portland, OR, USA). The characteristic absorbance peak of PS NPs at 230 nm was monitored to determine the concentration of the NPs.

2.7. Characterization

The crystal structure of the synthesized boehmite and electrospun SiO_2/PVDF membrane was analyzed using an X-ray diffractometer (Rigaku, Tokyo, Japan) with a Cu Kα (λ = 0.154 nm) beam source. Samples were scanned from 2θ = 10° to 90°, and the instrument was driven at the acceleration voltage of 45 kV and emission current of 200 mA. The morphology of the synthesized boehmite and the cross-section of the electrospun SiO_2/PVDF membrane were analyzed by field-emission transmission electron microscopy (FE-TEM, Tecnai F20, FEI, Hillsboro, OR, USA). A boehmite solution was drop cast on a copper TEM grid, and the grid was thoroughly dried in a vacuum oven before the TEM measurement. To measure the cross-section of the electrospun SiO_2/PVDF membrane, the fabricated membrane was cross-sectioned using a Cryo-Ultramicrotome (RMC-PTPC + CR-X), and the prepared sample was put on a copper TEM grid for measurements. Dynamic laser scattering (DLS, Malvern Zetasizer, Malvern, UK) analysis was used to determine the zeta potential of the samples at different pH values and at the isoelectric point (IEP). The synthesized boehmite and 0.20 μm PS particles were dispersed in ethyl alcohol to make a 100 ppm dispersion for the DLS measurements. Then, 0.1 mol/L NaOH and 0.1 mol/L HCl solutions were added to vary the pH of the dispersion, and the pH was monitored in situ using a pH meter (Orion Star A211, Thermo Scientific, Waltham, MA, USA). The morphologies of the electrospun SiO_2/PVDF and boehmite/SiO_2/PVDF composite membranes were analyzed by FE-scanning electron microscopy (SEM) (Nova NanoSEM450, FEI, Hillsboro, OR, USA) at an acceleration voltage of 10 kV. The atomic composition and presence of specific chemical bonds were studied using X-ray photoelectron spectroscopy (XPS) with an Al K-alpha X-ray source (K-Alpha X-ray Photoelectron Spectrometer System, Thermo Fisher Scientific, Waltham, MA, USA). For the survey scan, 200 eV pass energy was used, and the scanning was performed twice. For fine scans, 50 eV pass energy was used, and five scans were performed. All scans were executed with a flood gun to minimize the charge accumulation. The resultant spectra were analyzed using the Advantage software provided with the XPS instrument. To check the amount of boehmite detached from the boehmite/SiO_2/PVDF composite membrane during filtration, the filtrate was analyzed using inductively coupled plasma–optical emission spectroscopy (ICP–OES, iCAP 6000, Thermo Scientific, Loughborough, UK). Furthermore, 100, 300, and 500 mL of deionized water were filtered through the boehmite/SiO_2/PVDF composite membrane, and the aluminum contents of filtrates were measured. The contact angles of the membranes were determined using a contact angle and surface tension analyzer (Phoenix 300).

3. Results

3.1. Characterization of the Synthesized Boehmite (γ–AlOOH)

Figure 1a shows the TEM image of the synthesized boehmite. It reveals an assembly of nanorods, which is a characteristic feature of boehmite [14,21,22]. Furthermore, the X-ray diffraction (XRD) pattern of the sample (Figure 1b) showed diffraction peaks at 2θ = 14.50°, 28.24°, 38.44°, and 49.18°, which are ascribed to the (020), (120), (031), and (200) planes of boehmite, respectively [22], confirming the successful synthesis of boehmite nanorods. Finally, the synthesized boehmite exhibited a highly positive zeta potential ranging from 30 to 50 mV under acidic conditions (Figure 1c). Even at a neutral pH, the zeta potential was 32.8 mV. A highly positive zeta potential was maintained from low pH values to the IEP (pH 9), where the zeta potential became zero. Therefore, the sample would exhibit an electropositive effect for filtration in a wide range of environments.

3.2. Characterization of Electrospun SiO_2/PVDF Membrane

The morphology and nanofiber dimensions of the electrospun SiO_2/PVDF membrane were characterized by SEM. The average diameter of the SiO_2/PVDF nanofibers in the membrane was 134 ± 43 nm. Furthermore, a well-developed internetworking structure with mesopores was observed, indicating the formation of electrospun nanofibers.

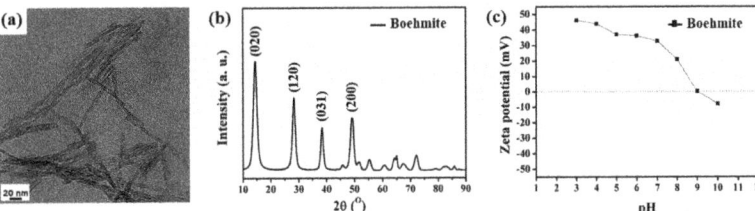

Figure 1. (a) Transmission electron microscopy (TEM) image of boehmite synthesized via hydrothermal treatment. (b) X-ray diffraction (XRD) pattern of as-prepared boehmite. (c) Zeta potential of the synthesized boehmite at pH 3–10.

Next, the atomic composition and available chemical bonds of the electrospun SiO_2/PVDF membrane surface were determined by XPS to investigate the presence of SiO_2. Si 2p and O 1s scans (Figure 2b,c) showed peaks at the binding energies of 103.7 and 533.2 eV, which were ascribed to the Si–O bond of SiO_2 [29], indicating SiO_2. In addition, the atomic ratio of Si:O decreased from 24.4:44.4 to 1:2 (Table 1), further indicating SiO_2 in the electrospun SiO_2/PVDF nanofiber. The XPS survey spectra (Figure 2a) showed Si 2p, C 1s, O 1s, and F 1s, which are atomic components of SiO_2 and PVDF. This shows that at an XPS penetration depth of up to 10–20 nm, PVDF and SiO_2 were present in the sample. However, as the combined atomic percentage of carbon and fluoride from PVDF is much lower than that of silicon and oxide from SiO_2, the outer portion of the electrospun SiO_2/PVDF fiber mainly consists of SiO_2 [30]. Thus, the electrospun SiO_2/PVDF fiber was cross-sectioned using ultramicrotomy, and the prepared sample was analyzed by TEM (Figure 3). Consistent with the XPS result, the outermost part of the electrospun SiO_2/PVDF fiber comprised SiO_2. Additionally, the TEM image revealed the formation of multicore structures, which was attributed to the distinct phase separation induced by unfavorable molecular interactions between the SiO_2 sol and PVDF.

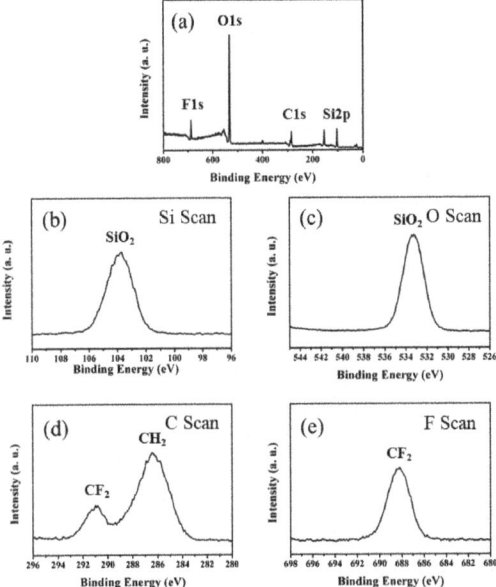

Figure 2. X-ray photoelectron spectroscopy (XPS) images of the electrospun SiO_2/PVDF membrane. (a) Wide scan spectra, (b) Si 2p spectra, (c) O 1s spectra, (d) C 1s spectra, and (e) F 1s spectra of the electrospun SiO_2/PVDF membrane.

Table 1. Atomic ratio of the electrospun SiO$_2$/PVDF membrane obtained from XPS.

	Binding Energy (eV)	Atomic Ratio (%)
Si 2p	103.7	24.44
C 1s	286.4	24.73
O 1s	533.2	44.41
F 1s	688.2	6.41

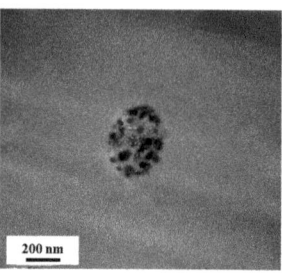

Figure 3. Cross-sectional TEM image of the multicore structure of the electrospun SiO$_2$/PVDF blend nanofiber.

3.3. Characterization of Boehmite/SiO$_2$/PVDF Composite Membrane

Electrospun SiO$_2$/PVDF membranes with highly electronegative SiO$_2$ skin can be effective in preparing electropositive membranes because highly electropositive boehmite can be easily attached to a membrane via electrostatic interactions [31]. Here, the electrospun SiO$_2$/PVDF membrane was simply submerged in a solution of the synthesized boehmite for 10 min to enable the attachment of the boehmite to the membrane via electrostatic attractions. Via a simple dipping process, boehmite was successfully attached to and fully covered the SiO$_2$/PVDF membrane, whereas in previous studies, intense hydrothermal treatment has been required to attach boehmite directly to the surfaces of host membranes. The SEM images of the SiO$_2$/PVDF membrane after the simple, short dipping process (Figure 4a,b) revealed the successful attachment of boehmite to the membrane. The SiO$_2$/PVDF fibers showed a smooth surface (panel a), whereas an irregular texture was observed on the fiber surface (panel b), confirming that boehmite coated the SiO$_2$/PVDF fibers. The increased membrane weight after the dipping process also indicated the successful attachment of boehmite to the membrane. On average, 12.65 wt% of boehmite was attached after dipping (Figure S1). Boehmite is mainly composed of aluminum. Although instant exposure to aluminum is not critical to human beings, prolonged exposure to aluminum can cause health issues [32]. Thus, the amount of detached boehmite after filtration was examined to ensure that no aluminum was introduced into the filtrate. First, 100, 300, and 500 mL deionized water samples were filtered through the electrospun boehmite/SiO$_2$/PVDF membrane, and the filtrates were analyzed by ICP–OES. The amount of aluminum in the three samples was undetectable, as it was less than 0.1 ppm (Figure 4c).

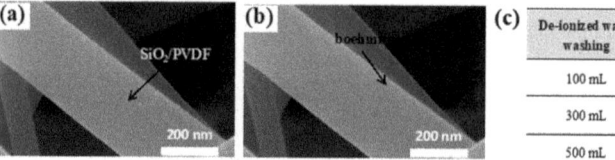

Figure 4. Scanning electron microscopy (SEM) images of (**a**) electrospun SiO$_2$/PVDF membrane and (**b**) boehmite-SiO$_2$/PVDF composite membrane fabricated by dipping. (**c**) Inductively coupled plasma–optical emission spectroscopy (ICP–OES) results of filtrates of deionized water samples.

3.4. Membrane-Filtration Performance

Pore size and pore size distribution, which are critical parameters that determine membrane retention capability, were evaluated using a capillary flow porometer (Figure 5a). The mean pore diameter of the electrospun SiO_2/PVDF membrane decreased from 0.48 to 0.35 μm upon the attachment of boehmite, which was attributed to the increased diameter of the nanofiber upon boehmite attachment. In addition, the mean pore size of Millipore GSWP, a commercial MF membrane with a smaller pore size, is 0.22 μm, which is smaller than that of the boehmite/SiO_2/PVDF composite. To determine the water flux efficiency of the boehmite/SiO_2/PVDF composite membrane, a dead-end cell system was assembled to measure the water flux. Millipore GSWP showed no considerable change in flux upon operation, whereas the flux of the electrospun SiO_2/PVDF and boehmite/SiO_2/PVDF composite membranes decreased upon operation, which was attributed to the water-flow-driven compression of the less densely packed electrospun membranes. The boehmite/SiO_2/PVDF membrane showed considerably improved water flux compared with the SiO_2/PVDF membrane (Figure 5b,c). The water flux through the SiO_2/PVDF membrane was 6463 $L \cdot m^{-2} \cdot h^{-1}$, whereas that of the boehmite/SiO_2/PVDF composite membrane was 18,827 $L \cdot m^{-2} \cdot h^{-1}$. This notable enhancement in water flux in the composite membrane can be attributed to the hydrophilicity of the membrane, which provides an additional driving force for water to easily pass through the capillary induced by membrane pores [29–31,33–41]. To verify our hypothesis, we measured the water contact angles of the boehmite/SiO_2/PVDF composite and SiO_2/PVDF membranes (Figure 5d). Due to the aforementioned hydrophilicity, the membrane with boehmite had a much lower contact angle than the SiO_2/PVDF membrane (37.92° vs. 127.85°). In the case of low-pressure conditions (2.5 psi), the water flux of the electrospun membranes was higher than that of Millipore GSWP.

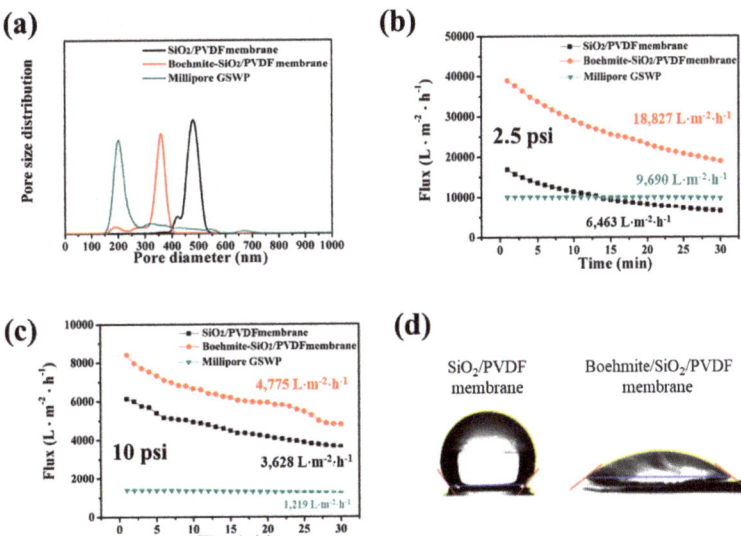

Figure 5. (**a**) Pore size distribution of the fabricated membranes (SiO_2/PVDF and boehmite/SiO_2/PVDF) and Millipore GSWP. (**b**,**c**) Water flux of each membrane at the pressures of 2.5 and 10 psi. (**d**) Water contact angles of SiO_2/PVDF and boehmite/SiO_2/PVDF membranes.

Finally, to evaluate the rejection ability of the boehmite/SiO_2/PVDF composite membrane, particle rejection tests were conducted using 0.20 ± 0.011 μm PS particles. The filtration efficiency of each membrane was measured for 30 min under a pressure of 2.5 psi by filtering a 100 ppm PS dispersion sample. The size of common bacteria was modeled by

0.20 μm particles. Additionally, the zeta potential of the PS particles was measured by DLS analysis (Figure 6a). The PS particles showed a highly negative zeta potential (−46.4 mV) at pH 7. Therefore, we concluded that the filtration of 0.20 μm PS NPs can effectively simulate bacteria rejection. The particle rejection percentage for the tested membranes increased with the filtration time due to the pore-blocking effect of the NPs. However, the boehmite/SiO$_2$/PVDF membrane with a large pore size and a considerably high water flux exhibited a high rejection rate similar to that of Millipore GSWP (Figure 6b). Considering the high water flux, which was twice that of Millipore GSWP, the boehmite/SiO$_2$/PVDF membrane exhibited better performance than Millipore GSWP. Therefore, the adoption of boehmite aids in more effective filtration via electrostatic attraction and enhancing interface interactions.

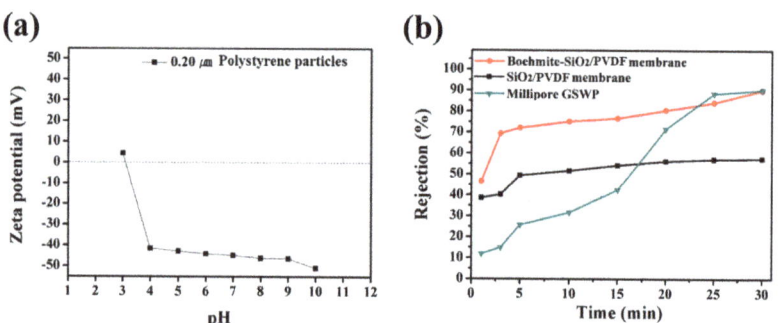

Figure 6. (a) Zeta potential of 0.20 μm polystyrene particles used for filtration tests at pH 3–10. (b) Rejection rates of polystyrene particles in SiO$_2$/PVDF, boehmite-SiO$_2$/PVDF composite, and Millipore GSWP membranes.

4. Conclusions

This study presents a simple, fast, cost-effective technique for fabricating electropositive membranes. Unlike conventional techniques that require intense hydrothermal reactions to attach electropositive boehmite to a membrane matrix, using the proposed technique, a dense layer of electropositive boehmite can be easily assembled on the surfaces of the electrospun membrane by simply dipping the membrane in a boehmite solution. Furthermore, the performance of the fabricated boehmite/SiO$_2$/PVDF membrane was evaluated and compared with that of electrospun membranes and Millipore GSWP filter. The water flux of the boehmite/SiO$_2$/PVDF membrane was twice that of the Millipore GSWP filter, which has a smaller pore size than the boehmite/SiO$_2$/PVDF membrane. The higher water flux of the boehmite/SiO$_2$/PVDF membrane was attributed to the improved interface interactions between the membrane and water upon the attachment of the hydrophilic boehmite. The boehmite/SiO$_2$/PVDF membrane showed a very high rejection rate comparable to that of Millipore GSWP. In existing surface treatment methods, pores on the treated surfaces are blocked, thereby lowering the filtration efficiency. The proposed technique overcomes this problem. The low water contact angle of the boehmite/SiO$_2$/PVDF increased the flow rate and reduced energy consumption. The boehmite/SiO$_2$/PVDF electropositive membrane also outperformed the commercial Millipore GSWP filters owing to the hydrophilicity of boehmite; thus, it is promising for applications in water purification and disinfection.

Supplementary Materials: The following supporting information can be downloaded at: https://www.mdpi.com/article/10.3390/polym15102270/s1, Figure S1. The attachment of boehmite onto the electrospun SiO$_2$/PVDF membrane.

Author Contributions: Formal analysis, C.H.L.; data curation, H.B.L.; writing—original draft preparation, D.C.; writing—review and editing, M.W.L.; supervision, S.M.J.; and funding acquisition, M.W.L. All authors have read and agreed to the published version of the manuscript.

Funding: This work was supported by the National Research Foundation of Korea (NRF) grant funded by the Korean Government (MSIT) (2020M3H4A3106354, 2020R1C1C1011817).

Institutional Review Board Statement: Not applicable.

Data Availability Statement: Not applicable.

Conflicts of Interest: The authors declare no conflict of interest.

References

1. Farahbakhsh, K.; Svrcek, C.; Guest, R.; Smith, D.W. A review of the impact of chemical pretreatment on low-pressure water treatment membranes. *J. Environ. Eng. Sci.* **2004**, *3*, 237–253. [CrossRef]
2. Wang, X.; Drew, C.; Lee, S.-H.; Senecal, K.J.; Kumar, J.; Samuelson, L.A. Electrospun nanofibrous membranes for highly sensitive optical sensors. *Nano Lett.* **2002**, *2*, 1273–1275. [CrossRef]
3. Wang, Y.; Hammes, F.; Düggelin, M.; Egli, T. Influence of size, shape, and flexibility on bacterial passage through micropore membrane filters. *Environ. Sci. Technol.* **2008**, *42*, 6749–6754. [CrossRef] [PubMed]
4. Mora, F.; Pérez, K.; Quezada, C.; Herrera, C.; Cassano, A.; Ruby-Figueroa, R. Impact of membrane pore size on the clarification performance of grape marc extract by microfiltration. *Membranes* **2019**, *9*, 146. [CrossRef]
5. Balamurugan, R.; Sundarrajan, S.; Ramakrishna, S. Recent trends in nanofibrous membranes and their suitability for air and water filtrations. *Membranes* **2011**, *1*, 232–248. [CrossRef] [PubMed]
6. Gopal, R.; Kaur, S.; Ma, Z.; Chan, C.; Ramakrishna, S.; Matsuura, T. Electrospun nanofibrous filtration membrane. *J. Membr. Sci.* **2006**, *281*, 581–586. [CrossRef]
7. Panteleev, A.; Aladushkin, S.; Kasatochkin, A.; Shilov, M.; Chudova, Y.V. Choice of filter material for a mechanical filter for purification of water after liming. *Therm. Eng.* **2020**, *67*, 492–495. [CrossRef]
8. Yang, C.; Wu, H.; Cai, M.; Li, Y.; Guo, C.; Han, Y.; Zhang, Y.; Song, B. Valorization of food waste digestate to ash and biochar composites for high performance adsorption of methylene blue. *J. Clean. Prod.* **2023**, *397*, 136612. [CrossRef]
9. El-Mehalmey, W.A.; Safwat, Y.; Bassyouni, M.; Alkordi, M.H. Strong interplay between polymer surface charge and MOF cage chemistry in mixed-matrix membrane for water treatment applications. *ACS Appl. Mater. Interfaces* **2020**, *12*, 27625–27631. [CrossRef]
10. Cao, X.-L.; Yan, Y.-N.; Zhou, F.-Y.; Sun, S.-P. Tailoring nanofiltration membranes for effective removing dye intermediates in complex dye-wastewater. *J. Membr. Sci.* **2020**, *595*, 117476. [CrossRef]
11. Bland, H.A.; Centeleghe, I.A.; Mandal, S.; Thomas, E.L.; Maillard, J.-Y.; Williams, O.A. Electropositive nanodiamond-coated quartz microfiber membranes for virus and dye filtration. *ACS Appl. Nano Mater.* **2021**, *4*, 3252–3261. [CrossRef]
12. Lv, Y.; Du, Y.; Chen, Z.-X.; Qiu, W.-Z.; Xu, Z.-K. Nanocomposite membranes of polydopamine/electropositive nanoparticles/polyethyleneimine for nanofiltration. *J. Membr. Sci.* **2018**, *545*, 99–106. [CrossRef]
13. Xiang, J.; Li, H.; Hei, Y.; Tian, G.; Zhang, L.; Cheng, P.; Zhang, J.; Tang, N. Preparation of highly permeable electropositive nanofiltration membranes using quaternized polyethyleneimine for dye wastewater treatment. *J. Water Process Eng.* **2022**, *48*, 102831.
14. Tepper, F. Novel nanofibre filter medium attracts waterborne pathogens. *Filtr. Sep.* **2002**, *39*, 16–19. [CrossRef]
15. Mandal, S.; Shaw, G.; Williams, O.A. Comparison of nanodiamond coated quartz filter with commercial electropositive filters: Zeta potential and dye retention study. *Carbon* **2022**, *199*, 439–443. [CrossRef]
16. Malaisamy, R.; Bruening, M.L. High-flux nanofiltration membranes prepared by adsorption of multilayer polyelectrolyte membranes on polymeric supports. *Langmuir* **2005**, *21*, 10587–10592. [CrossRef] [PubMed]
17. Malinowska, E.; Górski, Ł.; Meyerhoff, M.E. Zirconium(IV)-porphyrins as novel ionophores for fluoride-selective polymeric membrane electrodes. *Anal. Chim. Acta* **2002**, *468*, 133–141. [CrossRef]
18. Mazurkow, J.M.; Yüzbasi, N.S.; Domagala, K.W.; Pfeiffer, S.; Kata, D.; Graule, T. Nano-sized copper (oxide) on alumina granules for water filtration: Effect of copper oxidation state on virus removal performance. *Environ. Sci. Technol.* **2019**, *54*, 1214–1222. [CrossRef]
19. Ramasami, A.K.; Ravishankar, T.; Sureshkumar, K.; Reddy, M.; Chowdari, B.; Ramakrishnappa, T.; Balakrishna, G.R. Synthesis, exploration of energy storage and electrochemical sensing properties of hematite nanoparticles. *J. Alloys Compd.* **2016**, *671*, 552–559. [CrossRef]
20. Ikner, L.A.; Soto-Beltran, M.; Bright, K.R. New method using a positively charged microporous filter and ultrafiltration for concentration of viruses from tap water. *Appl. Environ. Microbiol.* **2011**, *77*, 3500–3506. [CrossRef]
21. Tepper, F.; Kaledin, L. A high-performance electropositive filter. *BioProcess Int.* **2006**, *4*, 64–68.
22. Nagai, N.; Mizukami, F. Properties of boehmite and Al_2O_3 thin films prepared from boehmite nanofibres. *J. Mater. Chem.* **2011**, *21*, 14884–14889. [CrossRef]

23. Karim, M.R.; Rhodes, E.R.; Brinkman, N.; Wymer, L.; Fout, G.S. New electropositive filter for concentrating enteroviruses and noroviruses from large volumes of water. *Appl. Environ. Microbiol.* **2009**, *75*, 2393–2399. [CrossRef] [PubMed]
24. Miao, Y.-E.; Wang, R.; Chen, D.; Liu, Z.; Liu, T. Electrospun self-standing membrane of hierarchical $SiO_2@\gamma$-AlOOH (Boehmite) core/sheath fibers for water remediation. *ACS Appl. Mater. Interfaces* **2012**, *4*, 5353–5359. [CrossRef] [PubMed]
25. Cashdollar, J.L.; Dahling, D.R. Evaluation of a method to re-use electropositive cartridge filters for concentrating viruses from tap and river water. *J. Virol. Methods* **2006**, *132*, 13–17. [CrossRef]
26. Ding, W.; Tong, Y.; Shi, L.; Li, W. Superhydrophilic PVDF Membrane Modified by Norepinephrine/Acrylic Acid via Self-Assembly for Efficient Separation of an Oil-in-Water Emulsion. *Ind. Eng. Chem. Res.* **2021**, *61*, 130–140. [CrossRef]
27. Luo, L.; Gao, Z.; Zheng, Z.; Zhang, J. "Polymer-in-Ceramic" Membrane for Thermally Safe Separator Applications. *ACS Omega* **2022**, *7*, 35727–35734. [CrossRef] [PubMed]
28. Liang, S.; Kang, Y.; Tiraferri, A.; Giannelis, E.P.; Huang, X.; Elimelech, M. Highly hydrophilic polyvinylidene fluoride (PVDF) ultrafiltration membranes via postfabrication grafting of surface-tailored silica nanoparticles. *ACS Appl. Mater. Interfaces* **2013**, *5*, 6694–6703. [CrossRef] [PubMed]
29. Jones, D.; Jiménez-Jiménez, J.; Jiménez-López, A.; Maireles-Torres, P.; Olivera-Pastor, P.; Rodriguez-Castellón, E.; Rozière, J. Surface characterisation of zirconium-doped mesoporous silica. *Chem. Commun.* **1997**, *5*, 431–432. [CrossRef]
30. Kumar, A.; Malik, A.K.; Tewary, D.K.; Singh, B. A review on development of solid phase microextraction fibers by sol–gel methods and their applications. *Anal. Chim. Acta* **2008**, *610*, 1–14. [CrossRef]
31. Cinibulk, M.K. Deposition of oxide coatings on fiber cloths by electrostatic attraction. *J. Am. Ceram. Soc.* **1997**, *80*, 453–460. [CrossRef]
32. Klotz, K.; Weistenhöfer, W.; Neff, F.; Hartwig, A.; van Thriel, C.; Drexler, H. The health effects of aluminum exposure. *Dtsch. Ärzteblatt Int.* **2017**, *114*, 653. [CrossRef] [PubMed]
33. Drew, C.; Liu, X.; Ziegler, D.; Wang, X.; Bruno, F.F.; Whitten, J.; Samuelson, L.A.; Kumar, J. Metal oxide-coated polymer nanofibers. *Nano Lett.* **2003**, *3*, 143–147. [CrossRef]
34. Abdal-hay, A.; Barakat, N.A.; Lim, J.K. Influence of electrospinning and dip-coating techniques on the degradation and cytocompatibility of Mg-based alloy. *Colloids Surf. A Physicochem. Eng. Asp.* **2013**, *420*, 37–45. [CrossRef]
35. Shin, Y.; Hohman, M.; Brenner, M.P.; Rutledge, G. Electrospinning: A whipping fluid jet generates submicron polymer fibers. *Appl. Phys. Lett.* **2001**, *78*, 1149–1151. [CrossRef]
36. Taylor, G.I. Disintegration of water drops in an electric field. *Proc. R. Soc. Lond. Ser. A Math. Phys. Sci.* **1964**, *280*, 383–397. [CrossRef]
37. Reneker, D.H.; Chun, I. Nanometre diameter fibres of polymer, produced by electrospinning. *Nanotechnology* **1996**, *7*, 216. [CrossRef]
38. Beamson, G.; Briggs, D. *High Resolution XPS of Organic Polymers*; Wiley: New York, NY, USA, 1992.
39. Wang, X.; Chen, X.; Yoon, K.; Fang, D.; Hsiao, B.S.; Chu, B. High flux filtration medium based on nanofibrous substrate with hydrophilic nanocomposite coating. *Environ. Sci. Technol.* **2005**, *39*, 7684–7691. [CrossRef]
40. Hester, J.; Banerjee, P.; Mayes, A. Preparation of protein-resistant surfaces on poly(vinylidene fluoride) membranes via surface segregation. *Macromolecules* **1999**, *32*, 1643–1650. [CrossRef]
41. Rana, D.; Matsuura, T. Surface modifications for antifouling membranes. *Chem. Rev.* **2010**, *110*, 2448–2471. [CrossRef]

Disclaimer/Publisher's Note: The statements, opinions and data contained in all publications are solely those of the individual author(s) and contributor(s) and not of MDPI and/or the editor(s). MDPI and/or the editor(s) disclaim responsibility for any injury to people or property resulting from any ideas, methods, instructions or products referred to in the content.

Article

Optimization of Lead and Diclofenac Removal from Aqueous Media Using a Composite Sorbent of Silica Core and Polyelectrolyte Coacervate Shell

Irina Morosanu [1], Florin Bucatariu [1,2], Daniela Fighir [1], Carmen Paduraru [1], Marcela Mihai [1,2,*] and Carmen Teodosiu [1,*]

1 Department of Environmental Engineering and Management, Gheorghe Asachi Technical University of Iasi, 73 D. Mangeron Street, 700050 Iasi, Romania; morosanu.irina@gmail.com (I.M.); fbucatariu@icmpp.ro (F.B.); daniela.arsene@ch.tuiasi.ro (D.F.); cpadur2005@yahoo.com (C.P.)
2 Petru Poni Institute of Macromolecular Chemistry, 41A Grigore Ghica Voda Alley, 700487 Iasi, Romania
* Correspondence: marcela.mihai@icmpp.ro (M.M.); cteo@ch.tuiasi.ro (C.T.)

Citation: Morosanu, I.; Bucatariu, F.; Fighir, D.; Paduraru, C.; Mihai, M.; Teodosiu, C. Optimization of Lead and Diclofenac Removal from Aqueous Media Using a Composite Sorbent of Silica Core and Polyelectrolyte Coacervate Shell. *Polymers* **2023**, *15*, 1948. https://doi.org/10.3390/polym15081948

Academic Editor: George Z. Kyzas

Received: 16 March 2023
Revised: 17 April 2023
Accepted: 17 April 2023
Published: 19 April 2023

Copyright: © 2023 by the authors. Licensee MDPI, Basel, Switzerland. This article is an open access article distributed under the terms and conditions of the Creative Commons Attribution (CC BY) license (https:// creativecommons.org/licenses/by/ 4.0/).

Abstract: The modification of inorganic surfaces with weak cationic polyelectrolytes by direct deposition through precipitation is a fast approach to generating composites with high numbers of functional groups. The core/shell composites present very good sorption capacity for heavy metal ions and negatively charged organic molecules from aqueous media. The sorbed amount of lead ions, used as a model for priority pollutants such as heavy metals, and diclofenac sodium salt, as an organic contaminant model for emerging pollutants, depended strongly on the organic content of the composite and less on the nature of contaminants, due to the different retention mechanisms (complexation vs. electrostatics/hydrophobics). Two experimental approaches were considered: (i) simultaneous adsorption of the two pollutants from a binary mixture and (ii) the sequential retention of each pollutant from monocomponent solutions. The simultaneous adsorption also considered process optimization by using the central composite design methodology to study the univariate effects of contact time and initial solution acidity with the purpose of enabling further practical applications in water/wastewater treatment. Sorbent regeneration after multiple sorption-desorption cycles was also investigated to assess its feasibility. Based on different non-linear regressions, the fitting of four isotherms (Langmuir, Freundlich, Hill, and Redlich–Peterson models) and three kinetics models (pseudo-first order (PFO), pseudo-second order (PSO), and two-compartment first order (TC)) has been carried out. The best agreement with experiments was found for the Langmuir isotherm and the PFO kinetic model. Silica/polyelectrolytes with a high number of functional groups may be considered efficient and versatile sorbents that can be used in wastewater treatment processes.

Keywords: silica; poly(ethyleneimine); diclofenac sodium salt; lead; sorption; optimization

1. Introduction

Different statistical experimental design methods, e.g., the central composite design or the Box–Behnken design, were applied for investigating the simultaneous effect of several operating conditions on the adsorption efficiency with a minimum number of experiments [1,2]. Hiew et al. [3] studied the kinetics, equilibrium, and thermodynamics of sodium diclofenac (DCF-Na) sorption onto graphene oxide. The single and interaction effects of different parameters, such as sorbent dosage, contact time, initial concentration, and temperature, were analyzed by central composite design (CCD). Masoumi et al. [4] reported four process factors, namely temperature, pH, and initial concentrations of lead, nickel, and cadmium, respectively, in which the heavy metals adsorption was optimized by the CCD method of the response surface methodology (RSM). Optimization of DCF-Na sorption on eucalyptus wood biochar was carried out using the Box–Behnken design, considering the following variables: stirring rate, drug concentration, and sorbent dosage [5]. The interaction impact of various parameters, i.e., DCF-Na concentration, initial pH, sorbent dose, temperature, and contact time, was evaluated by the CCD of RSM for the prediction of the response and optimization of DCF-Na sorption on graphene oxide decorated with a zeolitic imidazolate framework [6]. A central composite design was preferred among other RSM techniques for the optimization of DCF-Na sorption on a quaternized mesoporous silica, considering initial pH, sorbent dosage, reaction time, and initial concentration [7].

In our previous study [8], the composite sorbent obtained by direct deposition on an inorganic silica core (IS) of a coacervate based on poly(ethyleneimine) (PEI)/poly(acrylic acid) (PAA) and PEI/poly(sodium methacrylate) (PMAA), which underwent a strong crosslinking, presented the highest sorption capacity for Cd^{2+} ions as compared to layer-by-layer polyelectrolyte composite sorbents. This paper focused on the potential of the same composite sorbent, obtained by direct deposition of a coacervate, to remove lead ions (Pb^{2+}) and DCF-Na from aqueous solutions. Two main approaches were considered: (i) simultaneous adsorption of the two pollutants from a binary mixture and (ii) the alternative retention of each pollutant from monocomponent solutions. The first approach considered the optimization of the sorption process and the univariate effects of contact time and initial solution acidity. In this sense, a popular response surface experimental design based on CCD was used to model the simultaneous adsorption of Pb^{2+} ions and DCF-Na from aqueous solutions. The novelty of this paper lies in reporting the use of the CCD methodology for the optimization of the operating conditions for the simultaneous adsorption of these two priority pollutants. It is of high interest to determine the impact of different ratios between the two pollutants in solution by experimental design and to optimize the operational process conditions. Therefore, the experimental design was composed of twenty batch experimental runs, which were used to evaluate the influence of important operating parameters and their interactions upon the bicomponent adsorption system. Sorbent regeneration after multiple sorption-desorption cycles was also investigated to assess the feasibility of the sorbent. The second approach is sequential adsorption, with the focus on the possibility of functionalizing the sorbent with metal ions for improved DCF-Na sorption.

2. Materials and Methodology

2.1. Materials

All chemical reagents, i.e., lead nitrate (Fluka, Buchs, Switzerland), diclofenac sodium salt (98%, Acros Organics, Geel, Belgium), methanol Uvasol (Merck, Bucuresti, Romania), hydrochloric acid (Merck, Bucuresti, Romania), sodium hydroxide (Merck, Bucuresti, Romania), and ethylenediaminetetraacetic acid disodium salt (EDTA, Merck, Bucuresti, Romania), were of analytical purity and used as received. Silica particles of 40–60 microns in diameter were purchased from Daiso Co. (Osaka, Japan). Poly(sodium methacrylate) (M_w = 1800 g/mol) and branched poly(ethyleneimine) (M_w = 25,000 g/mol) were acquired from Aldrich (Redox, Otopeni, Romania) and used as is.

IS/(PEI/PMAA)$_c$ composite sorbent was prepared as described in our previous study [8]. Briefly, 5.0 g of silica microparticles (spherical shape) were first dispersed into 1 mol L^{-1} PEI (12 mL), and then the addition of 1 mol L^{-1} PMAA (6 mL) was made dropwise under vigorous stirring with a glass rod until a transparent solution was obtained. Then, PEI crosslinking with glutaraldehyde (2.5% w/v) was performed, obtaining a crosslinking ratio of [−CHO]:[−NH$_2$] = 1:1. Finally, the core/shell composite microparticles were treated with 1 mol L^{-1} of NaOH for PMAA extraction.

2.2. Batch Sorption and Desorption Studies

The preliminary investigation consisted of verifying the influence of solution acidity (4 < pH < 6) and contact time (1–8 h) on the simultaneous sorption of Pb^{2+} ions and DCF-Na on the IS/(PEI/PMAA)$_c$ composite sorbent. The pH of the solution was adjusted with 0.01 N HNO$_3$. A sorption equilibrium was considered to be attained when the sorption capacities of two successive samples did not vary significantly with time. The effect of the pollutants initial concentrations was studied considering the concentration ranges from the experimental design. All univariate experiments were conducted utilizing 2 g of composite per liter of aqueous solution, at room temperature.

The DCF-Na concentration was determined by a direct spectrophotometric method at a λ_{max} = 276 nm using Analytik Jena Specord 210 Plus (Analytik-Jena, Bucuresti, Romania) equipment. The concentration of Pb^{2+} ions in water solutions was quantified using an Analytik Jena 800 atomic absorption spectrometer (Analytik-Jena, Bucuresti, Romania). The calibration curves were validated by a certified reference material, namely diclofenac sodium 1 mg/mL in methanol from LGC and a multi-element metal standard solution from Merck. In addition, the analysis methods were verified for any interference from other components of the binary mixture. For a lead concentration of up to 71 mg/L, no interference could be detected on the DCF-Na determination. Additionally, no interference was observed for DCF-Na up to 65 mg/L in the analysis of lead ions.

The mass of Pb^{2+} ions or DCF-Na retained on IS/(PEI/PMAA)$_c$ (r = 1.0) was evaluated from the mass balance of the pollutant, which is represented by the following equation:

$$q_i = (C_0 - C_i)V/m \qquad (1)$$

where q_i (mg/g polymer) is the sorption capacity at a certain time t (q_t) or at equilibrium (q_e), C_0 (mg/L) is the initial concentration, C_i (mg/L) is the concentration at a certain time t (C_t) or at equilibrium (C_e), m (g) is the sorbent amount, and V (L) is the volume of the solution. The pollutant removal efficiency (RE, %) was calculated as follows:

$$RE = (C_0 - C_i)\,100/C_0. \qquad (2)$$

The regeneration of the composite sorbent was achieved by repeatedly washing the material with EDTA (0.1 mol/L) and HCl (1 mol/L), followed by activation with NaOH (1 mol/L) [8].

The sequential sorption tests were carried out by contacting the composite sorbent with one mono-element solution of each of the pollutants at a time [9] in order to investigate the possibility of sorbent functionalization and sorption efficiency.

2.3. Adsorption Kinetics and Isotherm Models

The sorption data obtained at different contact times have been described by pseudo-first order, pseudo-second order, two-compartment first-order kinetics models, as well as a liquid film diffusion equation. The sorption process was investigated through the following equilibrium isotherms: Langmuir, Freundlich, Hill, and Redlich–Peterson models.

The following objective functions were minimized in turn, and the optimum model parameters were chosen by the minimum sum of normalized errors:

$$\text{sum of the square error (SSE)}, \sum_{j=1}^{x}\left(q_{j,exp} - q_{j,model}\right)^2, \qquad (3)$$

$$\text{chi} - \text{square test } (\chi^2), \frac{\sum_{j=1}^{x}\left(q_{j,exp} - q_{j,model}\right)^2}{q_{i,exp}}, \qquad (4)$$

$$\text{average relative error (ARE)}, \frac{100}{x}\sum_{j=1}^{x}\left|\frac{q_{j,exp} - q_{j,model}}{q_{j,exp}}\right|. \qquad (5)$$

The best-fit model was selected based on the smallest value of the Hannan–Quinn information criterion [10].

2.4. Design of Experiments

The influence of the initial priority pollutant concentrations and the adsorption temperature were studied using the central composite design along with the coded independent variables shown in Table 1. The experimental design was composed of six center points, six axial points, and eight factorial points, which were obtained using Design Expert software, version 10.0.1 (DX10, Stat-Ease, US). The design matrix is indicated in Table 2, along with the two response functions considered—the sorption capacity for each of the priority pollutants involved and their predicted values.

Table 1. Experimental design levels of factors for Pb^{2+} ions and DCF-Na simultaneous adsorption on $IS/(PEI/PMAA)_c$ (r = 1.0).

Factor	Units	Level				
		−2	−1	0	+1	+2
A: $C_{0\ Pb}$	mg/L	5	15	25	35	45
B: $C_{0\ DCF-Na}$	mg/L	6	14	22	30	38
C: T	°C	4	12	20	28	36

Table 2. CCD matrix with experimental (*exp*) and predicted (*pred*) responses.

Run	A	B	C	Y1: $q_{exp,\ Pb}$ (mg/g Polymer)	$q_{pred,\ Pb}$ (mg/g Polymer)	Residual ($q_{exp,\ Pb}$ − $q_{pred,\ Pb}$)	Y2: $q_{exp,\ DCF-Na}$ (mg/g Polymer)	$q_{pred,\ DCF-Na}$ (mg/g Polymer)	Residual ($q_{exp,\ DCF-Na}$ − $q_{pred,\ DCF-Na}$)
1	35	30	12	104.39	104.71	−0.32	81.80	79.07	2.73
2	25	22	20	78.98	77.95	1.02	58.44	57.20	1.24
3	45	22	20	123.27	122.53	0.74	61.27	60.90	0.37
4	35	14	12	102.23	102.57	−0.34	39.60	37.81	1.79
5	25	22	20	77.35	77.95	−0.60	57.36	57.20	0.16
6	25	22	4	78.98	79.07	−0.084	52.46	55.98	−3.52
7	5	22	20	15.55	14.80	0.76	56.37	53.50	2.87
8	25	22	20	76.82	77.95	−1.13	55.37	57.20	−1.83
9	35	14	28	98.81	100.01	−1.20	39.02	39.03	−0.011
10	25	22	36	79.08	77.49	1.59	59.67	58.42	1.26
11	15	30	12	47.58	47.88	−0.30	76.81	75.37	1.44
12	15	14	28	47.92	49.10	−1.18	32.58	35.33	−2.75
13	35	30	28	101.12	102.25	−1.13	79.25	80.29	−1.03
14	25	22	20	78.81	77.95	0.85	56.50	57.20	−0.70
15	25	38	20	78.57	77.86	0.71	95.33	98.45	−3.13
16	25	22	20	78.30	77.95	0.34	57.70	57.20	0.50
17	15	14	12	47.84	48.21	−0.37	33.91	34.11	−0.21
18	25	22	20	78.98	77.95	1.02	59.00	57.20	1.80
19	25	6	20	76.74	75.95	0.79	14.97	15.95	−0.97
20	15	30	28	47.71	48.87	−1.17	76.58	76.59	-9.537×10^{-3}

The behavior of the simultaneous adsorption of Pb^{2+} ions and DCF-Na on the $IS/(PEI/PMAA)_c$ (r = 1.0) composite sorbent was mathematically modeled by a second-order polynomial equation (Equation (6)), which considers the interaction between the independent process variables:

$$q_{pred} = \beta_0 + \sum_{i=1}^{n} \beta_i x_i + \sum_{i=1}^{n} \beta_{ii} x_i^2 + \sum_{i=1}^{n-1} \sum_{j=i+1}^{n} \beta_{ij} x_i x_j + \varepsilon, \qquad (6)$$

where q_{pred} is the predicted sorption capacity (mg/g polymer), x_i, x_i^2, and $x_i x_j$ are the individual variables, the quadratic effects, and the interactions between variables, respectively; β_0 is a constant; β_i, β_{ii}, and β_{ij} are the linear, quadratic, and interaction coefficients, respectively; and ε is the random error. The total number of variables n = 3.

The regressed models were optimized by the analysis of variance (ANOVA) with the DX10 software. ANOVA indicated the statistical significance and influence of the independent model parameters and their interactions. For this purpose, a probability value (p-value), the Fischer's test value (F-value) at a 95% confidence level, the coefficient of determination (R^2), adjusted R^2 (R^2_{adj}), and predicted R^2 (R^2_{pred}), as well as normalized residue plots, were used.

3. Results and Discussion

3.1. Univariate Simultaneous Experiments

The influence of solution acidity on the efficiency of the simultaneous sorption process is illustrated in Figure 1. The experiments were carried out at an initial concentration of 47 mg of Pb^{2+}/L and 30 mg of DCF-Na/L. The sorption capacity of both pollutants increases with an increase in the initial pH. The highest values of the sorption capacity were obtained at a pH of 5, reaching a q_{Pb} = 138 mg/g polymer and a $q_{DCF\text{-}Na}$ = 87.6 mg/g polymer. Hence, the optimum pH was considered 5, and the subsequent tests were performed using this value.

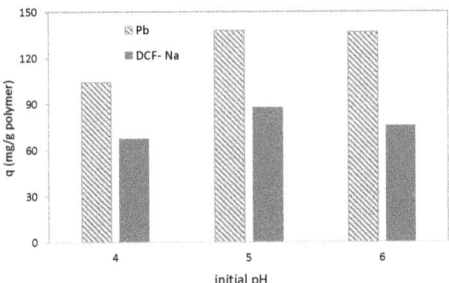

Figure 1. Effect of solution acidity on the simultaneous sorption process.

At pH = 5, near the pK_a of DCF-Na (~4.2) and the point of zero charge of the sorbent (~4.3), the DCF-Na molecules with the carboxylic groups interacted electrostatically with protonated amino groups inside the composite shell. Moreover, π-stacking, hydrogen bonds, and hydrophobics could favor DCF-Na retention as possible secondary forces inside the shell. The Pb^{2+} ions interacted, inside the cross-linked shell, with the amino groups of the PEI by forming coordinative bonds. Due to different types of pollutant entity interactions, coordinative for Pb^{2+} and electrostatic/H-bonds/hydrophobics for DCF-Na, we can conclude that the sorption could be additive at the same sorption site inside the cross-linked organic part of the composite. Thus, the DCF-Na sorbed amount is not influenced drastically by the Pb^{2+} sorbed amount. DCF-Na can form a complex with lead ions (molar ratio 2:1) [11] and further interact with Pb^{2+} as a counter ion, replacing nitrate ions.

The variation with time of the sorption capacity for the two pollutants, illustrated in Figure 2, shows that the sorption process occurs in two stages. The first stage is characterized by the rapid migration of the pollutants from the liquid phase to the sorbent's surface. For lead ions, this happens in the first hour of contact between the solid and liquid phases, when a sorption capacity of 103.3 mg/g polymer is achieved. After 3 h, the increment in the sorption capacity is small because the process advances slowly. For the organic pollutant, the initial adsorption stage takes place after around 2 h and after 6 h, equilibrium is reached. Due to the slower sorption kinetics of the sodium diclofenac, the optimum contact time between the binary mixture and IS/(PEI/PMAA)$_c$ (r = 1.0) was considered to be 6 h. Hence, all experiments, including the CCD tests, were performed within this time interval.

The kinetics parameters for the pseudo-first order (PFO) [12], pseudo-second order (PSO), and two-compartment first-order (TC) [13] models are presented in Table 3. The PFO equation assumes that the sorption rate is proportional to the distance from equilibrium, whereas the PSO equation describes that the rate is proportional to the second power of the distance from equilibrium [14].

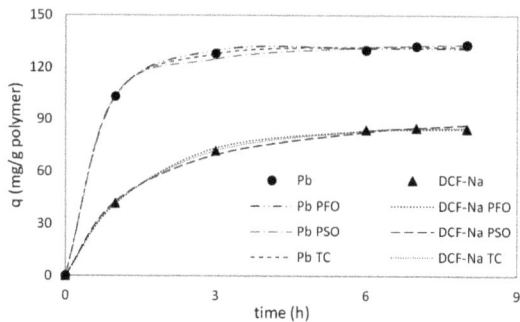

Figure 2. The influence of contact time on the sorption capacity of IS/(PEI/PMAA)$_c$ (r = 1.0) on the lead ions and diclofenac sodium from a bi-component mixture.

Table 3. Kinetics parameters for Pb^{2+} ions and DCF-Na simultaneous sorption on IS/(PEI/PMAA)$_c$ (r = 1.0).

		Pb^{2+} Ions	DCF-Na
Pseudo-first order $q_t = q_{e,1}(1 - \exp(-k_1 t))$	q_e (mg/g)	131.27	85.18
	k_1 (min^{-1})	0.02562	0.01112
	R^2	0.984	0.997
	SSE	9.82	3.72
	χ^2	0.08	0.05
	ARE	0.97	0.96
	HQC	5.28	0.43
Pseudo-second order $q_t = \frac{k_2 q_{e,2}^2 t}{1 + k_2 q_{e,2} t}$	q_e (mg/g)	139.06	102.06
	k_1 (min^{-1})	0.000356	0.000118
	R^2	0.980	0.991
	SSE	12.71	12.52
	χ^2	0.10	0.17
	ARE	0.96	1.89
	HQC	6.57	6.49
Two-compartment $q_t = q_e F_{fast}\left(1 - e^{-tk_{fast}}\right) + q_e F_{slow}\left(1 - e^{-tk_{slow}}\right)$, $F_{fast} + F_{slow} = 1$	q_e (mg/g)	132.02	86.16
	F_{fast}	0.425	0.084
	k_{fast} (min^{-1})	0.247	0.212
	F_{slow}	0.575	0.916
	k_{slow} (min^{-1})	0.016	0.010
	R^2	0.993	0.999
	SSE	4.64	0.86
	χ^2	0.04	0.01
	ARE	0.53	0.41
	HQC	4.39	−4.06
Film diffusion $\ln\left(1 - \frac{q_t}{q_e}\right) = -k_{fd} t$	k_{fd} (min^{-1})	0.0083	0.0126
	R^2	0.899	0.996

Analyzing the results for the kinetics of lead ions sorption displayed in Table 1, it can be observed that the error functions of the PFO and PSO models show close values for R^2, χ^2, and *ARE*, but lesser values for *SSE*, and hence, for the *HQ* information criterion. In the case of DCF-Na sorption kinetics, this is noticeable for PFO and the two-compartment model. According to Azizian [15], a high initial concentration of solute determines a better fit with the pseudo-first-order kinetics, while the sorption kinetics follows the PSO model better when the initial concentration is not too high. The experimental equilibrium sorption capacity was 133.2 mg/g for the Pb^{2+} ions and 84.7 mg/g for the DCF-Na. All models estimated $q_{e,\,Pb}$ values close to the experimental ones, but the estimation of $q_{e,\,DCF\text{-}Na}$ for the PSO model was 20% higher than the experimental one. It was reported that sorption kinetics often follow the PFO equation when the process occurs through diffusion through an interface [16].

Based on the smaller values for the HQ criterion, the degree of goodness-of-fit decreases in the following order: two-compartment, PFO, and then PSO for both pollutants. The TC model assumes that the sorption process takes place as a two-phase process [13]. The TC equation provided the best fit with the experimental data, giving the best estimations of the experimental sorption capacities on IS/(PEI/PMAA)$_c$ (r = 1.0). The constants mass fraction F_{fast} and first-order rate k_{fast}, corresponding to the rapid initial sorption stage, are larger than those of the slow compartment, suggesting that the fast sorption stage predominated during the pollutants uptake by the composite sorbent.

The diffusion mechanism was analyzed by applying the film diffusion mass transfer rate model [17]. A linear plot of $-\ln(1 - q_t/q_e)$ vs. t with an intercept of zero indicates that the sorption kinetics are governed by the diffusion through the liquid film surrounding the sorbent microparticles. The plots depicted in Figure 3 show a deviation from linearity for lead ion sorption, while the curve for DCF-Na presents very good linearity, with an intercept very close to zero. When the line for DCF-Na is forced to go through the origin, the coefficient R^2 becomes 0.9925. Hence, it can be considered that the curve for DCF-Na meets the conditions for the film diffusion to be rate-controlled. On the other hand, diffusion through the liquid film is not the main determining step in the kinetics of lead ion uptake by IS/(PEI/PMAA)$_c$ (r = 1.0). The sorption isotherms of lead (Figure 4a) and diclofenac sodium (Figure 4b) are presented in Figure 4.

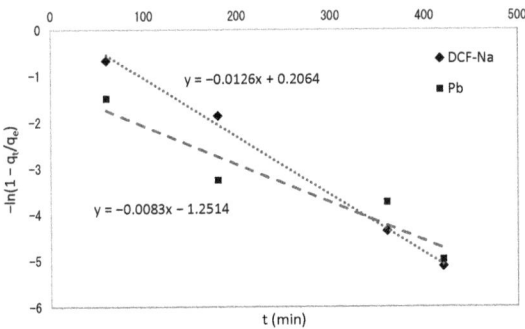

Figure 3. Film diffusion plot for the simultaneous sorption process.

Relevant parameters of the four isotherm equations, i.e., Langmuir [18], Freundlich [19], Hill [20], and Redlich–Peterson [21], are presented in Table 4. According to the HQ information criterion, the Langmuir model describes best the uptake of the pollutants on the composite sorbent, followed by Redlich–Peterson, Hill, and Freundlich for Pb^{2+} ion sorption and by Hill, Freundlich, and Redlich–Peterson for DCF-Na sorption.

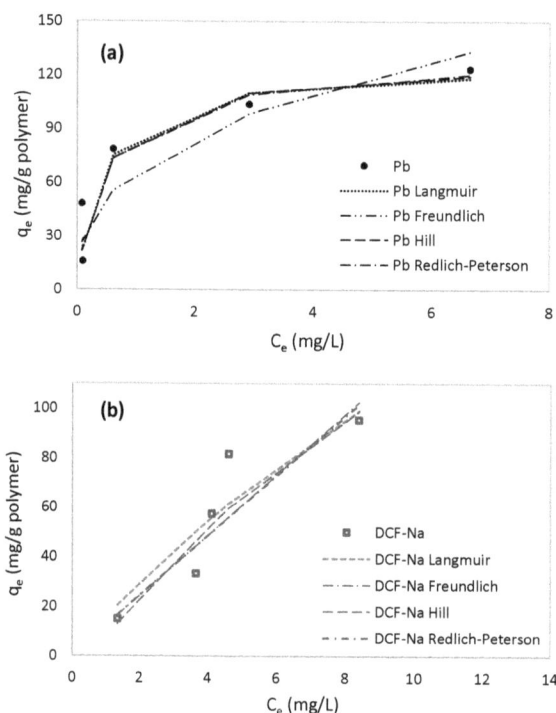

Figure 4. Isotherms of lead (**a**) and diclofenac sodium (**b**) for simultaneous sorption process.

The Langmuir model describes the equilibrium condition of a monolayer coverage of a homogeneous surface having identical adsorption sites [18]. Saturation sorption capacities of 125.68 mg/g polymer and 388.35 mg/g polymer were determined for Pb^{2+} and DCF-Na, respectively. A larger K_L equilibrium coefficient for lead ions suggests a stronger affinity of the sorbent towards this contaminant. In addition, the Langmuir isotherm and HQC indicate that the Redlich–Peterson and Hill models also describe the Pb^{2+} ion equilibrium data fairly well. The Redlich–Peterson model is an empirical isotherm, incorporating both the Langmuir and Freundlich models. As such, it presents a constant (K_{RP}) quantifying the linear affinity of concentration in the numerator and an exponent in the denominator to explain sorption saturation over a wide range of concentrations, providing the opportunity to apply the model to homogeneous and heterogeneous sorption processes [4]. The Hill isotherm model can explain the sorption of contaminants as a cooperative binding process, where the pollutant species bind to the active sites on a homogeneous substrate, producing an effect on the binding at other sites in the vicinity [22]. As noticed in Table 4, both isotherm exponents, β and n_H, are close to unity and thus reduce the models to the Langmuir equation. In this case, the closeness of the maximum sorption capacity between the three models is verified: for Langmuir—125.68 mg/g polymer; for Hill—129.11 mg/g polymer; and for Redlich–Peterson—$q_m \approx K_{RP}/\alpha$ = 118.51 mg/g polymer.

In the case of DCF-Na sorption on the composite surface, the highest R^2 was 0.837 for the Langmuir isotherm, indicating that at most 83.7% of the data can be explained by this model. The maximum sorption capacity indicated by this model is 388.35 mg/g polymer, which is higher than the DCF-Na individual sorption of 128.01 mg/g polymer (Hill model, R^2 = 0.996, data not shown). This could be explained by the overestimation of the Langmuir isotherm in the case of simultaneous sorption due to the enhancement of DCF-Na sorption by Pb^{2+}. This fact could be attributed to the ionic exchange of the NO_3^- ion being replaced by the carboxylate group of ionized DCF-Na.

Table 4. Isotherm parameters for the adsorption of DCF-Na and Pb^{2+} ions.

		Pb^{2+} Ions	DCF-Na
Langmuir $q_e = \frac{q_{m,L} K_L C_e}{1+K_L C_e}$	$q_{m,L}$ (mg/g)	125.68	388.35
	K_L (L/mg)	2.36	0.04
	R^2	0.907	0.837
	SSE	868.38	739.89
	χ^2	20.31	15.79
	ARE	24.90	23.69
	HQC	27.69	26.89
Freundlich $q_e = K_F C_e^{\frac{1}{n_F}}$	n_F	2.69	0.99
	K_L (mg/g)(mg/L)$^{1/n_F}$	65.84	12.05
	R^2	0.871	0.810
	SSE	1262.84	878.98
	χ^2	27.29	13.25
	ARE	33.39	18.84
	HQC	29.56	27.75
Hill $q_e = \frac{q_{m,H} C_e^{n_H}}{K_S + C_e^{n_H}}$	$q_{m,H}$ (mg/g)	129.11	177.18
	K_H	0.49	20.05
	n_H	0.94	1.52
	R^2	0.909	0.849
	SSE	849.74	692.16
	χ^2	20.25	11.86
	ARE	25.39	18.16
	HQC	28.53	27.51
Redlich–Peterson $q_e = \frac{K_{RP} C_e}{1+\alpha C_e^\beta}$	K_{RP} (L/g)	314.28	13.89
	α (L/mg)$^\beta$	2.652	0.135
	β	0.96	0.04
	R^2	0.911	0.811
	SSE	839.61	873.70
	χ^2	20.22	13.26
	ARE	25.20	18.99
	HQC	28.47	28.67

A comparison of the maximum sorption capacity of the IS/(PEI/PMAA)$_c$ (r = 1.0) sorbent with other materials reported for the sorption of lead ions and/or diclofenac sodium is presented in Table 5.

Table 5. Literature comparison of sorbents for Pb^{2+} ions and DCF-Na removal.

Sorbent	Sorption Solution	Maximum Sorption Capacity (mg/g) for Pb	Isotherm	Maximum Sorption Capacity (mg/g) for DCF-Na	Isotherm	Reference
Eucalyptus wood biochar	monocomponent	-	-	33.13	Sips	[5]

Table 5. *Cont.*

Sorbent	Sorption Solution	Maximum Sorption Capacity (mg/g) for Pb	Isotherm	Maximum Sorption Capacity (mg/g) for DCF-Na	Isotherm	Reference
Hyper-cross-linked polymer (HCP)	multicomponent: Pb, Ni, and Cd	173.10	RSM optimization	-	-	[4]
Magnetic GO/ ZIF-8/γ-AlOOH	monocomponent	-	-	2594.3	Langmuir	[6]
Magnetic amine-functionalized chitosan	monocomponent	-	-	469.48	Langmuir	[23]
Peanut hull-g-methyl methacrylate biopolymer	monocomponent	370.40	Langmuir	-	-	[24]
Activated carbon from cocoa pod husk	monocomponent	-	-	5.67	Experimental	[25]
Activated carbon	monocomponent	-	-	180	Langmuir	[26]
Cross-linked chitosan beads grafted with polyethylenimine	monocomponent	-	-	253.32	Langmuir	[27]
Polyethyleneimine (PEI)-modified chitosan magnetic hydrogel	multicomponent: Pb, Ni, and Cu	100.32	Langmuir	-	-	[28]
Magnetic iron oxide-silica shell nanocomposite	monocomponent	17.1	Langmuir	-	-	[29]
Graphene oxide-silica-chitosan adsorbent	monocomponent	256.41	Langmuir	-	-	[30]
Ordered mesoporous silica nanoparticles of MCM-41 type	monocomponent	22.2	Sips	-	-	[31]
Chemically modified silica monolith	monocomponent	574.71	Langmuir	-	-	[32]
IS/(PEI/PMAA)$_c$ (r = 1.0)	multicomponent: Pb and DCF-Na	125.68	Langmuir	388.35	Langmuir	This study

3.2. Multivariate/Multi-Objective Simultaneous Process Modeling and Optimization

3.2.1. Statistical Analysis of the Process Model

The experimental data from Table 2 were subjected to multiple regression analysis to fit the quadratic model (Equation (6)) for the two responses: Y_1—the sorption capacity of lead ions (q_{Pb}) and Y_2—the sorption capacity of diclofenac sodium (q_{DCF-Na}). The mathematical expressions, in actual units, are given below:

$$q_{Pb} = -6.65936 + 3.90003 C_{0,Pb} + 0.038835 C_{0,DCF-Na} + 0.16015T + 0.007723 C_{0,Pb} C_{0,DCF-Na} - 0.010766 C_{0,Pb} T + 0.0004017 C_{0,DCF-Na} T - 0.023225 C_{0,Pb}^2 - 0.0040925 C_{0,DCF-Na}^2 + 0.0012756 T^2, \quad (7)$$

$$q_{DCF-Na} = -5.67172 + 0.18502 C_{0,Pb} + 2.57842 C_{0,DCF-Na} + 0.076041T. \quad (8)$$

The goodness-of-fit and lack-of-fit test results obtained by ANOVA are presented in Table 6. A second-order polynomial model was developed to predict the influence of parameters on lead ion sorption on the IS/(PEI/PMAA)$_c$ (r = 1.0) composite sorbent, while a linear model was obtained for diclofenac sodium sorption. In the latter, no interaction between the parameters could be observed. As seen in Table 6, both models are statistically significant based on a determined p-value of less than 0.0001 and high values for the F-value. The linear and quadratic effects of the initial concentration of metallic ions were significant on the sorption behavior of lead ions (p-value < 0.05). The sorption capacity of DCF-Na is significantly impacted (p-value < 0.05) by the initial concentration of both organic and inorganic pollutants in the binary system. The lack-of-fit test indicates if the model is adequate by measuring the error arising from a deficiency in the model: if the

F-value of the lack-of-fit error is large and the corresponding error probability is small, the model is a poor fit to the experimental data [33]. The ANOVA summary shows the sorption of the two pollutants in this paper as a non-significant lack-of-fit for the regression models. The lack-of-fit F-values of 2.63 and 2.87, respectively, suggest that the lack-of-fit error is not significant in comparison with the pure error.

Table 6. ANOVA summary for the CCD models for the sorption of Pb^{2+} ions and DCF-Na by the IS/(PEI/PMAA)c (r = 1.0) composite sorbent.

	q_{Pb}					q_{DCF-Na}				
Source	Sum of Squares	df	Mean Square	F-Value	p-Value	Sum of Squares	df	Mean Square	F-Value	p-Value
Model	11,766.72	9	1307.41	856.07	<0.0001	6868.48	3	2289.49	577.59	<0.0001
A-C0 Pb	11,606.63	1	11,606.63	7599.82	<0.0001	54.77	1	54.77	13.82	0.0019
B-C0 DCF-Na	3.67	1	3.67	2.40	0.1521	6807.79	1	6807.79	1717.47	<0.0001
C-T	2.47	1	2.47	1.62	0.2319	5.92	1	5.92	1.49	0.2393
AB	3.05	1	3.05	2.00	0.1877					
AC	5.93	1	5.93	3.89	0.0770					
BC	5.288×10^{-3}	1	5.288×10^{-3}	3.462×10^{-3}	0.9542					
A^2	135.62	1	135.62	88.80	<0.0001					
B^2	1.72	1	1.72	1.13	0.3129					
C^2	0.17	1	0.17	0.11	0.7473					
Residual	15.27	10	1.53			63.42	16	3.96		
Lack of Fit	11.06	5	2.21	2.63	0.1561	54.75	11	4.98	2.87	0.1273
Pure Error	4.21	5	0.84			8.68	5	1.74		
R^2	0.9987					0.9909				
Adjusted R^2	0.9975					0.9891				
Predicted R^2	0.9920					0.9835				
Adequate precision	123.287					92.668				

The goodness-of-fit test, given by the R^2 coefficient, showed that over 99% of the variation in the observed data is explained by the generated models. In both cases, the predicted R^2 was in agreement with the adjusted R^2, with a difference of less than 0.2. This shows a good correlation between the experimental and predicted sorption capacities. The signal-to-noise ratio, expressed by the adequate precision value, should be higher than four for a good response signal and for the model to be suitable for use, as can be seen in Table 6.

The models' adequacy was further analyzed by diagnostic plots. The plots of the normal distribution of studentized residuals shown in Figure 5 are characterized by good linearity, with no large variation from the straight line for both responses, q_{Pb} and q_{DCF-Na}. To check the constant error, the plot of the residuals relative to the predicted responses was used. A uniform distribution of the residuals, as depicted in Figure 6, suggests that the variance is constant. Finally, the accuracy of the models is confirmed by the closeness between the experimental sorption capacities and the values obtained from the models, as seen in Figure 7 and Table 2.

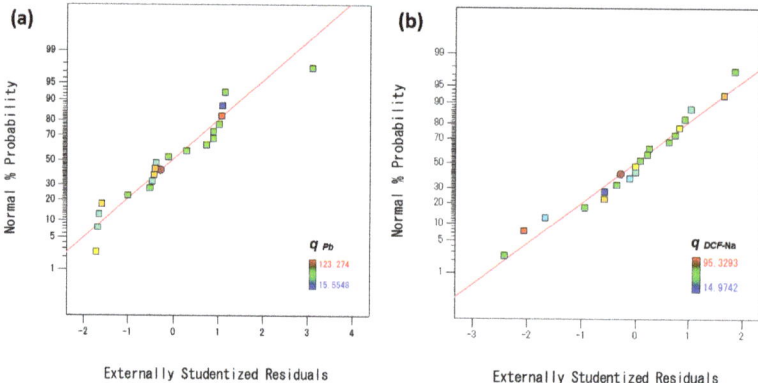

Figure 5. Normal plots of residuals for (**a**) q_{Pb} and (**b**) q_{DCF-Na}.

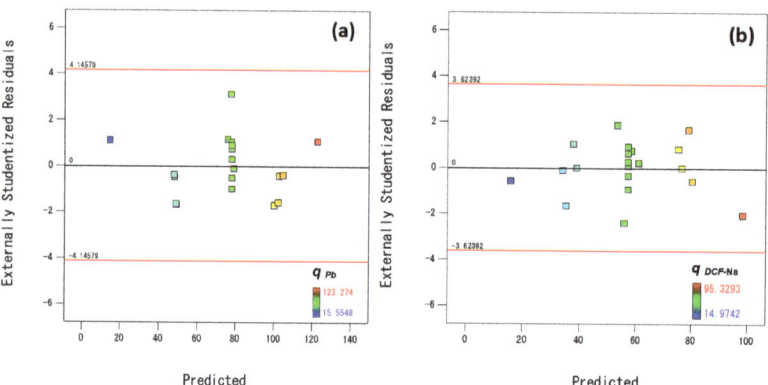

Figure 6. Diagnostic plots of residuals versus predicted responses for (**a**) q_{Pb} and (**b**) q_{DCF-Na}.

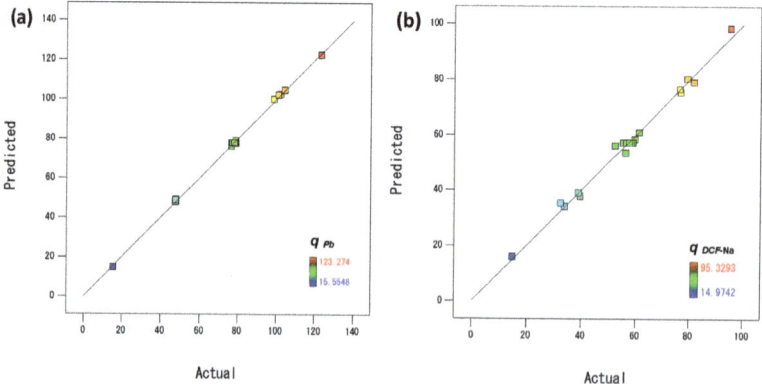

Figure 7. Plots of experimental versus predicted responses for q_{Pb} (**a**) and q_{DCF-Na} (**b**).

Based on the results of the analysis of variance and the diagnostic plots, the developed models could be used to describe the sorption of Pb^{2+} ions and DCF-Na on IS/(PEI/PMAA)$_c$ (r = 1.0) at a 95% confidence level.

3.2.2. Effects of Factors

The degree of influence and the corresponding effect of each independent variable on the process responses, q_{Pb} and q_{DCF-Na}, can be determined from the sign and values of the models' coefficients, as shown in Table 7. The negative sign of temperature, the $C_{0,Pb}$-T interaction, and the quadratic effect of the pollutant's initial concentration indicated the negative impact on the Pb^{2+} ions' sorption capacity. On the other hand, all parameters presented a positive influence on DCF-Na sorption, as the coefficients' signs were positive. The magnitudes of $C_{0,Pb}$ and $C_{0,DCF-Na}$ coefficients were higher than the other terms, suggesting these variables have more influence on the process responses.

Table 7. Factor coefficients of the models generated by CCD.

Factor	q_{Pb}		q_{DCF-Na}	
	Coefficient Estimate	Standard Error	Coefficient Estimate	Standard Error
Intercept	77.95	0.49	57.20	0.45
A - $C_{0, Pb}$	26.93	0.31	1.85	0.50
B - $C_{0, DCF-Na}$	0.48	0.31	20.63	0.50
C - T	−0.39	0.31	0.61	0.50
AB	0.62	0.44		
AC	−0.86	0.44		
BC	0.026	0.44		
A^2	−2.32	0.25		
B^2	−0.26	0.25		
C^2	0.082	0.25		

One aspect of this study is to explore the effect of the pollutant initial concentration on the lead and DCF-Na simultaneous adsorption from aqueous media. As seen until now, the CCD method was useful in determining if an interaction between these variables existed or not. Moreover, the 3D response surface plots were used to examine the interaction between the two variables, while the other factors were maintained at fixed values. The interaction between $C_{0,Pb}$ and $C_{0,DCF-Na}$ on the sorption capacity q_{Pb} is illustrated in Figure 8a. The response q_{Pb} varies between 15.6 and 123.3 mg/g polymer and increases with the initial concentration of Pb^{2+} ions due to a higher driving force between the liquid phase and the sorbent surface. The presence of DCF-Na with an initial concentration between 14 and 30 mg/L does not affect the metal ion sorption from the binary mixture. Looking back at the experimental results, at Pb:DCF-Na molar ratios between 1.0 and 6.41 ($C_{0,Pb}$ = 25 mg/L), the capacity q_{Pb} only varies slightly (about 1.2%).

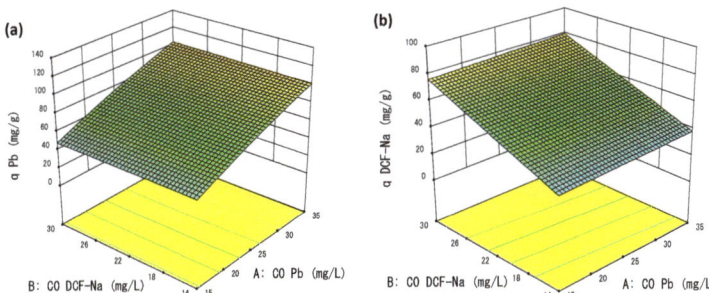

Figure 8. Three-dimensional response surface of the effect of lead ions and DCF-Na initial concentrations (at a temperature of 20 °C) on the simultaneous sorption process: (a) q_{Pb} and (b) q_{DCF-Na}.

In the current study, the initial concentration of DCF-Na varied from 6 to 38 mg/L. An increase from 15 to 95.3 mg/g polymer in the q_{DCF-Na} values was noticed with an increase in the organic moiety in the binary sorption system. A closer inspection of Figure 8b shows a small increase in q_{DCF-Na} at a high concentration of lead ions. Indeed, the experimental data indicates an augmentation of 8% in the sorption capacity of DCF-Na from a Pb:DCF-Na molar ratio of 0.34–3.14 ($C_{0,DCF-Na}$ = 22 mg/L).

The third factor considered in this study, e.g., the adsorption temperature, was found to have a weak influence on both lead ions and DCF-Na sorption from the binary mixture. This proves that the IS/(PEI/PMAA)$_c$ (r = 1.0) composite sorbent is not a thermosensitive material for ionic pollutants separation from aqueous media.

3.2.3. Optimization of the Adsorption Process

The sorption of Pb^{2+} ions and DCF-Na from binary mixtures on IS/(PEI/PMAA)$_c$ (r = 1.0) is optimized by the desirability function based on the pollutants sorption capacities as a response. The highest value of the desirability function was obtained for the following conditions: $C_{0,Pb}$ = 35 mg/L, $C_{0,DCF-Na}$ = 30 mg/L, and T = 12 °C, when the sorption capacities were q_{Pb} = 104.71 mg/g polymer and q_{DCF-Na} = 79.069 mg/g polymer. Sorption tests were performed in quadruplicate according to the mentioned conditions, and the experimental results indicated a deviation of −2.27% and 0.23% from the optimized q values for Pb^{2+} ions and DCF-Na, respectively. At a temperature of 20 °C, for the same initial pollutant concentrations (a desirability of 0.810), the experimental values deviated by −1.81% and 2.76% for q_{Pb} and q_{DCF-Na}, respectively, as against the optimized responses. These results demonstrate the accuracy of the developed models.

3.3. Desorption and Sorbent Regeneration

Experiments of pollutants' desorption were carried out at room temperature, with initial optimized concentrations from the RSM-CCD design of experiments. The simultaneous sorption efficiency for the two contaminants is presented in Figure 9. The removal efficiency (RE) of Pb^{2+} ions varied from a high of 89.2% to a low of 72%, while for DCF-Na the RE remained almost constant after five cycles of sorption-desorption.

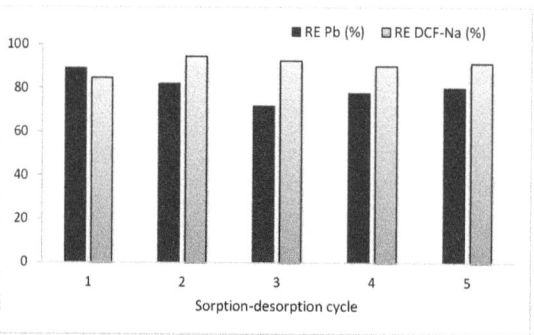

Figure 9. Removal efficiency (RE) of the pollutants in consecutive adsorption-desorption cycles.

3.4. Sequential Sorption

The sequential sorption experiments were carried out at room temperature for 7 h, with $C_{0,Pb}$ = 43 mg/L and $C_{0,DCF-Na}$ = 30 mg/L. The results, presented in Figure 10, indicate a significant decrease of lead ions RE from 70.4% to 29.2%.

This fact may be explained by the saturation of the sorbent surface as the available active sites, especially amino groups, for dative bond formation decrease. Meanwhile, a high DCF-Na is maintained even in stage six, when it reaches a value of 88.2%. At the same time, a leaching of Pb^{2+} ions was observed experimentally in stages four (16%) and six (20.9%), possibly due to the formation of Pb(DCF)$_2$ in solution by ionic exchanges. The physico-chemical interactions are tremendously important because they dictate the sorbent capacity for pollutant detection, retention, concentration in the solid phase, and subsequent release in the desorption step. The processes of retention of inorganic (Pb^{2+}) or organic (DCF-Na) pollutants are driven by coordinative bond formation, electrostatic interactions, H-bonding, hydrophobic interactions, or ionic exchange interactions.

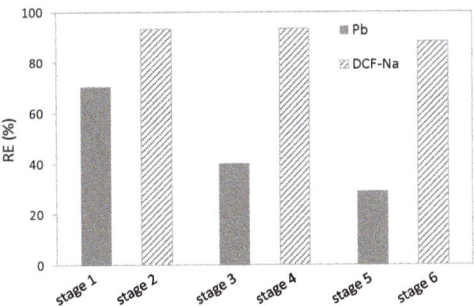

Figure 10. DCF-Na and Pb^{2+} ion removal efficiencies in sequential sorption experiments.

The schematic representation of possible interactions between pollutants and the composite surface, presented in Scheme 1, gave a clear overview of the advantages of using core/shell composite particles with different functional groups on the surface, which can act as active sites for the removal by sorption of different types of pollutants.

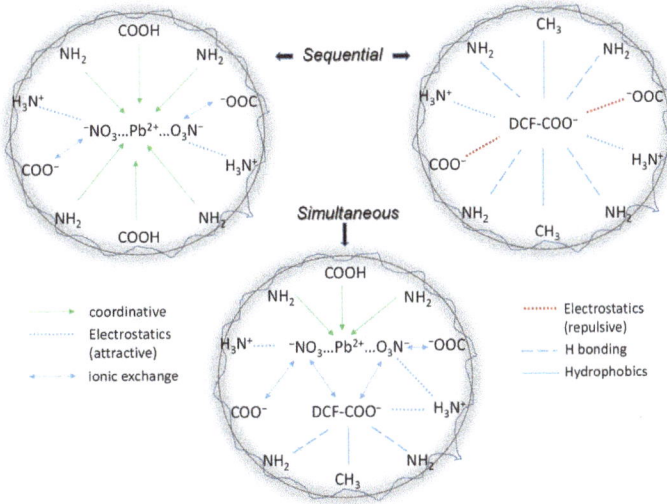

Scheme 1. Different types of interactions between Pb^{2+} and DCF-Na with composites based on PEI/PMAA chains on the surface.

4. Conclusions

This study proposed to investigate the sorption processes in batch conditions of two model pollutants, Pb^{2+} and DCF-Na, onto a core/shell composite based on PEI and PMAA. The synthesis strategy of the sorbent has been previously reported; in this approach, only one support (IS/(PEI/PMAA)$_c$), which contains the highest deposited amount of polyelectrolytes (~20% organic), has been tested as a sorbent. The sequential and simultaneous sorption cycles of lead ions and DCF-Na molecules, carried out in batch conditions, showed an additive capacity of the composite towards two different types of pollutants, with an important amount of DCF-Na being sorbed together with Pb^{2+} ions. This additive sorption demonstrated that the chemical (coordinative bonds) and physical (electrostatic, H-bonding, and hydrophobic) interactions are the main driving forces for pollutants retention at the same active sorption sites inside the cross-linked organic shell. The process optimization used the central composite design to model the simultaneous adsorption of Pb^{2+} ions and DCF-Na from aqueous solutions for various operating conditions. The experimental design was composed of twenty batch experimental runs and was used to evaluate the impact of different ratios between the two priority pollutants in solution and find optimum conditions for simultaneous adsorption. This study proved that this type of composite, obtained by the direct deposition of polyelectrolytes, may be used for the removal from wastewater of two main classes of priority pollutants: heavy metals and anionic organic molecules.

Author Contributions: Conceptualization, I.M., C.T. and M.M.; data curation, I.M., F.B., D.F. and C.P.; formal analysis, I.M. and F.B.; funding acquisition, C.T.; investigation, I.M., F.B., D.F. and C.P.; methodology, I.M., F.B. and C.P.; project administration, C.T.; resources, C.T. and M.M.; software, I.M.; supervision, C.T. and M.M.; validation, visualization, roles/writing—original draft, I.M. and F.B.; writing—review and editing, C.T., F.B. and M.M. All authors have read and agreed to the published version of the manuscript.

Funding: This work was supported by a grant from the Ministry of Research, Innovation, and Digitization, CNCS/CCCDI–UEFISCDI, project number PN-III-P4-ID-PCE-2020-1199, within PNCDI III, contract PCE 56/2021, "Innovative and sustainable solutions for priority and emerging pollutants removal through advanced wastewater treatment processes" (SUSTINWATER).

Institutional Review Board Statement: Not applicable.

Data Availability Statement: Not applicable.

Conflicts of Interest: The authors declare no conflict of interest.

References

1. Nair, A.T.; Makwana, A.R.; Ahammed, M.M. The Use of Response Surface Methodology for Modelling and Analysis of Water and Wastewater Treatment Processes: A Review. *Water Sci. Technol.* **2014**, *69*, 464–478. [CrossRef]
2. Anfar, Z.; Ait Ahsaine, H.; Zbair, M.; Amedlous, A.; Ait El Fakir, A.; Jada, A.; El Alem, N. Recent Trends on Numerical Investigations of Response Surface Methodology for Pollutants Adsorption onto Activated Carbon Materials: A Review. *Crit. Rev. Environ. Sci. Technol.* **2020**, *50*, 1043–1084. [CrossRef]
3. Hiew, B.Y.Z.; Lee, L.Y.; Lee, X.J.; Gan, S.; Thangalazhy-Gopakumar, S.; Lim, S.S.; Pan, G.-T.; Yang, T.C.-K. Adsorptive Removal of Diclofenac by Graphene Oxide: Optimization, Equilibrium, Kinetic and Thermodynamic Studies. *J. Taiwan Inst. Chem. Eng.* **2019**, *98*, 150–162. [CrossRef]
4. Masoumi, H.; Ghaemi, A.; Gilani Ghanadzadeh, H. Elimination of Lead from Multi-Component Lead-Nickel-Cadmium Solution Using Hyper-Cross-Linked Polystyrene: Experimental and RSM Modeling. *J. Environ. Chem. Eng.* **2021**, *9*, 106579. [CrossRef]
5. Treméa, R.; Quesada, H.B.; Bergamasco, R.; de Jesus Bassetti, F. Influence of Important Parameters on the Adsorption of Diclofenac Sodium by an Environmentally Friendly Eucalyptus Wood Biochar and Optimization Using Response Surface Methodology. *Desalin. WATER Treat.* **2021**, *230*, 384–399. [CrossRef]
6. Arabkhani, P.; Javadian, H.; Asfaram, A.; Ateia, M. Decorating Graphene Oxide with Zeolitic Imidazolate Framework (ZIF-8) and Pseudo-Boehmite Offers Ultra-High Adsorption Capacity of Diclofenac in Hospital Effluents. *Chemosphere* **2021**, *271*, 129610. [CrossRef]
7. Kang, J.-K.; Kim, Y.-G.; Lee, S.-C.; Jang, H.-Y.; Yoo, S.-H.; Kim, S.-B. Artificial Neural Network and Response Surface Methodology Modeling for Diclofenac Removal by Quaternized Mesoporous Silica SBA-15 in Aqueous Solutions. *Microporous Mesoporous Mater.* **2021**, *328*, 111497. [CrossRef]
8. Morosanu, I.; Paduraru, C.; Bucatariu, F.; Fighir, D.; Mihai, M.; Teodosiu, C. Shaping Polyelectrolyte Composites for Heavy Metals Adsorption from Wastewater: Experimental Assessment and Equilibrium Studies. *J. Environ. Manag.* **2022**, *321*, 115999. [CrossRef] [PubMed]
9. Morosanu, I.; Teodosiu, C.; Coroaba, A.; Paduraru, C. Sequencing Batch Biosorption of Micropollutants from Aqueous Effluents by Rapeseed Waste: Experimental Assessment and Statistical Modelling. *J. Environ. Manag.* **2019**, *230*, 110–118. [CrossRef]
10. Hannan, E.J.; Quinn, B.G. The Determination of the Order of an Autoregression. *J. R. Stat. Soc. Ser. B (Methodol.)* **1979**, *41*, 190–195. [CrossRef]

11. Refat, M.S.; Mohamed, G.G.; Ibrahim, M.Y.S.; Killa, H.M.A.; Fetooh, H. Synthesis and Characterization of Coordination Behavior of Diclofenac Sodium Drug Toward Hg(II), Pb(II), and Sn(II) Metal Ions: Chelation Effect on Their Thermal Stability and Biological Activity. *Synth. React. Inorg. Met. Nano-Metal Chem.* **2014**, *44*, 161–170. [CrossRef]
12. Lagergren, S.Y. Zur Theorie Der Sogenannten Adsorption Gelöster Stoffe. *K. Sven. Vetenskapsakad. Handl.* **1989**, *24*, 1–39.
13. Zhou, Y.; Liu, X.; Xiang, Y.; Wang, P.; Zhang, J.; Zhang, F.; Wei, J.; Luo, L.; Lei, M.; Tang, L. Modification of Biochar Derived from Sawdust and Its Application in Removal of Tetracycline and Copper from Aqueous Solution: Adsorption Mechanism and Modelling. *Bioresour. Technol.* **2017**, *245*, 266–273. [CrossRef] [PubMed]
14. Salvestrini, S. Analysis of the Langmuir Rate Equation in Its Differential and Integrated Form for Adsorption Processes and a Comparison with the Pseudo First and Pseudo Second Order Models. *React. Kinet. Mech. Catal.* **2018**, *123*, 455–472. [CrossRef]
15. Azizian, S. Kinetic Models of Sorption: A Theoretical Analysis. *J. Colloid Interface Sci.* **2004**, *276*, 47–52. [CrossRef]
16. Sahoo, T.R.; Prelot, B. Adsorption Processes for the Removal of Contaminants from Wastewater. In *Nanomaterials for the Detection and Removal of Wastewater Pollutants*; Elsevier: Amsterdam, The Netherlands, 2020; pp. 161–222.
17. Boyd, G.E.; Adamson, A.W.; Myers, L.S. The Exchange Adsorption of Ions from Aqueous Solutions by Organic Zeolites. II. Kinetics 1. *J. Am. Chem. Soc.* **1947**, *69*, 2836–2848. [CrossRef]
18. Langmuir, I. The Constitution and Fundamental Properties of Solids and Liquids. Part I. Solids. *J. Am. Chem. Soc.* **1916**, *38*, 2221–2295. [CrossRef]
19. Freundlich, H. Über Die Adsorption in Lösungen. *Zeitschrift für Phys. Chemie* **1907**, *57U*, 385–470. [CrossRef]
20. Hill, A. The Possible Effects of the Aggregation of the Molecules of Haemoglobin on Its Dissociation Curves. *J. Physiol.* **1910**, *40*, iv–vii. [CrossRef]
21. Redlich, O.; Peterson, D.L. A Useful Adsorption Isotherm. *J. Phys. Chem.* **1959**, *63*, 1024. [CrossRef]
22. Al-Ghouti, M.A.; Da'ana, D.A. Guidelines for the Use and Interpretation of Adsorption Isotherm Models: A Review. *J. Hazard. Mater.* **2020**, *393*, 122383. [CrossRef]
23. Liang, X.X.; Omer, A.M.; Hu, Z.; Wang, Y.; Yu, D.; Ouyang, X. Efficient Adsorption of Diclofenac Sodium from Aqueous Solutions Using Magnetic Amine-Functionalized Chitosan. *Chemosphere* **2019**, *217*, 270–278. [CrossRef] [PubMed]
24. Chaduka, M.; Guyo, U.; Zinyama, N.P.; Tshuma, P.; Matsinha, L.C. Modeling and Optimization of Lead (II) Adsorption by a Novel Peanut Hull-g-Methyl Methacrylate Biopolymer Using Response Surface Methodology (RSM). *Anal. Lett.* **2020**, *53*, 1294–1311. [CrossRef]
25. de Luna, M.D.G.; Murniati; Budianta, W.; Rivera, K.K.P.; Arazo, R.O. Removal of Sodium Diclofenac from Aqueous Solution by Adsorbents Derived from Cocoa Pod Husks. *J. Environ. Chem. Eng.* **2017**, *5*, 1465–1474. [CrossRef]
26. Salvestrini, S.; Fenti, A.; Chianese, S.; Iovino, P.; Musmarra, D. Diclofenac Sorption from Synthetic Water: Kinetic and Thermodynamic Analysis. *J. Environ. Chem. Eng.* **2020**, *8*, 104105. [CrossRef]
27. Lu, Y.; Wang, Z.; Ouyang, X.; Ji, C.; Liu, Y.; Huang, F.; Yang, L.-Y. Fabrication of Cross-Linked Chitosan Beads Grafted by Polyethylenimine for Efficient Adsorption of Diclofenac Sodium from Water. *Int. J. Biol. Macromol.* **2020**, *145*, 1180–1188. [CrossRef] [PubMed]
28. Chen, Z.; Wang, Y.-F.; Zeng, J.; Zhang, Y.; Zhang, Z.-B.; Zhang, Z.-J.; Ma, S.; Tang, C.-M.; Xu, J.-Q. Chitosan/Polyethyleneimine Magnetic Hydrogels for Adsorption of Heavy Metal Ions. *Iran. Polym. J.* **2022**, *31*, 1273–1282. [CrossRef]
29. Nicola, R.; Costişor, O.; Ciopec, M.; Negrea, A.; Lazău, R.; Ianăşi, C.; Picioruş, E.-M.; Len, A.; Almásy, L.; Szerb, E.I.; et al. Silica-Coated Magnetic Nanocomposites for Pb^{2+} Removal from Aqueous Solution. *Appl. Sci.* **2020**, *10*, 2726. [CrossRef]
30. Azizkhani, S.; Mahmoudi, E.; Abdullah, N.; Ismail, M.H.S.; Mohammad, A.W.; Hussain, S.A. Synthesis and Characterisation of Graphene Oxide-Silica-Chitosan for Eliminating the Pb(II) from Aqueous Solution. *Polymers* **2020**, *12*, 1922. [CrossRef]
31. Putz, A.-M.; Ivankov, O.I.; Kuklin, A.I.; Ryukhtin, V.; Ianăşi, C.; Ciopec, M.; Negrea, A.; Trif, L.; Horváth, Z.E.; Almásy, L. Ordered Mesoporous Silica Prepared in Different Solvent Conditions: Application for Cu(II) and Pb(II) Adsorption. *Gels* **2022**, *8*, 443. [CrossRef]
32. Ali, A.; Alharthi, S.; Ahmad, B.; Naz, A.; Khan, I.; Mabood, F. Efficient Removal of Pb(II) from Aqueous Medium Using Chemically Modified Silica Monolith. *Molecules* **2021**, *26*, 6885. [CrossRef] [PubMed]
33. Ekpenyong, M.; Antai, S.; Asitok, A.; Ekpo, B. Response Surface Modeling and Optimization of Major Medium Variables for Glycolipopeptide Production. *Biocatal. Agric. Biotechnol.* **2017**, *10*, 113–121. [CrossRef]

Disclaimer/Publisher's Note: The statements, opinions and data contained in all publications are solely those of the individual author(s) and contributor(s) and not of MDPI and/or the editor(s). MDPI and/or the editor(s) disclaim responsibility for any injury to people or property resulting from any ideas, methods, instructions or products referred to in the content.

Article

Delamination of Novel Carbon Fibre-Based Non-Crimp Fabric-Reinforced Thermoplastic Composites in Mode I: Experimental and Fractographic Analysis

Muhammad Ameerul Atrash Mohsin [1,2,*], Lorenzo Iannucci [1] and Emile S. Greenhalgh [1]

1 Department of Aeronautics, Imperial College London, Exhibition Road, London SW7 2AZ, UK
2 Empa, Swiss Federal Laboratories for Materials Science and Technology, Überland Str. 129, 8600 Dübendorf, Switzerland
* Correspondence: m.mohsin14@imperial.ac.uk or atrash.mohsin@empa.ch

Abstract: Delamination, a form of composite failure, is a significant concern in laminated composites. The increasing use of out-of-autoclave manufacturing techniques for automotive applications, such as compression moulding and thermoforming, has led to increased interest in understanding the delamination resistance of carbon-fibre-reinforced thermoplastic (CFRTP) composites compared to traditional carbon-fibre-reinforced thermosetting (CFRTS) composites. This study evaluated the mode I (opening) interlaminar fracture toughness of two non-crimp fabric (NCF) biaxial (0/90°) carbon/thermoplastic composite systems: T700/polyamide 6.6 and T700/polyphenylene sulphide. The mode I delamination resistance was determined using the double cantilever beam (DCB) specimen. The results were analysed and the Mode I interlaminar fracture toughness was compared. Additionally, the fractographic analysis (microstructure characterisation) was conducted using a scanning electron microscope (SEM) to examine the failure surface of the specimens.

Keywords: carbon-fibre-reinforced polymer (CFRP); thermoplastic composites; high performance composites; delamination resistance; non-crimp fabric (NCF)

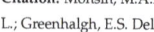

Citation: Mohsin, M.A.A.; Iannucci, L.; Greenhalgh, E.S. Delamination of Novel Carbon Fibre-Based Non-Crimp Fabric-Reinforced Thermoplastic Composites in Mode I: Experimental and Fractographic Analysis. *Polymers* **2023**, *15*, 1611. https://doi.org/10.3390/polym15071611

Academic Editors: Tomasz Makowski and Sivanjineyulu Veluri

Received: 26 January 2023
Revised: 21 March 2023
Accepted: 22 March 2023
Published: 23 March 2023

Copyright: © 2023 by the authors. Licensee MDPI, Basel, Switzerland. This article is an open access article distributed under the terms and conditions of the Creative Commons Attribution (CC BY) license (https://creativecommons.org/licenses/by/4.0/).

1. Introduction

Delamination is a critical failure mode in composite materials, particularly in laminated composites. These materials are made up of layers of different materials that are bonded together, and delamination occurs when the layers separate from one another. In recent years, there has been extensive research dedicated to enhancing the delamination resistance of composites [1–7]. The resistance to delamination is often referred to as the Mode I interlaminar fracture toughness and is represented by the G_{Ic} values, which are sometimes quoted in material supplier's data sheets. However, in the past, most of the research performed in regards to the interlaminar fracture toughness of laminated composites has revolved around thermosetting composites [1,8–16]. To the author's knowledge, very few studies can be found in the open literature on the delamination of laminated non-crimp fabric composite systems [15,17–20], and these are mostly thermoset systems. Furthermore, most research on the delamination of thermoplastic composites revolved around other types of fibre architecture, e.g., unidirectional laminates [21–23].

One of the most useful real-world applications of composites is low-velocity impact, as laminated composites have been proven to be capable of absorbing higher impact energy compared to conventional metals and metal alloys [24,25]. Mohsin et al. [24] have proven that thermoplastic composites are superior to most thermosetting composites. As the delamination of composite structures can be induced by the impact of low velocity, it is one of the major challenges that draws the most attention from a safety perspective. Strength, toughness, and fatigue life are all reduced as a result of the harm brought on by the formation of such a delamination [26].

The mode I interlaminar stiffness is often tested and extensively researched using double-cantilever beam experiments [1,7,10,20,27]. To measure the Mode I delamination resistance of a material, the double cantilever beam (DCB) specimen is commonly used. The DCB specimen is a rectangular-shaped piece of material with a notch cut into it, which simulates a crack. The resistance to delamination is measured by loading the specimen until the layers of the composite separate. After the experimental results are collected and data are analysed, the Mode I interlaminar fracture toughness of the material is determined. To understand the cause of failure, the fractographic analysis is conducted using a scanning electron microscope (SEM) to characterise the microstructure of the failed specimens. This allows researchers to understand the underlying mechanisms of delamination and develop strategies to improve the delamination resistance of composites.

The choice of polyamide 6.6 (PA6.6) and polyphenylene sulphide (PPS) as the two matrices presented in this study was driven by the UK-DATACOMP [28] and UK-THERMOCOMP [29] project due to the interest in the UK automotive and aerospace industry. The PA6.6 polymer was considered as a reasonable option for the automotive industry due its relatively low cost. The PPS was of interest due to its typically higher tensile strength and lower processing temperatures (300–345 °C) when compared to other well-known aerospace thermoplastics such as PEEK (350–400 °C) [30,31]. Additionally, this study was aimed at establishing and enhancing the materials database of the industrial partners of the project. Hence, there was a need to characterise the Mode I delamination of these novel NCF biaxial CFRTP systems.

2. Materials and Methods

2.1. Material System and Laminate Preparation

This research used two types of CFRTP material systems: (i) NCF biaxial (0°/90°) T700 (continuous) carbon fibres pre-impregnated with polyamide 6.6 veils (T700/PA6.6) and PA6.6 stitching, and (ii) NCF biaxial (0°/90°) T700 (continuous) carbon fibres pre-impregnated with polyphenylene sulphide veils (T700/PPS) and Kevlar® (DuPont, Wilmington, DE, USA) stitching. These materials were supplied by partners in the THERMOCOMP project [29]. The T700/PA6.6 material system has also previously been reported and discussed in Mohsin et al. [24,32].

The material (unreinforced and reinforced) and mechanical properties of the constituent materials of the composite are shown in Table 1. However, since the material system is proprietary, the mechanical properties of the laminates were obtained by the author using a series of standardised and non-standardised tests listed in Table 2 and described in [24,33].

Table 1 shows the mechanical properties of neat PA6.6, PPS, and T700 fibre. Table 2 shows the Quasi-static mechanical properties of T700/PA6.6 (FVF = 52%) and T700/PPS (FVF = 61%) [34].

Table 1. Mechanical properties of neat PA6.6, PPS, and T700 fibre.

	PA6.6	PPS	T700 Fibre
References	[33,35,36]	[33,37,38]	[39]
Density, ρ (kg/m^3)	1170	1310	1800
Tensile strength, ultimate (MPa)	71	111	4900
Tensile modulus (GPa)	0.2–3.8 *	2.6–6.1 *	230
Elongation at break (%)	53.9	13.9	2.1
Mode I fracture toughness, G_{IC} (kJ/m^2)	0.2	0.5	-

* Note that the mechanical properties of the polymers were not characterised/measured in-house. Based on vendor's data and the grade of the of matrix, the estimated tensile modulus of the PA6.6 and PPS are ~3.5 GPa and ~3 GPa, respectively.

Table 2. Quasi-static mechanical properties of T700/PA6.6 (FVF = 52%) and T700/PPS (FVF = 61%) [24,33,34,40].

Mechanical Properties	Material: T700/PA6.6	T700/PPS
Tensile Young's modulus (GPa)	65	60 [1]
Compressive Young's modulus (GPa)	69	47 [1]
Tensile strength (MPa)	918	852
Compressive strength (MPa)	461	265
In-plane shear modulus (GPa)	3.2	3.3
In-plane shear stress at 5% (MPa)	52	73

[1] The measured tensile and compression Young's modulus of the T700/PPS, which has a higher fibre volume fraction (61%) than the T700/PA6.6 (52%), is lower due to the influence of the stitching material of the former (Kevlar®), causing undulation in the laminate. This reduces the modulus.

To ensure accuracy and eliminate any effects of moisture, the densities of both CFRTP material systems were measured using a pycnometer after being stored in an oven at 40 °C for three days. The densities of the T700/PA6.6 and T700/PPS systems were measured to be 1485 kg/m^3 and 1553 kg/m^3, respectively.

Additionally, the fibre–volume–fraction (FVF) of the laminates produced was measured via a thermogravimetric analysis (TGA) (T700/PA6.6 = 52% and T700/PPS = 61%). The process has been detailed in [34]. The FVF measurement is important to understand the composite material properties and its behaviour under a load. The accurate measurement of FVF is essential to make sure that the design is robust enough and to select the appropriate composite material for a specific application.

2.2. Manufacturing Process

The laminates were created by layering T700/PA6.6 with a layup sequence of $(0/90)_{12s}$ using a hand lay-up method. The hand lay-up method is a process where the layers of material are manually placed and arranged on the mould before being shaped with a thermoforming method. The thermoforming method used in this case was a laboratory hydraulic press at 275 °C. This process of layering and shaping the T700/PA6.6 material creates a laminate that is strong and durable.

- The recommended processing parameters for T700/PA6.6 are:
- Dwell time: 10 min;
- Processing temperature: 275 °C;
- Heating rate: 15 °C/min;
- Pressure: 1.5 MPa;
- Demoulding temperature: 25–35 °C.

The average thickness of the T700/PA6.6 panel was measured to be 4.1 ± 0.28 mm. This measurement is important, as it ensures that the final product meets the desired thickness specifications.

For the T700/PPS, the manufacturer's recommended processing parameters are as follows:

- Dwell time: 10 min;
- Processing temperature: 315 °C;
- Heating rate: 15 °C/min;
- Pressure: 2.5 MPa;
- Demoulding temperature: <100 °C.

These parameters are similar to the ones for T700/PA6.6, with the main difference being the processing temperature of 315 °C and the pressure of 2.5 MPa. The demoulding temperature is also slightly different, with T700/PPS having a more flexible range of simply < 100 °C.

In summary, the laminates were prepared by layering and shaping T700/PA6.6 and T700/PPS materials using a hand lay-up method and a laboratory hydraulic press, with specific processing parameters to ensure the optimal bonding and curing of the material.

The final product meets the desired thickness specifications and has a good strength and durability.

3. Experimental Setup and Data Reduction

3.1. Experimental Setup

Measuring a laminated composite material's resistance against the interlaminar fracture, known as the Mode I interlaminar fracture toughness (G_{Ic}), is critical in the process of material selection and design for various structural applications. The reason for this is that vulnerability to delamination has always been considered as one of the most vital weaknesses of advanced laminated composite structures. Therefore, the knowledge of a laminated composite material's resistance against the interlaminar fracture is always valuable in the process of material selection and design for various structural applications.

One of the most commonly used methods to determine G_{Ic} values is the DCB test. The DCB test allows for the determination of the Mode I interlaminar fracture toughness (critical energy release rate), G_{Ic}. The DCB test performed on both NCF biaxial carbon-fibre-reinforced thermoplastic (CFRTP) systems (T700/polyamide 6.6 (PA6.6) and T700/polyphenylene sulphide (PPS)) in this study is partially in accordance to the existing standardised test outlined in the American Society for Testing and Materials (ASTM) D5528-13 [41], which was originally tailored for unidirectional fibre-reinforced plastic composites.

The advantage of using G_{Ic} values is that they are independent of the specimen or method of load introduction. This makes G_{Ic} values useful for determining the design allowable and damage tolerance, specifically, the delamination failure criteria of composite structures manufactured from a specific composite material system. This makes the DCB test a valuable tool for engineers and designers in order to select the appropriate composite material for a specific application and to make sure that the design is robust enough to avoid a delamination failure.

The typical the DCB specimen types are shown in Figure 1. It is important to note that the specification of the specimen dimensions and the preparation of the specimen are crucial for the accuracy of the results obtained from the DCB test. Therefore, it is recommended to follow the standardized procedures outlined in ASTM D5528-13 in order to obtain accurate results.

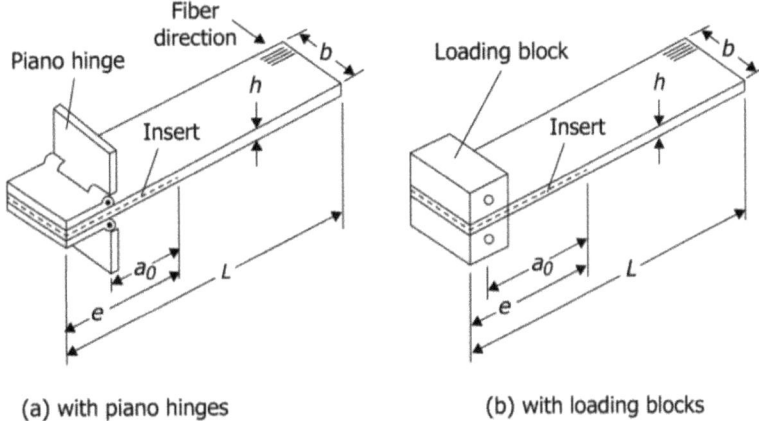

Figure 1. Standardised DCB specimens according to the ASTM D5528-13 [41]: (**a**) with piano hinges and (**b**) with loading blocks. Reprinted, with permission, from ASTM D5528-13 [41] Standard Test Method for Mode I Interlaminar Fracture Toughness of Unidirectional Fibre-Reinforced Polymer Matrix Composites, copyright ASTM International. A copy of the complete standard may be obtained from www.astm.org.

The test specimen is a rectangular laminated composite material with a length of 170 mm and a width of 20 mm. It should have a uniform thickness of 4 ± 0.2 mm and a 60 mm non-adhesive insert in the middle that serves as the point of delamination initiation (Figure 2). The specimen is attached to loading blocks or hinges, which are bonded to one end of the specimen using an epoxy glue, and opening forces are applied to these points. The width and thickness of each specimen were measured and averaged across its length. During the test, the specimen is opened by controlling the opening displacement while measuring the delamination length and load. The DCB specimens were coated in white before the test to improve visibility (Figure 3) and the experimental setup of the test can be observed in Figure 4, which includes a travelling microscope used to record the delamination length and monitor crack growth. The test machine used is an INSTRON® testing machine (Instron Corporation, Norwood, MA, USA) and the cross-head displacement rate for each test was set to 0.5 mm/min.

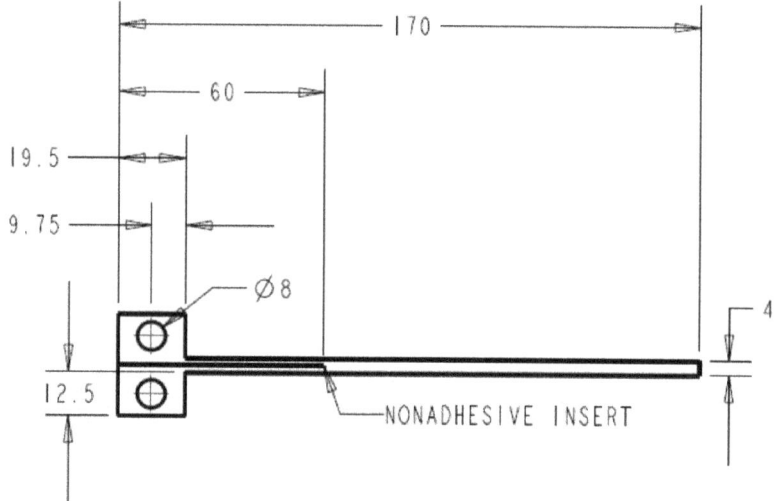

Figure 2. Exact dimensions of the DCB specimen (in mm).

3.2. Data Reduction

The graph of load versus cross-head displacement is generated by plotting the data obtained from the test. The Mode I interlaminar fracture toughness is then calculated using the modified beam theory (MBT) method [42]. The Mode I interlaminar fracture toughness, also known as the strain energy release rate, is a measure of the resistance of the material to the crack propagation and is represented by G_{Ic}. This value can be calculated using Equation (1).

$$G_{Ic} = \frac{3P\delta}{2b(a+|\Delta|)} \quad (1)$$

where P is the load, δ is the cross-head displacement, b is the specimen width, a is the delamination length, and Δ is a correction term applied to the delamination length.

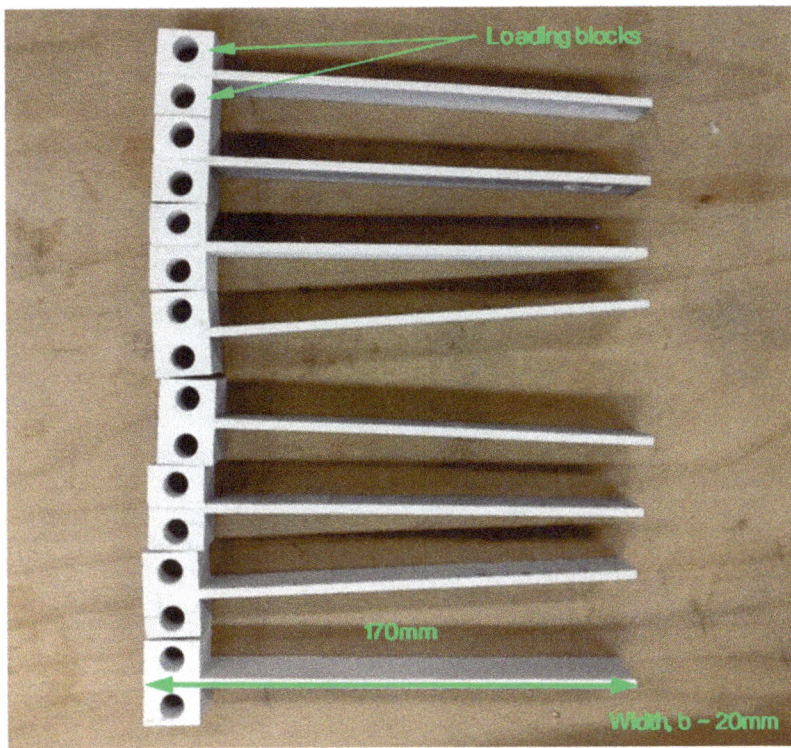

Figure 3. The DCB specimens that have been painted white prior to being marked to enhance visibility.

Figure 4. DCB test setup with a travelling microscope, which is a microscope that is mounted on a slider that can be moved along a scale.

The correction term, Δ, is determined from the experimental data by generating a least square plot of the cubic root of compliance, $C^{1/3}$, as a function of delamination length,

a. This is conducted to account for any variations in the delamination length that may have occurred during the test. The correction term, Δ, is the value that is added to the delamination length to make the plot pass through the origin. The compliance, C, can be calculated using the Equation (2).

$$C = \frac{\delta}{P} \tag{2}$$

This equation is used to determine the relationship between the displacement and the load applied on the specimen. The results of these calculations are then used to determine the mode I interlaminar fracture toughness of the material.

4. Results and Discussion

The DCB test, which was performed on both T700/PA6.6 and T700/PPS specimens, revealed a stable crack growth throughout the duration of the test. The failure was determined by the first significant load drop in the load-displacement curve, as outlined in the appendices. During the test, it was observed that the crack propagated consistently along the midplane and length of the specimen, as the load gradually decreased with the increasing delamination length. The maximum delamination lengths of the T700/PA6.6 samples, before the DCB arms exhibited a bending failure, were found to be larger, at around 70–80 mm compared to the 40–50 mm observed in the T700/PPS samples.

Additionally, the R-curves for both specimens showed a positive slope, which is a common characteristic for these types of materials. This can be observed in Figures 5 and 6 for the T700/PA6.6 and T700/PPS samples, respectively. Furthermore, a representative R-curve is highlighted against the crack length in Figure 7, providing a clear visual representation of the results.

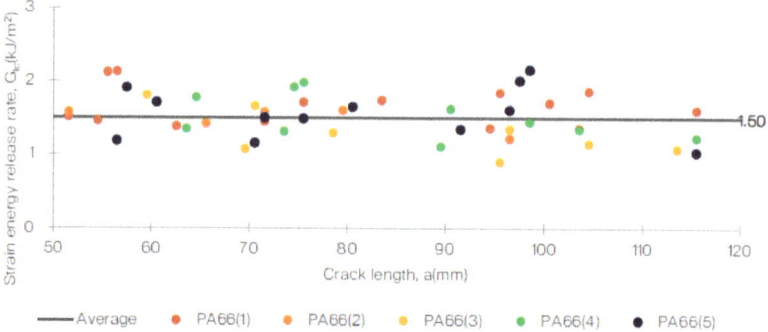

Figure 5. Delamination resistance curve (R-curve) from the DCB test of the T700/PA6.6 system.

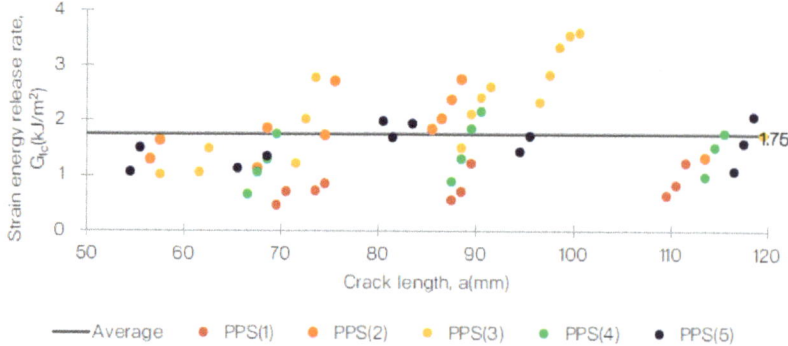

Figure 6. Delamination resistance curve (R-curve) from the DCB test of the T700/PPS system.

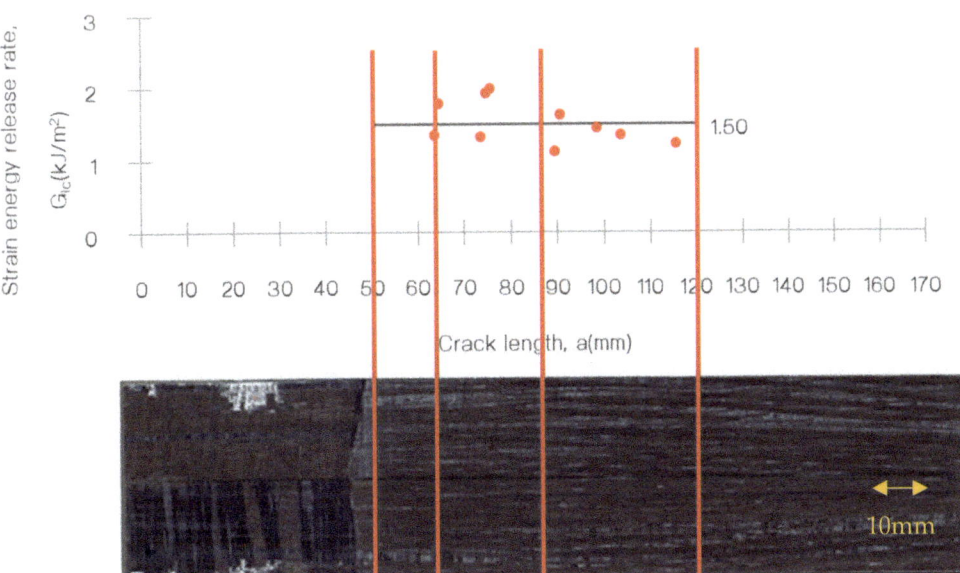

Figure 7. Representative R-curve (specimen #1 T700/PA6.6) against measured crack length.

Based on the R-curves obtained for both specimens, the Mode I fracture toughness, G_{Ic}, for both samples, was calculated and tabulated as shown in Table 3. The average G_{Ic} calculated for the T700/PA6.6 and T700/PPS were 1.50 kJ/m² (CV = 9.7%) and 1.75 kJ/m² (CV = 8.6%), respectively. These results demonstrate that the Mode I interlaminar fracture toughness of the T700/PPS is approximately 17% higher than the T700/PA6.6 system.

Table 3. Summary of the Mode I interlaminar fracture toughness of T700/PA6.6 and T700/PPS.

Study	Material System	Fibre Orient.	G_{Ic} (kJ/m²)	CV (%)
Mohsin et al. [33]	T700/PA6.6 (this study)	Bidirectional [1]	1.50	9.7
Mohsin et al. [33]	T700/PPS (this study)	Bidirectional [1]	1.75	8.6
Pret et al. [43]	AS4/PEEK	Unidirectional	1.46	
Ivanov et al. [44]	T300JB/PPS	Multidirectional [2]	0.81	
Ivanov et al. [44]	T300/BP-907	Multidirectional [2]	1.29	
Tijs et al. [45]	AS4D/PEKK	Unidirectional	0.70	
Reis et al. [46]	HS-Carbon/PA6	Unidirectional	2.20	

[1] NCF biaxial 0/90°; [2] Woven 0/90°.

In conclusion, the DCB test provided valuable information on the stable crack growth and delamination behaviour of the T700/PA6.6 and T700/PPS specimens. The results of the test, including the R-curves, the maximum delamination lengths, and the Mode I fracture toughness, have allowed for a comprehensive understanding of the mechanical properties of these materials. These findings can be used to inform future materials' development and application decisions.

5. Fractographic Analysis

Fractographic analysis is a powerful tool that allows for the characterisation of the microstructure of materials, particularly in the case of delamination failures. By using a scanning electron microscope (SEM), the morphologies of the fractured surfaces can be examined and interpreted as a Mode I fracture. Mode I fractures are characterised by rough surfaces, due to the presence of broken fibre ends, and can be further divided into two types: matrix cleavage and fibre bridging [47].

At the microstructural level, the delamination failure extends along the fibres and propagates into the surrounding matrix, as shown in Figures 8 and 9. The textured micro flow and corresponding river lines converge within the matrix next to the fibres, and are the key features of the matrix cleavage. The river lines' growth direction typically follows the direction of the global crack growth [11].

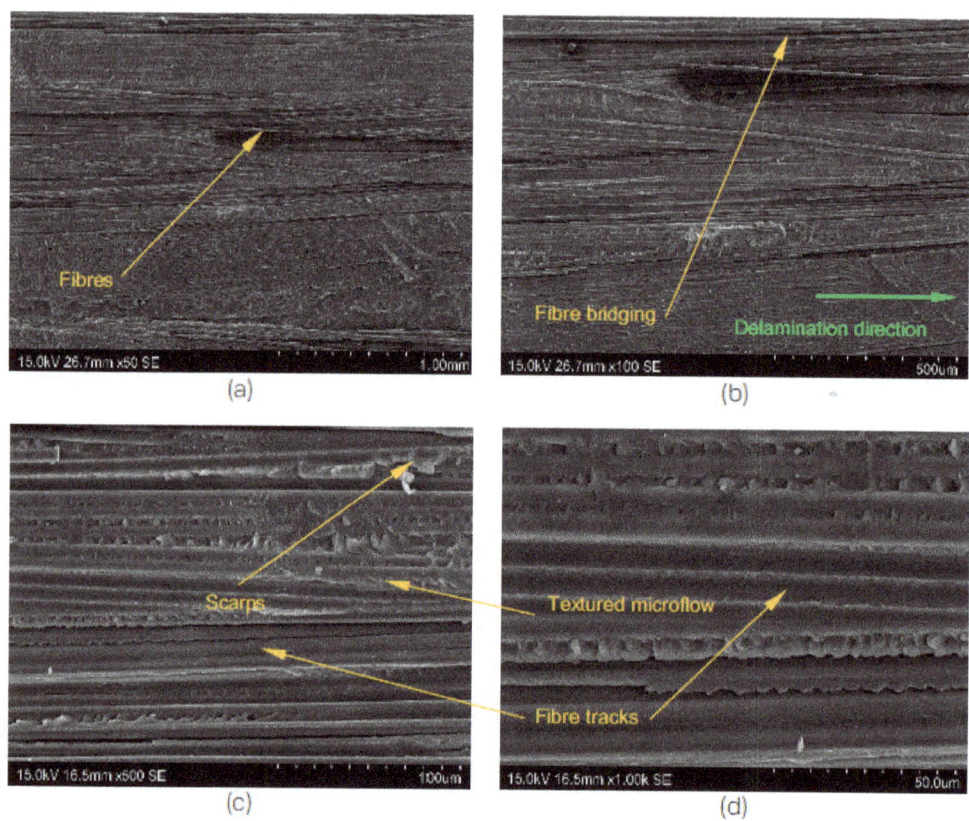

Figure 8. SEM view of the T700/PA6.6 DCB specimens after the test; scale bar is (**a**) 1 mm, (**b**) 500 μm, (**c**) 100 μm, and (**d**) 50 μm.

The degree of fibre bridging is often associated with the material and processing conditions of the laminated composites, and it tends to increase with increasing crack lengths [47]. Additionally, it can also contribute to a corrugated cross-section in the fracture surface, as shown in Figure 9.

The corrugated cross-section of the T700/PPS samples after the test, as shown in Figure 9, is the result of extensive fibre bridging and a combination of fibre-matrix debonding, matrix deformation, and void formation. This also contributes to the stick-slip crack growth behaviour that was observed during the experiment. The distinctively associated fracture morphology with thermoplastic composites, namely ductile drawing, could be observed in both systems, as shown in the micrographs in Figures 8 and 9.

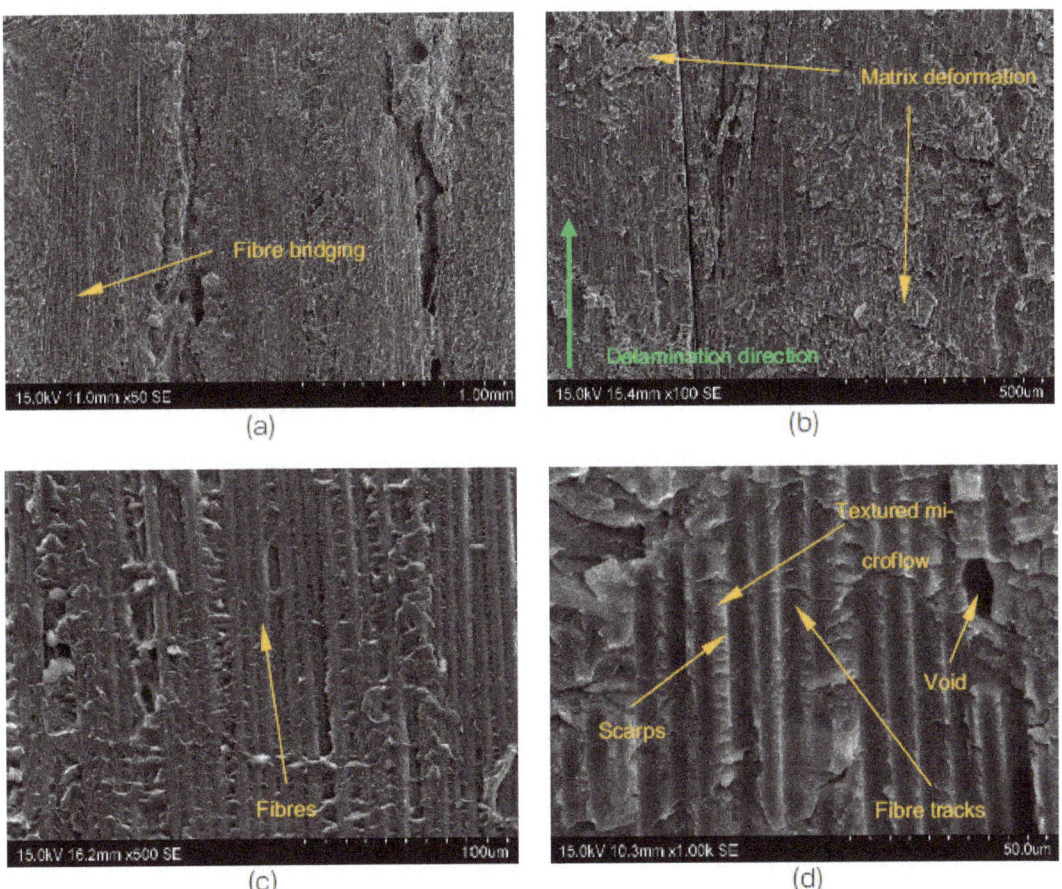

Figure 9. SEM view of the T700/PPS DCB specimens after the test; scale bar is (**a**) 1 mm, (**b**) 500 µm, (**c**) 100 µm, and (**d**) 50 µm.

These results were largely influenced by the through-thickness stitching of the NCF. The presence of stitching tends to enhance the Mode I delamination resistance. Based on Figure 10, the stitches of the T700/PPS are more apparent on the fracture surfaces compared to the T700/PA6.6. The stronger Kevlar® stitches on the former (compared to the Nylon® (DuPont, Wilmington, DE, USA) (stitching on the latter), to a certain extent, promote a greater crimp development of the tows. This results in an increase in the fracture toughness. Additionally, in materials with a high degree of crimp, stitches themselves can sometimes fail. When the stitches impose too much or add insufficient constraint to the tows, fibre waviness can develop. This misalignment leads to an increase in Mode I toughness and reduction in compression strength.

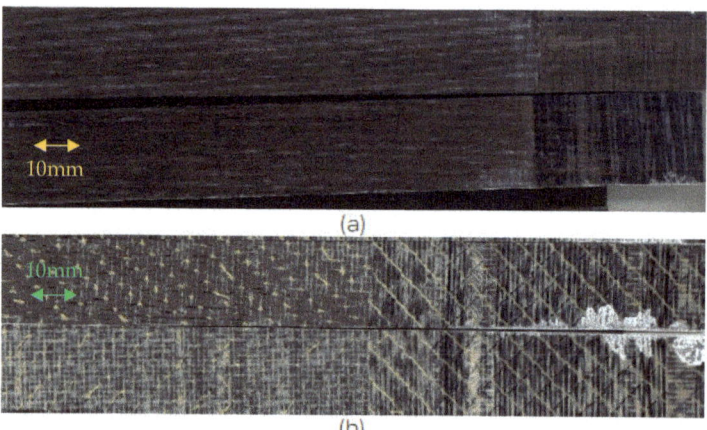

Figure 10. Mode I failure surfaces of NCF (**a**) T700/PA6.6 and (**b**) T700/PPS; crack growth from right to left.

6. Conclusions

The results of the study indicate that the performance of the composite laminate systems is highly dependent on the stitching material and the mechanical properties of the polymer used. This was evident from the microstructure characterisation of both systems, which was conducted using fractography. The analysis revealed that the crack propagation characteristics of both laminate systems were very comparable, with both systems displaying stable crack growth. Additionally, both systems showed stable delamination growth, which is an important factor to consider when evaluating the overall strength and durability of the composite materials.

One of the key findings of the study was that both CFRTP systems performed superiorly when compared to typical CFRTS systems. This is likely due to the fact that the CFRTP systems are able to maintain their structural integrity better under high stress and strain conditions, which is a result of the improved mechanical properties of the polymer used. This makes the CFRTP systems more suitable for applications that require high strength and durability, such as in aerospace and automotive industries.

Overall, the study provides valuable insights into the performance of composite laminate systems and the factors that influence their properties. The findings of the study can be used to develop more advanced composite materials that can meet the demanding requirements of various industries.

Author Contributions: Conceptualisation, M.A.A.M.; methodology, M.A.A.M.; software, L.I.; validation, M.A.A.M.; formal analysis, M.A.A.M.; investigation, M.A.A.M.; resources, L.I. and E.S.G.; data curation, M.A.A.M.; writing—original draft preparation, M.A.A.M.; writing—review and editing, M.A.A.M.; visualisation, M.A.A.M.; supervision, L.I. and E.S.G.; project administration, M.A.A.M.; funding acquisition, M.A.A.M. and L.I. All authors have read and agreed to the published version of the manuscript.

Funding: This research received funding from the Department of Aeronautics, Imperial College London and partners of the THERMOCOMP and DATACOMP project.

Institutional Review Board Statement: Not applicable.

Informed Consent Statement: Not applicable.

Data Availability Statement: Not applicable.

Acknowledgments: The author would like to express his deepest gratitude to Lorenzo Iannucci (Imperial College London) and Emile Greenhalgh (Imperial College London) for their guidance and contribution towards this research.

Conflicts of Interest: The authors declare no conflict of interest.

Nomenclature

ASTM	American Society for Testing and Materials
CAI	Compression after impact
CFRP	Carbon-fibre-reinforced polymer
CFRTP	Carbon-fibre-reinforced thermoplastic
CV	Coefficient of variation
DCB	Double cantilever beam
FVF	Fibre–volume–fraction
NCF	Non-crimp fabric
NFLS	Normal failure stress (Pa)
PA	Polyamide
PPS	Polyphenylene sulphide
SEM	Scanning electron microscope
SFLS	Shear failure stress (Pa)
TGA	Thermogravimetric analysis
UD	Unidirectional
A	Cross-sectional area (m^2)
C	Compliance
E	Young's modulus (Pa)
F	Force (N)
G_{IC}	Mode I fracture toughness (kJ/m^2)
K	Kinetic energy (J)
L	Length of specimen (m)
P	Load
T	Loading duration (s)
a	Delamination or crack length
b	Specimen width
Δ	Correction term
δ	Cross-head displacement
ρ	Density (kg/m^3)

References

1. Davies, P.; Moulin, C.; Kausch, H.H.; Fischer, M. Measurement of GIc and GIIc in Carbon/Epoxy Composites. *Compos. Sci. Technol.* **1990**, *39*, 193–205. [CrossRef]
2. Lee, K.Y.; Kwon, S.M. Interlaminar Fracture Toughness for Composite Materials. *Eng. Fract. Mech.* **1993**, *45*, 881–887.
3. Todo, M.; Jar, P.-Y.B. Study of Mode-I Interlaminar Crack Growth in DCB Specimens of Fibre-Reinforced Composites. *Compos. Sci. Technol.* **1998**, *58*, 105–118. [CrossRef]
4. Robinson, P.; Das, S. Mode I DCB Testing of Composite Laminates Reinforced with Z-Direction Pins: A Simple Model for the Investigation of Data Reduction Strategies. *Eng. Fract. Mech.* **2004**, *71*, 345–364. [CrossRef]
5. Del Saz-Orozco, B.; Ray, D.; Stanley, W.F. Effect of thermoplastic veils on interlaminar fracture toughness of a glass fiber/vinyl ester composite. *Polym. Compos.* **2017**, *38*, 2501–2508. [CrossRef]
6. Kuhtz, M.; Hornig, A.; Gude, M. Modification of Energy Release Rates in Textile Reinforced Thermoplastic Composites to Control Delamination Characteristics. In Proceedings of the 17th European Conference on Composite Materials (ECCM–17), Munich, Germany, 26–30 June 2016.
7. Haldar, S.; Lopes, C.S.; Gonzalez, C. Interlaminar and Intralaminar Fracture Behavior of Carbon Fiber Reinforced Polymer Composites. *Key Eng. Mater.* **2016**, *713*, 325–328. [CrossRef]
8. Arakawa, K.; Takahashi, K. Interlaminar Fracture Analysis of Composite DCB Specimens. *Int. J. Fract.* **1995**, *74*, 277–287. [CrossRef]
9. Caprino, G. The Use of Thin DCB Specimens for Measuring Mode I Interlaminar Fracture Toughness of Composite Materials. *Compos. Sci. Technol.* **1990**, *39*, 147–158. [CrossRef]

10. Bin Mohamed Rehan, M.S.; Rousseau, J.; Fontaine, S.; Gong, X.J. Experimental Study of the Influence of Ply Orientation on DCB Mode-I Delamination Behavior by Using Multidirectional Fully Isotropic Carbon/Epoxy Laminates. *Compos. Struct.* **2017**, *161*, 1–7. [CrossRef]
11. Mohsin, M.A.A.; Iannucci, L.; Greenhalgh, E.S. Mode I Interlaminar Fracture Toughness Characterisation of Carbon Fibre Reinforced Thermoplastic Composites. In *Proceedings of the American Society for Composites 2017*; DEStech Publications Inc.: Lancaster, PA, USA, 2017; Volume 1.
12. Carreras, L.; Bak, B.L.V.; Jensen, S.M.; Lequesne, C.; Xiong, H.; Lindgaard, E. Benchmark Test for Mode I Fatigue-Driven Delamination in GFRP Composite Laminates: Experimental Results and Simulation with the Inter-Laminar Damage Model Implemented in SAMCEF. *Compos. Part B Eng.* **2023**, *253*, 110529. [CrossRef]
13. Low, K.O.; Johar, M.; Sung, A.N.; Mohd Nasir, M.N.; Rahimian Koloor, S.S.; Petrů, M.; Israr, H.A.; Wong, K.J. Displacement Rate Effects on Mixed-Mode I/II Delamination of Laminated Carbon/Epoxy Composites. *Polym. Test.* **2022**, *108*, 107512. [CrossRef]
14. Yao, L.; Liu, J.; Lyu, Z.; Alderliesten, R.C.; Hao, C.; Ren, C.; Guo, L. In-Situ Damage Mechanism Investigation and a Prediction Model for Delamination with Fibre Bridging in Composites. *Eng. Fract. Mech.* **2023**, *281*, 109079. [CrossRef]
15. Vallons, K.; Behaeghe, A.; Lomov, S.V.; Verpoest, I. Impact and Post-Impact Properties of a Carbon Fibre Non-Crimp Fabric and a Twill Weave Composite. *Compos. Part A Appl. Sci. Manuf.* **2010**, *41*, 1019–1026. [CrossRef]
16. Chen, Q.; Wu, F.; Jiang, Z.; Zhang, H.; Yuan, J.; Xiang, Y.; Liu, Y. Improved Interlaminar Fracture Toughness of Carbon Fiber/Epoxy Composites by a Combination of Extrinsic and Intrinsic Multiscale Toughening Mechanisms. *Compos. Part B Eng.* **2023**, *252*, 110503. [CrossRef]
17. Tugrul Seyhan, A.; Tanoglu, M.; Schulte, K. Mode I and Mode II Fracture Toughness of E-Glass Non-Crimp Fabric/Carbon Nanotube (CNT) Modified Polymer Based Composites. *Eng. Fract. Mech.* **2008**, *75*, 5151–5162. [CrossRef]
18. Colin de Verdiere, M.; Skordos, A.A.; May, M.; Walton, A.C. Influence of Loading Rate on the Delamination Response of Untufted and Tufted Carbon Epoxy Non Crimp Fabric Composites: Mode I. *Eng. Fract. Mech.* **2012**, *96*, 11–25. [CrossRef]
19. Gouskos, D.; Iannucci, L. A Failure Model for the Analysis of Cross-Ply Non-Crimp Fabric (NCF) Composites under in-Plane Loading: Experimental & Numerical Study. *Eng. Fract. Mech.* **2022**, *271*, 108575. [CrossRef]
20. Drake, D.A.; Sullivan, R.W. Prediction of Delamination Propagation in Polymer Composites. *Compos. Part A Appl. Sci. Manuf.* **2019**, *124*, 105467. [CrossRef]
21. Quan, D.; Murphy, N.; Ivanković, A.; Zhao, G.; Alderliesten, R. Fatigue Delamination Behaviour of Carbon Fibre/Epoxy Composites Interleaved with Thermoplastic Veils. *Compos. Struct.* **2022**, *281*, 114903. [CrossRef]
22. Donough, M.J.; Farnsworth, A.L.; Phillips, A.W.; St John, N.A.; Prusty, B.G. Influence of Deposition Rates on the Mode I Fracture Toughness of In-Situ Consolidated Thermoplastic Composites. *Compos. Part B Eng.* **2023**, *251*, 110474. [CrossRef]
23. Zhou, J.; He, T.; Li, B.; Liu, W.; Chen, T. A Study of Mode I Delamination Resistance of a Thermoplastic Composite. *Compos. Sci. Technol.* **1992**, *45*, 173–179. [CrossRef]
24. Mohsin, M.A.A.; Iannucci, L.; Greenhalgh, E.S. Experimental and Numerical Analysis of Low-Velocity Impact of Carbon Fibre-Based Non-Crimp Fabric Reinforced Thermoplastic Composites. *Polymers* **2021**, *13*, 3642. [CrossRef] [PubMed]
25. Chen, Y.; Hou, S.; Fu, K.; Han, X.; Ye, L. Low-Velocity Impact Response of Composite Sandwich Structures: Modelling and Experiment. *Compos. Struct.* **2017**, *168*, 322–334. [CrossRef]
26. Brunner, A.J.; Murphy, N.; Pinter, G. Development of a Standardized Procedure for the Characterization of Interlaminar Delamination Propagation in Advanced Composites under Fatigue Mode I Loading Conditions. *Eng. Fract. Mech.* **2009**, *76*, 2678–2689. [CrossRef]
27. Chai, H. The Characterization of Mode I Delamination Failure in Non-Woven, Multidirectional Laminates. *Composites* **1984**, *15*, 277–290. [CrossRef]
28. UK-DATACOMP|CIC. Available online: http://the-cic.org.uk/uk-datacomp (accessed on 5 October 2015).
29. UK-THERMOCOMP|CIC. Available online: http://the-cic.org.uk/uk-thermocomp (accessed on 5 October 2015).
30. Jiang, Z.; Liu, P.; Chen, Q.; Sue, H.-J.; Bremner, T.; DiSano, L.P. The Influence of Processing Conditions on the Mechanical Properties of Poly(Aryl-Ether-Ketone)/Polybenzimidazole Blends. *J. Appl. Polym. Sci.* **2020**, *137*, 48966. [CrossRef]
31. Jiang, Z.; Liu, P.; Sue, H.-J.; Bremner, T. Effect of Annealing on the Viscoelastic Behavior of Poly(Ether-Ether-Ketone). *Polymer* **2019**, *160*, 231–237. [CrossRef]
32. Mohsin, M.A.A.; Iannucci, L.; Greenhalgh, E.S. On the Dynamic Tensile Behaviour of Thermoplastic Composite Carbon/Polyamide 6.6 Using Split Hopkinson Pressure Bar. *Materials* **2021**, *14*, 1653. [CrossRef]
33. Mohsin, M.A.A. *Manufacturing, Testing, Modelling and Fractography of Thermoplastic Composites for the Automotive Industry*; Imperial College: London, UK, 2019.
34. Mohsin, M.A.A.; Iannucci, L.; Greenhalgh, E.S. Fibre-Volume-Fraction Measurement of Carbon Fibre Reinforced Thermoplastic Composites Using Thermogravimetric Analysis. *Heliyon* **2019**, *5*, e01132. [CrossRef]
35. MatWeb Overview of Materials for Nylon 66, Unreinforced. Available online: http://www.matweb.com/search/DataSheet.aspx?MatGUID=a2e79a3451984d58a8a442c37a226107&ckck=1 (accessed on 12 October 2021).
36. Tohgo, K.; Hirako, Y.; Ishii, H.; Sano, K. Mode I Interlaminar Fracture Toughness in Carbon Fiber Reinforced Thermoplastic Laminate. *Nippon Kikai Gakkai Ronbunshu A Hen/Transactions Japan Soc. Mech. Eng. Part A* **1995**, *61*, 1273–1279. [CrossRef]
37. Overview of Materials for Polyphenylene Sulfide (PPS), Unreinforced, Extruded. Available online: http://www.matweb.com/search/DataSheet.aspx?MatGUID=f277b224f135406caa973d38d49104ca&ckck=1 (accessed on 29 June 2017).

38. Radlmaier, V.; Obermeier, G.; Ehard, S.; Kollmannsberger, A.; Koerber, H.; Ladstaetter, E. Interlaminar Fracture Toughness of Carbon Fiber Reinforced Thermoplastic In-Situ Joints. *AIP Conf. Proc.* **2016**, *1779*, 090003. [CrossRef]
39. Toray Composite Materials America Inc. T700S Standard Modulus Carbon Fiber. Available online: https://www.toraycma.com/wp-content/uploads/T700S-Technical-Data-Sheet-1.pdf.pdf (accessed on 12 October 2021).
40. Mohsin, M.A.A.; Iannucci, L.; Greenhalgh, E.S. Translaminar Fracture Toughness Characterisation of Carbon Reinforced Thermoplastic Composites. In Proceedings of the ECCM17—17th European Conference on Composite Materials, European Society for Composite Materials, Munich, Germany, 26–30 June 2016.
41. *ASTM D5528-13*; Standard Test Method for Mode I Interlaminar Fracture Toughness of Unidirectional Fiber-Reinforced Polymer Matrix Composites 1. ASTM International: West Conshohocken, PA, USA, 2017; pp. 1–13.
42. Hodgkinson, J.M. *Mechanical Testing of Advanced Fibre Composites*; CRC Press: Boca Raton, FL, USA, 2000; ISBN 1855733129.
43. Prel, Y.J.; Davies, P.; Benzeggagh, M.L.; Charentenay, F.-X. Mode I and Mode II Delamination of Thermosetting and Thermoplastic Composites. *Compos. Mater. Fatigue Fract.* **1989**, *2*, 251–269. [CrossRef]
44. Ivanov, S.G.; Beyens, D.; Gorbatikh, L.; Lomov, S.V. Damage Development in Woven Carbon Fibre Thermoplastic Laminates with PPS and PEEK Matrices: A Comparative Study. *J. Compos. Mater.* **2017**, *51*, 637–647. [CrossRef]
45. Tijs, B.H.A.H.; Abdel-Monsef, S.; Renart, J.; Turon, A.; Bisagni, C. Characterization and Analysis of the Interlaminar Behavior of Thermoplastic Composites Considering Fiber Bridging and R-Curve Effects. *Compos. Part A Appl. Sci. Manuf.* **2022**, *162*, 107101. [CrossRef]
46. Reis, J.P.; de Moura, M.F.S.F.; Moreira, R.D.F.; Silva, F.G.A. Pure Mode I and II Interlaminar Fracture Characterization of Carbon-Fibre Reinforced Polyamide Composite. *Compos. Part B Eng.* **2019**, *169*, 126–132. [CrossRef]
47. Greenhalgh, E.S. *Failure Analysis and Fractography of Polymer Composites (Google EBook)*; CRC Press LLC: Oxford, UK, 2009; ISBN 1845696816.

Disclaimer/Publisher's Note: The statements, opinions and data contained in all publications are solely those of the individual author(s) and contributor(s) and not of MDPI and/or the editor(s). MDPI and/or the editor(s) disclaim responsibility for any injury to people or property resulting from any ideas, methods, instructions or products referred to in the content.

Article

On the Addition of Multifunctional Methacrylate Monomers to an Acrylic-Based Infusible Resin for the Weldability of Acrylic-Based Glass Fibre Composites

Henri Perrin, Masoud Bodaghi *, Vincent Berthé and Régis Vaudemont

Luxembourg Institute of Science and Technology (LIST), 5, rue Bommel, L-4940 Hautcharage, Luxembourg
* Correspondence: masoud.bodaghi@list.lu; Tel.: +352-2758884575

Abstract: The melt strength of Elium® acrylic resin is an important factor to ensure limited fluid flow during welding. To provide Elium® with a suitable melt strength via a slight crosslink, this study examines the effect of two dimethacrylates, namely butanediol-di-methacrylate (BDDMA) and tricyclo-decane-dimethanol-di-methacrylate (TCDDMDA), on the weldability of acrylic-based glass fibre composites. The resin system impregnating a five-layer woven glass preform is a mixture of Elium® acrylic resin, an initiator, and each of the multifunctional methacrylate monomers in the range of 0 to 2 parts per hundred resin (phr). Composite plates are manufactured by vacuum infusion (VI) at an ambient temperature and welded by using the infrared (IR) welding technique. The mechanical thermal analysis of the composites containing multifunctional methacrylate monomers higher than 0.25 phr shows a very little strain for the temperature range of 50 °C to 220 °C. The quantity of 0.25 phr of both of the multifunctional methacrylate monomers in the Elium® matrix improves the maximum bound shear strength of the weld by 50% compared to those compositions without the multifunctional methacrylate monomers.

Keywords: adhesion; composites; crosslinking; IR welding; multifunctional methacrylate monomers

Citation: Perrin, H.; Bodaghi, M.; Berthé, V.; Vaudemont, R. On the Addition of Multifunctional Methacrylate Monomers to an Acrylic-Based Infusible Resin for the Weldability of Acrylic-Based Glass Fibre Composites. *Polymers* **2023**, *15*, 1250. https://doi.org/10.3390/polym15051250

Academic Editors: Tomasz Makowski and Sivanjineyulu Veluri

Received: 27 January 2023
Revised: 21 February 2023
Accepted: 26 February 2023
Published: 28 February 2023

Copyright: © 2023 by the authors. Licensee MDPI, Basel, Switzerland. This article is an open access article distributed under the terms and conditions of the Creative Commons Attribution (CC BY) license (https://creativecommons.org/licenses/by/4.0/).

1. Introduction

As they are lower in crosslink density, thermoplastic polymers show a generally lower strength and stiffness than thermoset ones [1]. Continuous fibre arrangements for fibre-reinforced thermoplastic composites are therefore a better choice than short fibres for structural applications due to their much higher modulus and strength [2]. Being able to melt matrix instead of cure can lead to the joining of continuous fibre-reinforced thermoplastic composites (continuous fibre TPCs) by fusion-bonding techniques [3,4]. This welding family involves two consecutive steps: (1) deconsolidation: heating and melting the surfaces of thermoplastic composites to be joined, and (2) consolidation: pressing the exposed surfaces together for consolidation. The welding technology can be divided into four groups based on alternative heat sources (Figure 1): friction welding, electromagnetic welding, bulk heating, and thermal welding [5].

Among the variants of fusion bonding (Figure 1), ultrasonic welding, induction welding, and resistance welding are more mature than infrared (IR) welding [6,7]. Similar to the others, the IR welding also offers short processing cycles, a potential for mass production, and a high degree of integration (Table 1).

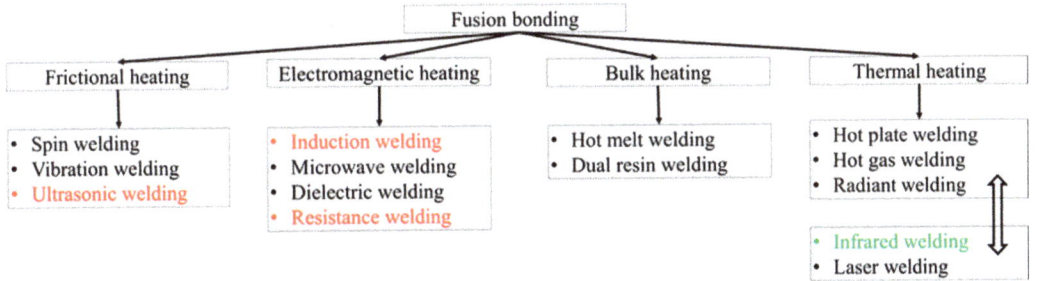

Figure 1. Variants of fusion-bonding methods for thermoplastic composites (adopted with permission from [5]. 2001, Elsevier).

Table 1. Important welding factors for fusion welding processes.

Fusion-Bonding Process	Heating Time (s)	Specific Parameters	Scale-Up Potential	Reference
Ultrasonic welding	3–4	Frequency: 20–40 kHz	A thermoplastic composite airframe panel	Palardy and Villegas [8]
		Amplitude: 10–100 μm		
Induction welding	10–3600	Frequency: 60–100 MHz	One-shot welding up to 600 cm	Williams et al. [9]
Resistance welding	30–3000	Power input: 30–160 kW/m^2	Double lap joints with the length of 1.2 m	McKnight et al. [10]
		Pressure: 0.1–1.4 MPa		
Infrared welding	10–30	Beam temperature: 650 °C	Welding of a surface area of 1.2 × 2.4 m^2	Swartz [11]

IR welding is a non-contact technique for bonding fibre-reinforced thermoplastic composites. The heat generation at the joint interface is achieved by absorption and conversion of electromagnetic radiation without physical contact, and hence the technique eliminates the occurrence of surface contamination. This technique has the ability of fast heating, and welding at high productivity rates in an automated system [12,13].

Only a few studies have examined the weld strength of thermoplastic composite plates by IR welding [12,14] (Table 2). The examined continuous fibre TPCs were manufactured based on melt processing, such as hot stamping. One of the main difficulties associated with the production of continuous fibre TPCs is the impregnation of fibre bundles and wetting of the fibre due to the high resin melt viscosity [6].

Table 2. Conducted research on IR welding of thermoplastics.

References	Polymer Composite	Interests of Study	Observations
Baere et al. [12,15]	5-harness satin weave carbon-reinforced polphenylene sulfide (PPS)	Heating time, contact pressure, consolidation time using lap shear experiments	Pre-consolidation of PPS resulted in reproducible results with IR process
Chateau et al. [14]	Glass fibre Polycarbonate/ polycarbonate composite	Welding temperature on the joining quality	A complete self-diffusion was not feasible due to inadequate welding temperature
Perrin et al. [13]	Carbon fibre-reinforced peek	Chemical surface modification on the mechanical performance	IR welding was proven to be an appropriate technique for welding of dissimilar composite matrix

A shift from melt processing towards the reactive processing of thermoplastic composites allows the injection of a low viscosity (i.e., 0.01 to 0.1 Pa·s) mono- or oligo-meric precursor, such as polyamides (PA6, PA12) and polybutylene terephthalates, into a fibre preform under vacuum or positive pressures and subsequent in situ polymerisation [16].

1.1. Infusible Thermoplastic Resin Welding Process

The reactive process is associated with the high processing temperature for in situ polymerisation and hence is not a proper choice for the large composite parts, such as wind turbine blades [17]. In a possible response to the low-temperature reactive processing solution, Arkema developed infusible thermoplastic resins (i.e., Elium®) from acrylic resins for reactive processing [18]. The components of Elium® acrylic resin are 2-propenoic acid, 2-methyl-, methyl ester, or methylmethacrylate monomer (MMA), and acrylic copolymer. For the in situ polymerisation, the resin is mixed with a compatible initiator system such as peroxide. The polymerisation depending on the initial compositions can be carried out from room temperature up to 90 °C [19].

Fusion joining of continuous fibre-reinforced Elium® composites opens a new field of study. Ultrasonic welding, resistance welding, and induction welding (highlighted in the red-font colours in Figure 1) have been investigated for this new infusible thermoplastic resin (Elium®) [20,21]. Murray, R.E. et al. [20] applied fusion-bonding techniques including resistance and induction welding for the glass fibre thermoplastic acrylic-based Elium®-188 composite for wind blade applications. They obtained an improvement of 30% in the lap shear weld joint strength of the plates, as compared to the ones bonded with adhesives. However, the scale-up of the resistance welding may be challenging due to the requirements of power and voltage for a longer weld line. In addition, the requirement of physical contact with the welded joints in resistance welding may pose challenges for the connection of the heating element to the power supply, particularly for the internal joints. In other study, the ultrasonic welding technique was applied for carbon/Elium®-150, and Bhudolia et al. [21,22] obtained a 23% improvement in the weld joint strength of plates compared to the adhesively bonded joints. Nevertheless, the application of ultrasonic welding is particularly limited to aerospace industries, where the high mechanical performance is often preferred rather than the associated cost.

The relative newcomer IR welding is still unknown for its effects on thermoplastic acrylic-based Elium® composites. Current research focuses on the first attempt to apply the IR technique and investigate the weldability of glass fibre composite plates reinforced with the addition of multifunctional methacrylate monomers into the Elium® matrix.

1.2. Problem Statement and Objectives

In all variants of fusion bonding, a polymer across the exposed surfaces is heated above 75% of its transition temperature (Elium®, which is an amorphous thermoplastic polymer, is above its glass transition temperature). Above this transition temperature, the melt viscosity of the polymer is reduced, and hence polymer chains flow and form the joint area. In the welding zone where the local temperature could be as high as 200 °C, the low melting temperature of Elium® [23,24] may lead to a poor bond. Without crosslinks, the melt strength is too low due to the low viscosity. The high local temperature results in limitations on the exposure heating time during the bonding process. When the heating time is too long, Elium® would melt if it has a linear chemical structure. A way to compensate these detrimental effects is boosting the formation of crosslinks in Elium® by using multifunctional methacrylate monomers. The crosslinks improve the melt strength of Elium® to ensure it does not flow during the welding. The quantity of the multifunctional methacrylate monomers must be optimised. The higher quantity of multifunctional methacrylate monomer could limit the welding by lowering the diffusion of chains [25].

Currently, the literature is lacking on the crosslinking of Elium®-based composites [26], and to the best of our knowledge, no studies have covered the welding of partially crosslinked Elium®-based structural composites. This study focuses on the addition of the multifunctional methacrylate monomers in 0 to 2 phr (parts per hundred resin) into Elium® resin. The key hypothesis here is that the differences in the phr may produce a significant impact on the Elium® composite's weldability. A set of experiments are then performed by IR welding of thermoplastic composite plates manufactured by a typical vacuum infusion. Finally, the sample preparations are performed to assess the thermomechanical properties of the welded joints. The experiment steps are shown in Figure 2.

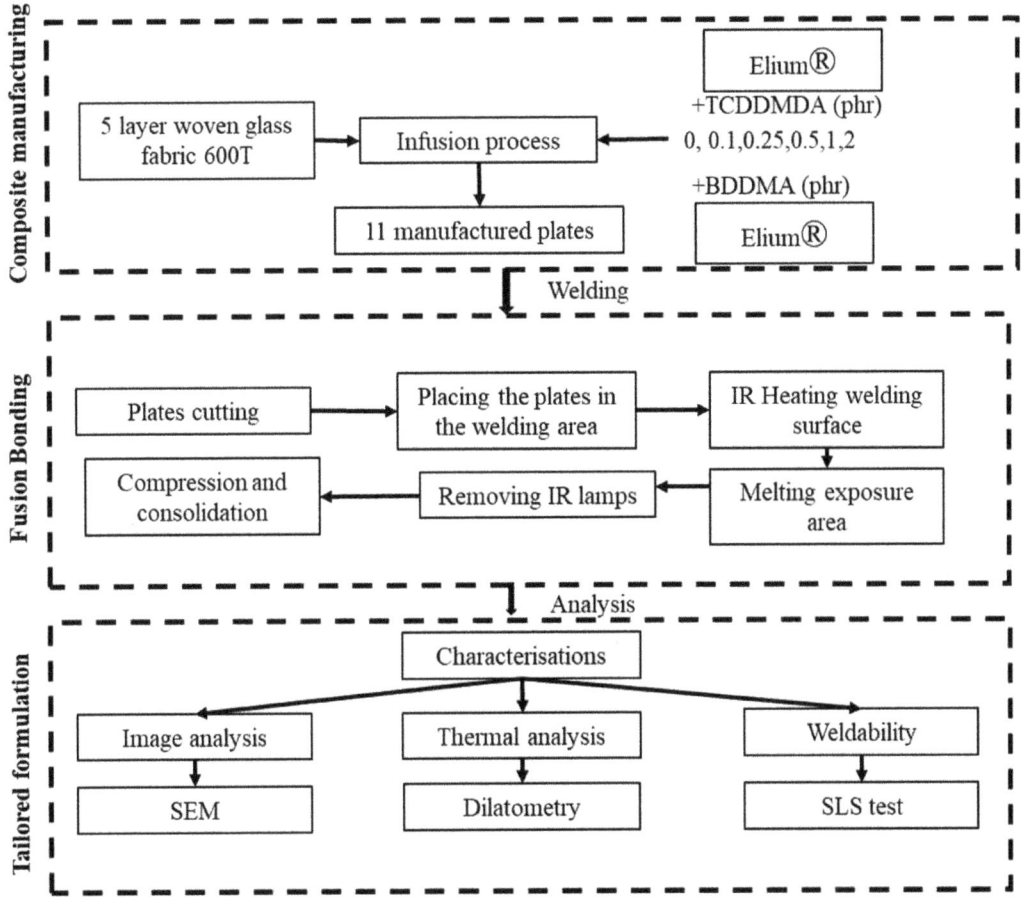

Figure 2. The experiment steps for the IR welding study of the modification of the Elium® formulation.

2. Experimental Details

2.1. Material Selection

2.1.1. Multifunctional Methacrylate Monomers

Two multifunctional methacrylate monomers, namely butanediol-di-methacrylate (BDDMA) and tricyclo-decane-dimethanol-di-methacrylate (TCDDMDA), were supplied by Sartomer, France. These monomers are aimed at reducing the volatile organic compounds, increasing heat resistance, and improving adhesion on glass fibre. Their different physical properties are summarised in Table 3.

Table 3. Physical properties of acrylic-based multifunctional methacrylate monomers.

Chemical Name	Commercial Name	Structure	Functionality	Viscosity at 25 °C (mPa.s)	Features	T_g (°C)
TCDDMDA	SR834		4	108	Toughness	High T_g
BDDMA	SR214		4	12	Hardness Excellent solubility	210

2.1.2. Constituents of Thermoplastic Composites

The reinforcement was a woven glass fabric 600 T from Chomarat with the aerial density of 600 g/m². The Elium® 188XO resin impregnating the fabric was supplied by ARKEMA, Colombes, France. The resin can easily be processed by vacuum infusion and resin transfer moulding (RTM) and assembly by welding. The viscosity of resin was 100 mPa.s at room temperature and the addition of benzoyl peroxide will initiate the polymerisation without a heat source. It should be noted that the resin viscosity after the addition of both multifunctional methacrylate monomers to the Elium® resin was not changed. The glass fibre preform was impregnated with the resin.

For the current study, the resin system for the reference samples (without multifunctional methacrylate monomers) was a mixture of the Elium® resin and the initiator with a mass ratio of 100:2, and each of the multifunctional methacrylate monomers in the range of 0 to 2 parts per hundred resin (phr). Depending on the Elium® composition, 11 composite plates were manufactured with a vacuum infusion process.

2.2. Composite Manufacturing

The vacuum infusion technique, which is one of the LCM variants, was shown to be a viable technique for manufacturing a 9 m fibre-reinforced Elium® wind turbine spar cap in [17]. Following this successful demonstration, the current study also applied the vacuum infusion technique. The process starts by placing a five-layer glass fibre preform and follows with the sealing of the mould by vacuum bag at 0.8 mbar. The resin system was degassed for 10 min at 150 mbar. Under vacuum of 160 mbar, the resin was pulled from the reservoir into the mould.

The material temperature and the polymerisation cycle were monitored from the exterior layer of the vacuum bag. Three heat flux sensors were placed on the area of interest, as shown in Figure 3. As the sensors are isolated from the composite part, non-intrusive heat flux sensors do not disturb the processing cycle. The sensors allowed us to measure the local heat flux and temperature versus time and distance from the resin injection point. In addition, the resin flow front can be tracked with a change in thermal conductivity between the dry and impregnated fibre preforms. As the current paper focuses on the weldability of Elium® composites, the data including heat flux, temperature, and resin flow front versus time are not presented.

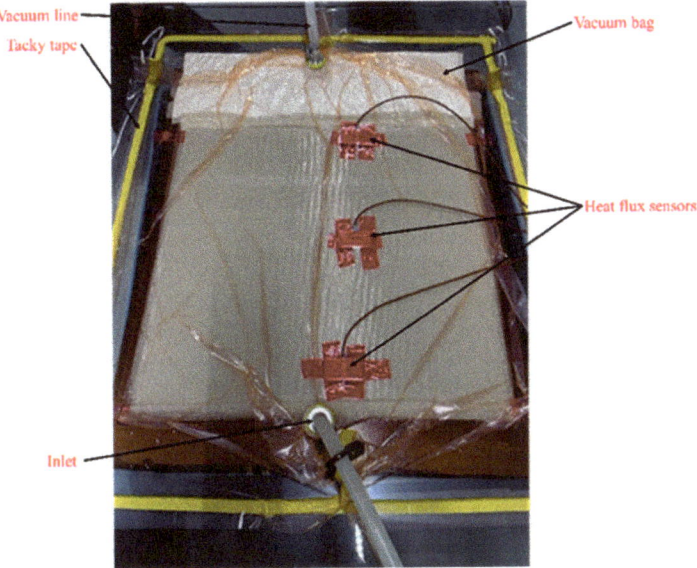

Figure 3. The position of heat flux sensors during vacuum infusion.

2.3. IR Welding Process

Infrared welding steps involve placing two thermoplastic composite parts on the upper and lower welding surfaces. Vacuum grippers will hold the parts in place (Figure 4-1). Subsequently, IR lamps on a mobile frame were inserted between the plates to homogenously melt a thin layer of plastic on the surface of each composite part (Figure 4-2). Then, the IR lamps were removed, and the parts were clamped to solidify the melted surface under pressure (Figure 4-3). Finally, after cooling, the joined composites were unloaded from the IR welding tool (Figure 4-4).

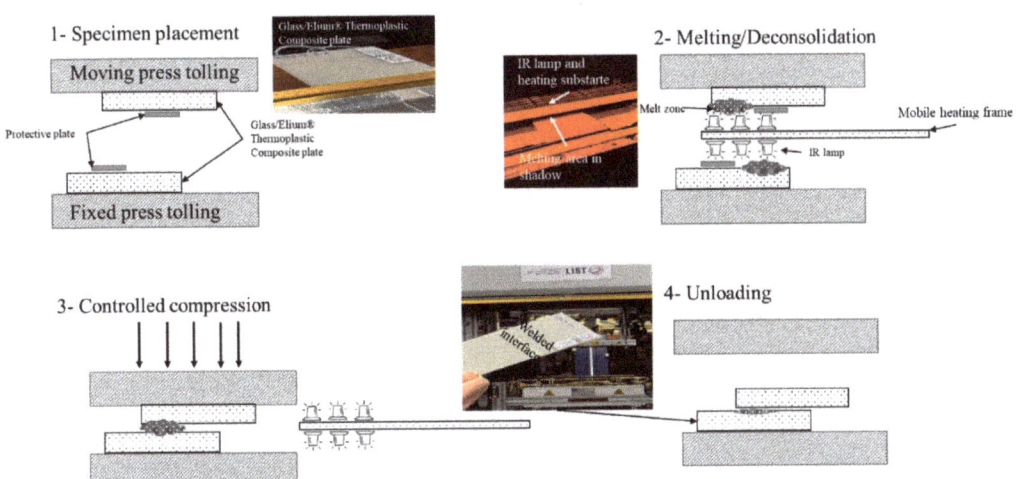

Figure 4. Descriptive IR welding process for Elium® thermoplastic welding: (**1**) specimen placement, (**2**) melting/deconsolidation, (**3**) controlled compression, and (**4**) unloading.

For the current study, an infrared welding machine (IR-V-ECO-800) from FRIMO [26], which is a lead supplier of infrared welding systems, was used. The composite plates were first cut from the middle along the resin infusion direction. Subsequently, the plates were welded together by the IR-V-ECO-800 welding machine, FRIMO, Lotte, Germany with a 25.3 mm overlap.

2.4. Analysis and Testing

The welding samples were cut into several specimens for the purpose of thermal and mechanical assessments.

2.4.1. Dilatometry

The thermal strain of samples was measured by using an optical sensor with a constant load of 0.2 N. The tests were conducted using a Netzsch DIL 402 Expedis dilatometer at a temperature range of 25–200 °C, with a heating rate of 5 °C/min. For both multifunctional methacrylate monomers in the range of 0 to 2 parts per hundred resin (phr), one composite sample of 5 mm × 5 mm was cut. The thermal analysis was carried out once per resin composition.

2.4.2. Single-Lap Shear (SLS)

According to the ASTM D5868-01, comparative shear strength data for joints were generated on the welded specimens with the dimensions of 170 mm for a welding line, and an overlap area of 25.3 mm × 25.3 mm. A universal Instron machine with a 1 kN load cell and a crosshead speed of 2 mm/min was used for conducting the tests. For both multifunctional methacrylate monomers in the range of 0 to 2 parts per hundred resin (phr), five composite samples were cut to produce SLS test samples. Thus, the SLS tests were repeated five times for both BDDMA and TCDDMDA samples.

2.4.3. Image Analysis

The fracture surfaces of joints were examined by visual observations and by using pressure-controlled FEI Quanta 200 FEG scanning electron microscopy (SEM), Hillsboro, OR, USA.

3. Results and Discussion

3.1. Thermal Analysis

When designing thermal stable thermoplastic composites, thermal expansion is an important property. Figure 5 shows a comparison of thermal expansions between the composite samples without and with multifunctional methacrylate monomers. The addition of the multifunctional methacrylate monomers to the Elium® matrix below 0.25 phr shows that the composites underwent strain as the temperature increased. This means that at quantities below 0.25 phr, the crosslinkers did not provide enough cross-linkage between the long linear Elium® matrix chain. This could be caused by the higher activity of the acrylate group in the Elium® matrix compared to the methacrylate group in multifunctional methacrylates, and hence three-dimensional crosslinks were not completely formed [27,28]. As the reactive quantity became higher than 0.25 phr in the Elium® compositions, the composites showed very little strain when increasing the temperature from 50 °C to 220 °C. This indicates that incorporation of multifunctional methacrylates forms a crosslinked structure on the long linear Elium® matrix chain. This crosslinked structure is associated with several bridges between the linear chain. Therefore, as phr increased, the number of bridges between the linear chain increased, forming a three-dimensional network that increased the rigidity of the Elium® matrix.

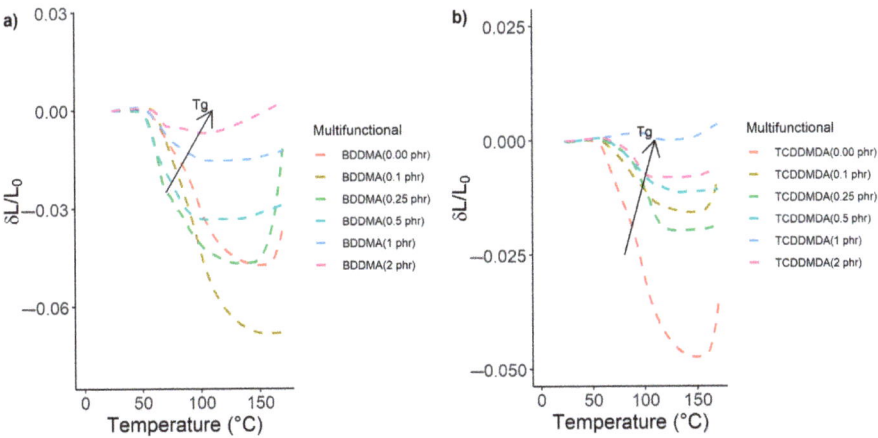

Figure 5. Dilatometry curves of Elium® composites for both multifunctional methacrylate monomers. Black arrows show the increasing trend in T_g: (**a**) butanediol-di-methacrylate (BDDMA) and (**b**) tricyclo-decane-dimethanol-di-methacrylate (TCDDMDA).

The shift to higher T_g also, as shown in Figure 5 (black arrows), demonstrates the increase in inflexibility of molecular chains [29] as the phr of multifunctional methacrylate monomers in the Elium® matrix increased. The correlation between T_g, multifunctional methacrylate monomer chain length, and its concentration is complex [30]. Different curing cycles may also change T_g up to 20 °C [31]. Apart from the TCDDMDA and BDDMA comonomers, certain other monomers have been added to MMA and their influence on the T_g has been examined [32]. For instance, T_g was increased by the addition of fluoro-monomers in MMA. Kubota et al.'s conclusion [33] is bound to the direct relationship between the C-chain length and the T_g. On the other hand, the addition of itaconate and nitro-monomers to MMA showed a reduction in T_g [34]. Therefore, T_g can be influenced by both C-chain length and the type of multifunctional methacrylate monomer added. However, Elium® is a formulated resin made of different acrylic compounds and is not like PMMA. The authors therefore believe these effects also need to be examined for Elium® and plan to address them in their future studies.

Figure 5 also shows an expanding trend for the thermoplastic composites without a multifunctional methacrylate monomer after 160 °C due to the deconsolidation of head-to-head linkage. However, the existence of multifunctional methacrylate monomers in the Elium® matrix reduced or inhibited the expansion. As the phr of multifunctional methacrylate monomers increased, it seems the methacrylate groups made the Elium® matrix less reactive, the radical polymerisation became slower, and the density of the crosslinked structure on the polymerised Elium® chain increased.

Figure 6 shows the colour changes of the weld area in the deconsolidated state during composite heating for the IR welding. The surface colouring of the reference sample without multifunctional methacrylate monomers changed to white due to the deconsolidation. The deconsolidation was caused by the decompaction of glass fibre reinforcements and the thermal expansion and viscoelastic behaviour of the Elium® matrix [35,36]. From 0 to 0.25 phr, their colouring changed to white. As the phr of multifunctional methacrylate monomers increased to 0.5 and beyond, the surface colouring remained unaffected by heat, as compared to those compositions below 0.25 phr, indicating a highly crosslinked structure.

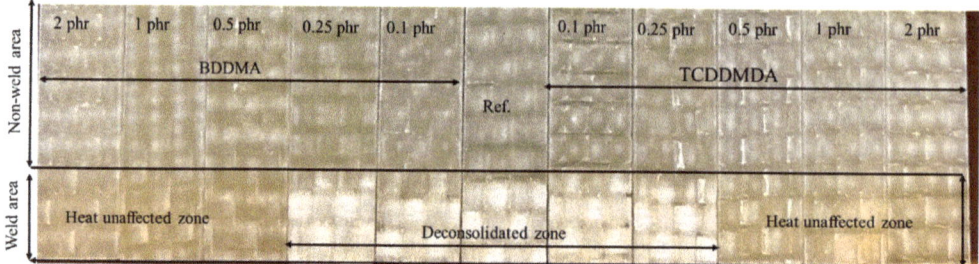

Figure 6. The deconsolidated area during heating the composites for IR welding. Ref. stands for the Elium® matrix without multifunctional methacrylate monomers.

3.2. Mechanical Properties

3.2.1. Single-Lap Shear (SLS)

The lap shear strength for the IR-welded samples varied depending on the phr of the multifunctional methacrylate monomers in the Elium® matrix. Figure 7 shows the scatter plot of average lap shear strengths of IR-welded specimens as a function of the multifunctional methacrylate monomers' phr in the Elium® matrix. The error bars correspond to the standard deviation of each set of characterised joints.

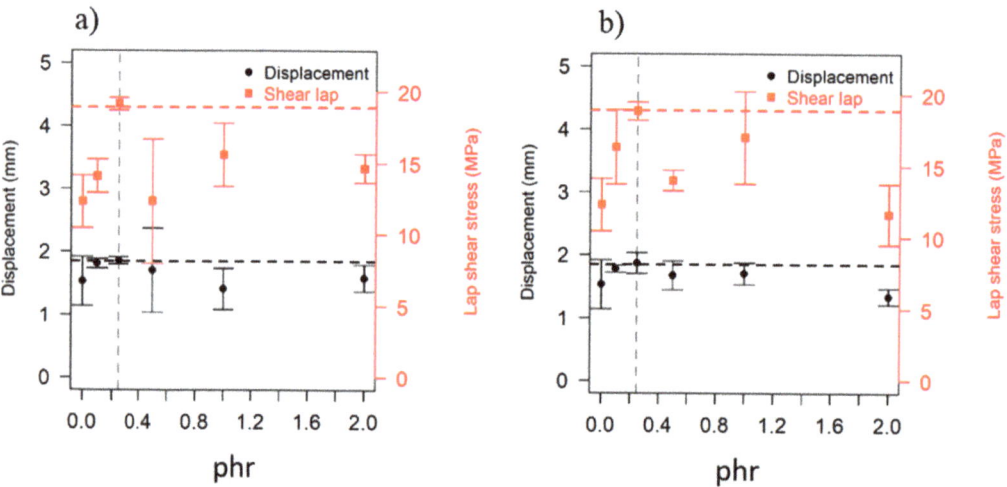

Figure 7. Lap shear strength of the IR-welded sample versus phr of each of the multifunctional methacrylate monomers: the red square points represent lap shear stress. The black circle points represent displacement during single-lap shear stress. (**a**) SLS test data with BDDMA and (**b**) SLS test data with TCDDMDA.

In Figure 7, the maximum lap shear strengths (red dashed line) of the specimens were obtained at 0.25 phr (grey dashed line) with the two multifunctional methacrylate monomers by up to 50% higher than the reference sample (0 phr). This increase in the lap shear strengths could be attributed to the increase in reactivity brought about by the monomers' addition. Notably, 0.25 phr could be an optimum point for high polymer chain mobility and high polymerisation reactivity. The increase beyond 0.25 phr could contribute to the further reduction in polymer chain mobility, and hence the lap shear strengths fell and could reach up to 40% below the maximum value at 2 phr for each of the multifunctional methacrylate monomers (as shown in Figure 7). Further studies are required to examine the gel content at different phr.

Figure 7 also provides additional information on the joint quality in terms of ductility, which is calculated from the joint displacement data during deformation under loading. The maximum displacement for the welding specimens with different concentrations of multifunctional methacrylate monomers is shown in Figure 7 in the black circle points. The 0.25 phr (vertical grey dashed line) had the highest displacement (horizontal black dashed line) as compared to the reference sample and those with different phr values. The specimens with 0.25 phr showed a better load distribution between the bonded plates, which will yield a robust and strong joint design [37]. The authors plan to address the fracture toughness characterisation to further explore the performance of IR-welded bonds with the modification of the Elium® matrix at different fibre volume fractions.

More scatters (higher coefficient of variation, as shown by the errors bar in Table 4) with lower or higher than 0.25 phr were observed in the bond strengths. There were a few sources of uncertainties for these wider scatters. First, the fabricated plates might have resin-starved zones on the surfaces, leading to the bonds failing at the glass fibres. To mitigate such effects, a controlled uniform resin filling process during the infusion process is necessary [38]. Additionally, for the phr values higher than 0.25, there is a high probability of occurrence of brittle debonding.

Table 4. The range of lap shear strengths for the fractography.

	Specimen with Multifunctional Methacrylate Monomer (phr)	SLS (MPa) ± Standard Deviation (MPa)	Coefficient of Variation (CoV) (Standard Deviation/Average) (%)
	Neat Elium (0)	12.27 ± 1.84	15
BDDMA	0.1	14.063 ± 1.76	12.5
	0.25	19.12 ± 0.48	2.5
	0.5	12.3 ± 4.34	35
	1	15.56 ± 2.22	14
	2	14.64 ± 1.00	6.83
TCDDMDA	0.1	16.32 ± 2.6	15.9
	0.25	18.84 ± 0.64	3.39
	0.5	13.99 ± 0.73	5.2
	1	17.00 ± 3.24	19.05
	2	11.63 ± 2.13	18.31

3.2.2. Comparison of Strengths with Other Welding Techniques

Average SLS results of the best bonding results obtained in the current study were compared with the results from Murray et al. [20] and Bhudolia et al. [21]. Although this study investigated the same family of infusible thermoplastic resins that were considered in the studies of Murray et al. [20] and Bhudolia et al. [21], the authors are fully aware that a direct comparison between the SLS results of the current study and those from [20,21] is not entirely possible. In fact, our results were generated based on the IR welding, whereas [20,21] applied other variants of the fusion-bonding technique (induction, resistance, and ultrasonic welding).

Table 5 summarises the technical information on manufacturing and welding and SLS results from the IR welding compared to those of SLS results from the indication, resistance, and ultrasonic welding [20,21]. The averages of the three data sources were very close and validated the effectiveness of multifunctional methacrylate monomers on the bonding strengths.

Table 5. SLS results of IR welding versus induction, resistance, and ultrasonic welding.

Parameters	Murray et al. [20] (2019)	Bhudolia et al. [21] (2020)	Current Study	
Welding method	-Induction welding -Resistance welding	Ultrasonic welding	IR welding	
Type of resin	Elium®188	Elium®150	Elium®188XO	
Ratio of resin:initiator	100:2	100:3	100:3	
Type of fabric	Fibreglast 3.5-oz plain-weave carbon fibre	Twill-weave dry carbon fibres	woven glass fabric	
Composite manufacturing technique	Vacuum infusion	Resin transfer moulding (RTM) and post-cured at 65 °C	Vacuum infusion	
Overlap area (mm^2)	645	645.16	640	
Variables	Comparing adhesives with resistance and induction-welded bonds	- weld time - weld pressure - amplitude - ED type	Multifunctional methacrylate monomer concentration	
Average shear strength of best bonding results (MPa)	Induction / resistance	18.68	BDDMA	TCDDMDA
	18.5 / 18.5		19.12	18.84

3.2.3. Fractography

This section investigates the fracture surfaces of lap shear specimens. From a macroscopic point of view (Figure 8), three major failure modes for welded bonds were recognisable. Interfacial failure is characterised by the failure between the adhered surfaces and was well-correlated to the lowest shear strength failures of the specimens containing 2 phr of both multifunctional methacrylate monomers, indicating that the composite surfaces were not well-welded to each other. The second macroscopic failure mode is cohesive failure, which is characterised by the failure within the resin to the resin bond. The mixed (interfacial and cohesive) failure, which is the third macroscopic failure mode, can be visualised at the resin-adhered joint interface. The specimens with 0.25 phr of both multifunctional methacrylate monomers, which showed the strongest IR-welded bonds (SLS tests), failed in a dominated cohesive failure mode. The cohesive failure can be correlated to the area of fibre imprints (red rectangular border in Figure 8). The higher fibre damage area represents stronger bonds. As the phr increased from 0 to 0.25, the fibre damage area increased, and at 0.25 phr it reached the highest area, representing the strongest bond. Beyond 0.25 phr, the fibre damage area decreased, and at 2 phr, the area approached to zero, representing the weakest IR bond, which is consistent with the SLS experiments.

To further explore the failure of Elium® composite bonds, Figure 9 shows the microscopic images of the failure surfaces captured by the scanning electron microscopy (SEM) for the specimens with the highest (0.25 phr) and the lowest values (2 phr) of the lap shear strengths.

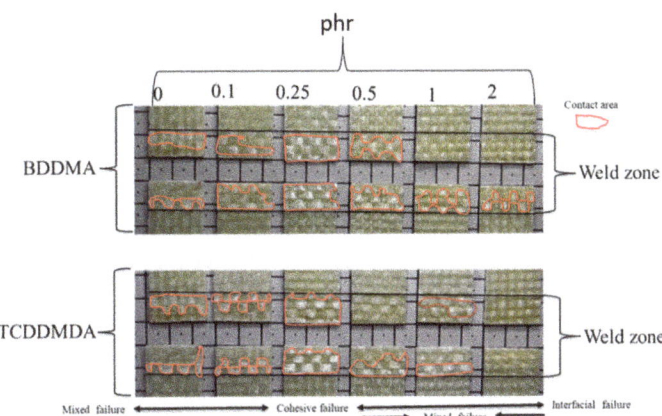

Figure 8. Macroscopic failure images of Elium® composites with varying phr values of multifunctional methacrylate monomers.

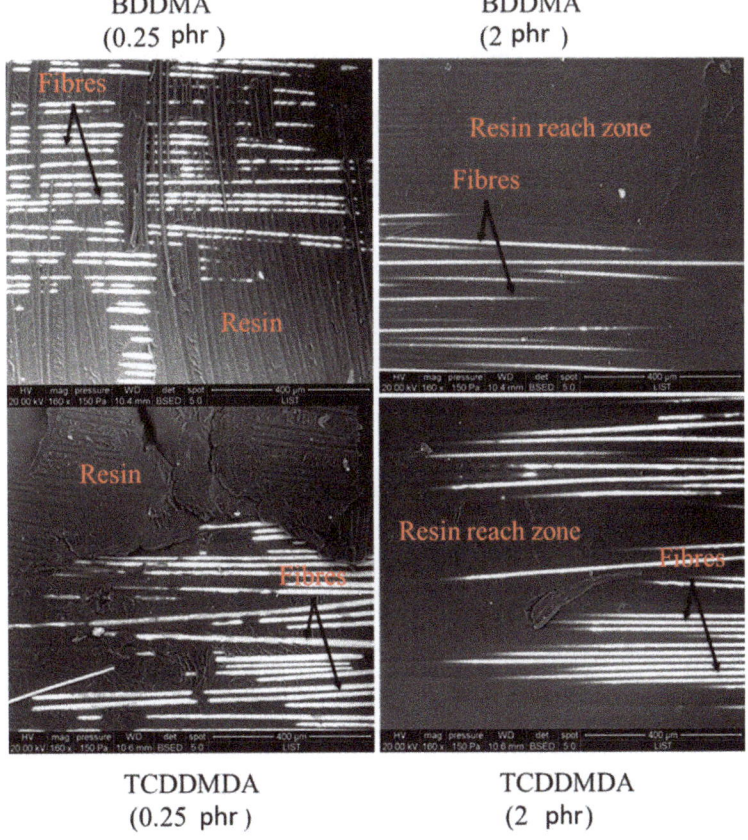

Figure 9. SEM images at the magnification of 160 times, showing the fracture surfaces of the welded composites plates with Elium® matrix containing 0.25 phr and 2 phr of each of the multifunctional methacrylate monomers, corresponding to the highest and lowest shear strengths.

As shown in Figure 9, two types of failure in shear were observable. The cohesive failure was specified by a highly rough surface with a large damage level throughout the welding line (Figure 9 BDDMA and TCDDMDA, with 0.25 phr), which confirmed that the adhesion between the thermoplastic and glass fibres was very good [39]. The second one is the interfacial failure, which was characterised by a smooth surface, leaving bare glass fibres on the surface (Figure 9 BDDMA and TCDDMDA, with 2 phr). One reason for the poor resin–fibre bonding (composite with 2 phr) was the composition of the matrix. With reference to dilatometry results, Elium® resin containing 2 phr of each of the multifunctional methacrylate monomers was highly cross-linked, and then deconsolidation was not expected to occur after 10 s of heating at 200 °C for the IR welding. The effects of the heat exposure time and its effect on the strengths of bond formation and the potential defects' formation, such as voids during the deconsolidation, are currently under investigation.

4. Conclusions

This study examined the addition of two dimethacrylates, namely butanediol-di-methacrylate (BDDMA) and tricyclo-decane-dimethanol-di-methacrylate (TCDDMDA), into the Elium® matrix on the weldability of vacuum-infused Elium® glass fibre composites. Our study provided the thermal data and strength data based on the effect of the parts per hundred resin (phr) of each of the multifunctional methacrylate monomers. Ten composite samples containing different phr values were manufactured. One composite sample without multifunctional methacrylate monomers was fabricated as a reference sample. The results of dilatometry and SLS tests showed that:

1. An appropriate melt strength for an acrylic-based glass fibre composite material was obtained with each of the multifunctional methacrylate monomers at 0.25 phr, indicating optimal welding conditions.
2. Compared to the reference sample, the addition of 0.25 phr of each of the multifunctional methacrylate monomers to the composite matrix offered the maximum bond shear strength of the weld, with an increase of 50%.
3. Both multifunctional methacrylate monomers showed similar behaviour in the composite matrix in terms of weldability, regardless of their phr values in the Elium® matrix.

Last but not least, it is necessary to explore important and remaining unknowns on the topic in future studies, focusing on the kinetics and macromolecular architecture development of the resin systems and their effects on the weldability of Elium® composites.

Author Contributions: H.P., conceptualization, methodology, supervision, validation, formal analysis, software, funding acquisition, visualization, investigation, review and editing; M.B., conceptualization, methodology, validation, formal analysis, visualization, investigation, writing—original draft, review and editing; V.B., review and editing, visualization; R.V., data curation, formal analysis, review and editing. All authors have read and agreed to the published version of the manuscript.

Funding: This research was funded by Luxembourg National Research Fund (FNR), for funding Structural composite material for 3d Printing, SAMIA-3D under the research grant BRIDGES18/MS/13321465.

Institutional Review Board Statement: Not applicable.

Informed Consent Statement: Not applicable.

Data Availability Statement: The data presented in this study are available on request from the corresponding author.

Acknowledgments: The authors greatly appreciate the support of Loïc Borghini and Sébastien Klein for the composite manufacturing, welding, and sample preparation, as well as Sébastien Gergen for Assembly Testing from Luxembourg Institute of Science and Technology, MRT Department, for the IR welding specimens.

Conflicts of Interest: The authors declare no conflict of interest.

References

1. Crosky, A.; Soatthiyanon, N.; Ruys, D.; Meatherall, S.; Potter, S. Thermoset matrix natural fibre-reinforced composites. In *Natural Fibre Composites*; Elsevier BV: Amsterdam, The Netherlands, 2014; pp. 233–270. [CrossRef]
2. Mallick, P.K. Thermoplastics and thermoplastic-matrix composites for lightweight automotive structures. In *Materials, Design and Manufacturing for Lightweight Vehicles*; Elsevier Ltd.: Amsterdam, The Netherlands, 2010; pp. 174–207.
3. Avenet, J.; Cender, T.A.; Le Corre, S.; Bailleul, J.-L.; Levy, A. Adhesion of High Temperature Thermoplastic Composites. *Procedia Manuf.* **2020**, *47*, 925–932. [CrossRef]
4. Avenet, J.; Cender, T.A.; Le Corre, S.; Bailleul, J.-L.; Levy, A. Experimental correlation of rheological relaxation and interface healing times in welding thermoplastic PEKK composites. *Compos. Part A Appl. Sci. Manuf.* **2021**, *149*, 106489. [CrossRef]
5. Ageorges, C.; Ye, L.; Hou, M. Advances in fusion bonding techniques for joining thermoplastic matrix composites: A review. *Compos. Part A Appl. Sci. Manuf.* **2001**, *32*, 839–857. [CrossRef]
6. Tao, W.; Su, X.; Wang, H.; Zhang, Z.; Li, H.; Chen, J. Influence mechanism of welding time and energy director to the thermoplastic composite joints by ultrasonic welding. *J. Manuf. Process.* **2018**, *37*, 196–202. [CrossRef]
7. Bhudolia, S.K.; Gohel, G.; Leong, K.F.; Islam, A. Advances in Ultrasonic Welding of Thermoplastic Composites: A Review. *Materials* **2020**, *13*, 1284. [CrossRef] [PubMed]
8. Palardy, G. Smart ultrasonic welding of thermoplastic composites. In Proceedings of the American Society for Composites- 31st Technical Conference, ASC 2016 DEStech Publications Inc, Williamsburg, VA, USA, 19–22 September 2016.
9. Williams, G.; Green, S.; McAfee, J.; Heward, C.M. Induction Welding of Thermoplastic Composites. In *FRC90–Proceedings*; Mechanical Engineering Publications: London, UK, 1990; pp. 133–136.
10. McKnight, S.H.; Holmes, S.T.; Gillespie, J.W.; Lambing, C.L.T.; Marinelli, J.M. Scaling issues in resistance-welded thermoplastic composite joints. *Adv. Polym. Technol.* **1997**, *16*, 279–295. [CrossRef]
11. Swartz, H.D.; Swartz, J.L. *Focused Infrared Melt Fusion: Another Option for Welding Thermoplastic Composites*; SME Technical Paper, Joining Composites; Society of Manufacturing Engineers: Southfield, Michigan, 1989; pp. 1–16.
12. De Baere, I.; Allaer, K.; Jacques, S.; Van Paepegem, W.; Degrieck, J. Interlaminar behavior of infrared welded joints of carbon fabric-reinforced polyphenylene sulfide. *Polym. Compos.* **2012**, *33*, 1105–1113. [CrossRef]
13. Perrin, H.; Mertz, G.; Senoussaoui, N.-L.; Borghini, L.; Klein, S.; Vaudemont, R. Surface functionalization of thermoset composite for infrared hybrid welding. *Funct. Compos. Mater.* **2021**, *2*, 1–15. [CrossRef]
14. Asseko, A.C.A.; Lafranche, É.; Cosson, B.; Schmidt, F.; Le Maoult, Y. Infrared welding process on composite: Effect of interdiffusion at the welding interface. *AIP Conf. Proc.* **2016**, *1769*, 020011. [CrossRef]
15. Allaer, K.; de Baere, I.; van Paepegem, W.; Degrieck, J. Infrared welding of carbon fabric reinforced thermoplastics. *Jec Compos. Mag.* **2012**, 44–47.
16. Van Rijswijk, K.; Bersee, H.E.N. Reactive Processing of Textile Fiber-Reinforced Thermoplastic Composites—An Overview. *Compos. Part A Appl. Sci. Manuf.* **2007**, *38*, 666–681. [CrossRef]
17. Murray, R.E.; Penumadu, D.; Cousins, D.; Beach, R.; Snowberg, D.; Berry, D.; Suzuki, Y.; Stebner, A. Manufacturing and Flexural Characterization of Infusion-Reacted Thermoplastic Wind Turbine Blade Subcomponents. *Appl. Compos. Mater.* **2019**, *26*, 945–961. [CrossRef]
18. ARKEMA. OLERIS® ACCOLA CAREFLEX® DICUP ELIUM® TEGOGLA. Available online: https://www.arkema.com/global/en/products/product-safety/disclaimer/ (accessed on 1 February 2023).
19. Bodaghi, M.; Park, C.H.; Krawczak, P. Reactive Processing of Acrylic-Based Thermoplastic Composites: A Mini-Review. *Front. Mater.* **2022**, *9*. [CrossRef]
20. Murray, R.E.; Roadman, J.; Beach, R. Fusion joining of thermoplastic composite wind turbine blades: Lap-shear bond characterization. *Renew. Energy* **2019**, *140*, 501–512. [CrossRef]
21. Bhudolia, S.K.; Gohel, G.; Kantipudi, J.; Leong, K.F.; Barsotti, R.J.B., Jr. Ultrasonic Welding of Novel Carbon/Elium® Thermoplastic Composites with Flat and Integrated Energy Directors: Lap Shear Characterisation and Fractographic Investigation. *Materials* **2020**, *13*, 1634. [CrossRef] [PubMed]
22. Gohel, G.; Bhudolia, S.K.; Kantipudi, J.; Leong, K.F.; Barsotti, R.J. Ultrasonic welding of novel Carbon/Elium®with carbon/epoxy composites. *Compos. Commun.* **2020**, *22*, 100463. [CrossRef]
23. Chebil, M.S.; Bouaoulo, G.; Gerard, P.; EL Euch, S.; Issard, H.; Richaud, E. Oxidation and unzipping in ELIUM resin: Kinetic model for mass loss. *Polym. Degrad. Stab.* **2021**, *186*, 109523. [CrossRef]
24. Hussein, M.A.; Albeladi, H.k.; AlRomaizan, A.N. Role of Cross-Linking Process on the Performance of PMMA. *Int. J. Biosens. Bioelectron.* **2017**, *3*, 279–284. [CrossRef]
25. Decker, C. UV-Radiation Curing of Adhesives. *Handb. Adhes. Sealants* **2006**, *2*, 303–353.
26. FRIMO Group GmbH. IR-Welding Innovations. Available online: https://trends.directindustry.com/frimo-group-gmbh/project-36161-133913.html (accessed on 1 February 2023).
27. Beckel, E.R.; Nie, J.; Stansbury, J.W.; Bowman, C.N. Effect of Aryl Substituents on the Reactivity of Phenyl Carbamate Acrylate Monomers. *Macromolecules* **2004**, *37*, 4062–4069. [CrossRef]

28. Maruo, Y.; Yoshihara, K.; Irie, M.; Nagaoka, N.; Matsumoto, T.; Minagi, S. Does Multifunctional Acrylate's Addition to Methacrylate Improve Its Flexural Properties and Bond Ability to CAD/CAM PMMA Block? *Materials* **2022**, *15*, 7564. [CrossRef] [PubMed]
29. Min, K.; Silberstein, M.; Aluru, N.R. Crosslinking PMMA: Molecular dynamics investigation of the shear response. *J. Polym. Sci. Part B: Polym. Phys.* **2013**, *52*, 444–449. [CrossRef]
30. Jerolimov, V.; Jagger, R.G.; Millward, P.J. Effect of Cross-linking Chain Length on Glass Transition of a Dough-moulded Poly (methylmethacrylate) Resins. *Acta Stomatol. Croat.* **1994**, *28*, 3–9.
31. Jerolimov, V.; Jagger, R.; Milward, P. Effect of the curing cycle on acrylic denture base glass transition temperatures. *J. Dent.* **1991**, *19*, 245–248. [CrossRef]
32. Ajay, R.; Suma, K.; Ali, S.A. Monomer modifications of denture base acrylic resin: A systematic review and meta-analysis. *J. Pharm. Bioallied Sci.* **2019**, *11*, S112–S125. [CrossRef] [PubMed]
33. Kubota, T.; Kobayashi, M.; Hayashi, R.; Ono, A.; Mega, J. Influence of Carbon Chain Length of Fluorinated Alkyl Acrylate on Mechanical Properties of Denture Base Resin. *Int. J. Oral-Medical Sci.* **2005**, *4*, 92–96. [CrossRef]
34. Spasojevic, P.; Panic, V.; Seslija, S.; Nikolic, V.; Popovic, I.; Velickovic, S. Poly(methyl methacrylate) denture base materials modified with ditetrahydrofurfuryl itaconate: Significant applicative properties. *J. Serbian Chem. Soc.* **2015**, *80*, 1177–1192. [CrossRef]
35. Brzeski, M.; Mitschang, P. Deconsolidation and Its Interdependent Mechanisms of Fibre Reinforced Polypropylene. *Polym. Polym. Compos.* **2015**, *23*, 515–524. [CrossRef]
36. Wan, Y.; Takahashi, J. Deconsolidation behavior of carbon fiber reinforced thermoplastics. *J. Reinf. Plast. Compos.* **2014**, *33*, 1613–1624. [CrossRef]
37. Pettersson, J. "Analysis and Design of an Adhesive Joint in Wind Turbine Blades," Lund University. 2016. Available online: https://www.byggmek.lth.se/fileadmin/byggnadsmekanik/publications/tvsm5000/web5217.pdf (accessed on 28 February 2023).
38. Bodaghi, M.; Costa, R.; Gomes, R.; Silva, J.; Curado-Correia, N.; Silva, F. Experimental comparative study of the variants of high-temperature vacuum-assisted resin transfer moulding. *Compos. Part A Appl. Sci. Manuf.* **2019**, *129*, 105708. [CrossRef]
39. Johnson, W.; Masters, J.; Wilson, D.; Melin, L.; Neumeister, J.; Pettersson, K.; Johansson, H.; Asp, L. Evaluation of Four Composite Shear Test Methods by Digital Speckle Strain Mapping and Fractographic Analysis. *J. Compos. Technol. Res.* **2000**, *22*, 161. [CrossRef]

Disclaimer/Publisher's Note: The statements, opinions and data contained in all publications are solely those of the individual author(s) and contributor(s) and not of MDPI and/or the editor(s). MDPI and/or the editor(s) disclaim responsibility for any injury to people or property resulting from any ideas, methods, instructions or products referred to in the content.

Enhanced Coating Protection of C-Steel Using Polystyrene Clay Nanocomposite Impregnated with Inhibitors

Aljawharah M. Alangari [1], Layla A. Al Juhaiman [1,*] and Waffa K. Mekhamer [1,2]

1 Chemistry Department, King Saud University, Riyadh 12372, Saudi Arabia
2 Department of Material Science, Institute of Graduate Studies, Alexandria University, Alexandria 5422004, Egypt
* Correspondence: ljuhiman@ksu.edu.sa

Abstract: Polymer–Clay Nanocomposite (PCN) coatings were prepared using the solution intercalation method. The raw Khulays clay was treated with NaCl to produce sodium clay (NaC). Thereafter, Cetyl Pyridinium Chloride (CPC) was used to convert NaC into the organic clay form (OC). PCN was prepared by adding polystyrene as the matrix to different weights of OC to prepare 1 wt.% and 3 wt.% PCN. To enhance the coating protection of C-steel in NaCl solution, PCN coatings were added to microcapsules loaded with some corrosion inhibitors PCN (MC). The microcapsules are prepared by the encapsulation of rare-earth metal Ce^{+3} ions and Isobutyl silanol into polystyrene via the Double Emulsion Solvent Evaporation (DESE) technique. Characterization techniques such as FTIR, X-Ray Diffraction (XRD), and Transmission Electron Microscopy (TEM) were employed. FTIR confirmed the success of the preparation, while XRD and TEM revealed an intercalated structure of 1 wt.% PCN while 3 wt.% PCN has a fully exfoliated structure. Electrochemical Impedance Spectroscopy (EIS), Electrochemical Frequency Modulation (EFM), and Potentiodynamic Polarization showed an enhanced protection efficiency of PCN (MC) coatings. The results demonstrated that the corrosion resistance (R_{Corr}) of 3% PCN (MC) coating was higher than all the formulations. These PCN (MC) coatings may provide corrosion protection for C-steel pipes in many industrial applications.

Keywords: carbon steel; local clay; polystyrene; polymer clay nanocomposites; microcapsules; enhanced coating protection

1. Introduction

Steel and its alloys are widely used in many industrial applications, like the oil industry and engineering structures, due to their high strength and ductility. Carbon steel (C-steel) pipes are mainly used in a corrosive medium where corrosion prevention is important. Corrosion of steel alloys may lead to degeneration in their properties, waste of resources, safety, and environmental problems [1]. Therefore, protecting steel from corrosion has acquired high importance in minimizing economic losses, cost reduction, and achieving acceptable performances. Thus, a new generation of high-performance protective coatings that can provide long-term protection is required [2]. The idea behind the application of different corrosion control strategies lies in the removal of an electrochemical cell component, such as the cathode, electron-conductive path, corrosive environment, or even by modifying the metal to be protected [1,2]. Protective coatings are one of the best strategies in corrosion protection. Through protective coatings, corrosion can be controlled by one or more of these mechanisms: barrier property, cathodic protection, and inhibition, including the principle of anodic protection. Coatings consist of major components, including binders, pigments, solvents, and additives as inhibitors, surface-active agents, thickeners, and coupling agents. Corrosive components, such as ions, oxygen, and water, can penetrate the coating to reach the interface between the metal and that coating, causing a hidden corrosion process in the metal. Nanotechnology, including nanocomposites, contributed significantly to enhancing anticorrosive coating efficiency. Various types of nanofillers have been used for the preparation of these nanocomposites. Recently, clay got the attention of many researchers because of its ability to give a stunning enhancement to composite coating efficiency. It is also eco-friendly and naturally abundant [2]. Among all the well-known silicate types, layered materials like smectite clay, such as bentonite and montmorillonite (MMT), play an important role in the preparation of different novel polymer–clay nanocomposites (PCNs), since it has various properties, including a huge cationic exchange capacity (CEC) [3].

Bentonite clay minerals are naturally formed phyllosilicates consisting mostly of montmorillonite with (2:1) layer composition. A single layer consists of two tetrahedral sheets (Silica, SiO_4), with one octahedral sheet (Alumina, Al_2O_3) sandwiched between them and can be expressed as (TOT). The gap between the layers is known as the interlayer or gallery or (d_{001} spacing), which can vary according to the cations present. The use of polymer nanocomposites mainly depends on their polymer and fillers, whether they are synthetic or natural with at least one dimension in a nanometer size scale, and they produce unusual properties that were not observed when using the same components separately [2].

Polymer clay nanocomposites (PCNs) have outstanding chemical, physical, mechanical, thermal, and barrier properties compared to polymers or conventional composites. PCN forms a maze-like pathway that retards the diffusion of various corrosive components. Protective coating prevents corrosive ions such as chloride, protons, or molecules like water and oxygen from penetrating through the film into the metal surface. However, resistance does not last for long, and it will be penetrated eventually, creating an internal pathway through which corrosive substances pass to the metal surface, causing corrosion of the metal [4–13]. Despite the high efficiency of organic and PCN protective coatings, long-term protection is not yet guaranteed. PCN form a rigid barrier, but once corrosive components permeate them, their properties begin to collapse, leading to corrosion at the interface between the metal and the corrosive medium [1,2].

Some previous studies highlighted that the organic coating could work as a reservoir for carrying corrosion inhibitors, which then promote the coatings' life [14–16]. Currently, several studies have been conducted to develop protective coatings impregnated, loaded, or doped with some eco-friendly new additives called "green corrosion inhibitors". These impregnated coatings not only provide enhanced barrier properties but also interact with the environment and protect the active sites on the metal surface whenever the corrosive components penetrate them. Recently, many studies investigating how the functionality and life of the coating can be improved have been carried out [14–16]. Publication on this line of study, i.e., smart coatings, has increased approximately tenfold from 2000–2010. Several criteria, including the type of material, type of polymer, functionality, applications, as well as stimuli (or triggers), can be used to classify smart coatings. There are many triggering mechanisms, including chemical mechanisms such as pH, humidity change, external triggers like UV radiation, temperature, a mechanical mechanism, and electrochemical potential. Smart coatings have a response to the surrounding environment, which in turn provides long-term protection through different—self-healing or self-repairing mechanisms, and the healing process may have more than one triggering mechanism associated with each other; for example, a chemical–mechanical mechanism.

Not long ago, smart anticorrosive coatings used were based on chromium compounds, which are known for their high toxicity, prompting many researchers to develop alternative coatings using ceramics, metals, or composites [14,15]. Often, smart coatings contain cerium, sodium salt of tetra thiomolybdate (VI), and organofluorine and organosilane compounds. Organosilane compounds are widely used in smart coatings as coupling agents since they can present various functionalities, including the inhibition of active corrosion sites on the metal substrate and the self-healing of the coating resin. The general formula of organosilanes is $R-(CH_2)-Si-X_3$, which include a silicon atom, a hydrolyzable alkoxy group (X), and a non-hydrolyzable group (R), such as amino, epoxy, vinyl, and methacrylate.

Microcapsules (MCs) play an important role in smart protective coatings. These capsules are produced on a micrometer scale, the commercial ones usually range from

3–800 µm, and they consist of two main parts; a loaded material to be coated, which can be a solid, liquid or gas (known as Core) and an inert solid outer material that encapsulates the core (Shell) [5,14–16]. There are several techniques for encapsulation. The employed technique will depend on some factors, such as the hydrophobicity and hydrophilicity of polymers and loaded corrosion inhibitors, the self-healing mechanism, the matrix in the coating formulation, and the nature of the polymer used as a shell.

In our previous study [13], polystyrene clay nanocomposite was prepared using Khulays organoclay (OC). Differential Scanning Calorimetry (DSC) was applied for 1–10% PCN. The glass transition temperature decreased from 98.14 °C for pure PS to lower values for 1–3% PCN, then increased for 5% PCN to reach a higher Tg of 104.22 °C for 10% PCN. The exfoliated structure of the low clay-loading (1–3%) PCN leads to the movement of the polymer chains between the organoclay layers, and the PCN film becomes ductile and does not break when used as a coating material. In another study of our research group [6], we used 1, 3, 5, and 10% PCN as a protective coating of the C-steel. It was observed that the extent of protection provided by PCN coatings depended upon the loading of the organoclay, as found by other researchers [2,8]. At a higher clay loading than 3% PCN, the corrosion protection performance decreased, as proved by the electrochemical data [6].

In the present study, we will prepare the PCN with low clay loading at 1 and 3% PCN to obtain a PCN with an exfoliated structure to decrease the coating permeability. Moreover, it will improve the coating barrier properties explained by the concept of tortuous paths where the permeating molecules are forced to follow a wiggle path which will improve the metal coating protection. In order to enhance the protective properties of these PCN coatings, we aim to prepare a polystyrene/organoclay nanocomposite impregnated with an inhibitor prepared by double-emulsion solvent evaporation (DESE). In this study, several characterization techniques are applied, such as FT-IR, XRD, SEM, and TEM. Evaluation of the protection efficiency of C-steel in NaCl solution was done by performing three electrochemical methods, which are Electrochemical Impedance Spectroscopy (EIS), Electrochemical Frequency Modulation (EFM), and Potentiodynamic Polarization.

2. Materials and Methods

2.1. Materials

The raw clay (RC) was collected by a certified geologist from Khulays region, Jeddah, Saudi Arabia. It was previously found that this RC belongs to bentonite clay minerals [13]. The RC contains montmorillonite (35.22%), kaolinite (13.33%), mica (22.80%), quartz (8.57%), feldspars (6.66%), ilmenite (5.71%), dolomite (3.81%), and gypsum (3.81%). The exact composition of this clay mineral was indicated by X-ray Fluorescence (XRF) [13] to be:

$$Na_{0.67}K_{0.13}Ca_{0.02}Ba_{0.04}(Si_{7.47}Al_{0.53})(Al_{2.59}Fe_{0.78}Ti_{0.14}Mg_{0.44}Cr_{0.04})O_{20}(OH)_4$$

A local polystyrene ((M = 259,000 g/mol) was used from the SABIC company in Saudi Arabia. Carbon steel (C-steel) rods were provided by ODS Co., Schleswig-Holstein, Germany. To ensure the exposed cross-sectional area remained constant, the outer wall of the C-steel rod was covered by an epoxy resin. The chemical composition of the C-steel is shown in Table 1. The materials are summarized in Table 2.

Table 1. The chemical composition of the carbon steel (C-steel).

Element	C	Mn	Cr	Si	Ni	Cu	Al	P	S
Weight (%)	0.46	0.6	0.18	0.18	0.04	0.03	0.023	0.013	0.006

Table 2. The material used in this work.

Product Name	Source
RC	Khulays clay—Saudi Arabia
C-steel (Grade 1046)	ODS Co.—Berlin, Germany
Polystyrene (PS125)	SABIC—Riyadh, Saudi Arabia
Sodium Chloride (AR grade)	Win lab Company, Queensland, Australia
Ethanol 96%—AR grade	Avonchem—Waterloo, UK
Cetyl pyridinium chloride (CPC)	BDH Co., Istanbul, Turkey
Polyvinyl Alcohol (PVA)	LOBAChemie Co., Mumbai, India
Glacial acetic acid 99.5%—Extra Pure	Sigma-Aldrich Co., Saint Louis, MI, USA
Dichloromethane 99.5%—AR\ACS grade	Sigma-Aldrich Co.
Isobutyl Trimethoxy Silane 97% (IBTMS)	Alfa Aeasar Co., Ward Hill, MA, USA
$Ce(NO_3)_3 \cdot 6H_2O$—AR	LOBAChemie Co.

2.2. Clay Separation, Saturation, and Organic Modification

Most previous studies used pretreated clay (obtained from the market) in the preparation of polymer clay nanocomposites. In our lab, we ground the local Khulays clay from the raw rocks to fine grain size in the micrometer range, as discussed in [13]. Our strategy of preparing PCN started with the modification of RC. Fifty grams of RC was dispersed in 500 mL of distilled water and shaken for two hours, then left overnight. The raw clay was centrifuged and then washed with distilled water, and this process was repeated for three days to remove any impurities in RC. Then, 300 mL of 0.5 M NaCl solution was added to the clay, and the suspension was mixed by shaking for 2 h and then left overnight. This process was repeated three times to produce sodium clay NaC. The cationic surfactant Cetyl pyridinium Chloride (CPC) was used in the ion exchange treatment to convert the NaC into the organic clay form (OC). It is known that the highest cation exchange capacity (CEC) of clay is about 85 meq/100 g [3]. Five grams of NaC was dispersed in 500 mL of distilled water. To ensure that the organic cations in CPC replace all the cations in the NaC structure, the amount of CPC corresponding to two times the CEC of clay was dissolved in 100 mL of distilled water. The CPC solution was added to the NaC suspension at a slow rate, and the mixture was stirred for 24 h at room temperature. The resulting organoclay (OC) was separated by centrifugation and washed with distilled water several times until no chloride ions were detected by 0.1 M AgNO3 solution. Finally, the OC was dried at 60 °C for 24 h in the oven.

2.3. Polystyrene-OC Nanocomposites (PCNs) Preparation

Anticorrosive coatings for C-steel based on polystyrene/organoclay nanocomposite (PCN) were successfully prepared using polystyrene (PS) as a matrix with different percentage weights of OC to prepare 1% PCN and 3% PCN, respectively [6,13]. In a 50 mL flask, a certain mass of OC (0.02 and 0.06 g) was added to 10 mL of toluene, and this suspension was stirred magnetically overnight at room temperature. Afterward, 2 g of PS was added, and the mixture was magnetically stirred for 6 h and then in an ultrasonic machine for 10 min. For characterization purposes, the PCN solution was cast into a Petri dish to allow the toluene to evaporate. The coating adhesion test was performed for the interface between C-steel and PCN coatings using a cross-hatch cutter instrument (Sheen Instruments, Cambridge, UK) according to ASTM D-3359-02.

2.4. Polystyrene Microcapsules (MCs) Preparation

Polystyrene microcapsules (MCs) loaded with Isobutylsilanol and Ce^{+3} ions as corrosion inhibitors were successfully prepared with the Double-Emulsion Solvent Evaporation (DESE) method following the procedures from a previous report [5]. The preparation included three steps, namely, the preparation of the aqueous inhibitor solution, first emulsion, and double emulsion labeled (W_1), (W_1/O), and ($W_1/O/W_2$), respectively. First, in a 50 mL flask, equal volumes of distilled water and 96 wt.% ethanol were added, and the pH was adjusted to pH 5 with glacial acetic acid. Cerium ions from Ce(NO$_3$)$_3$.6H$_2$O salt were added to the solution with a concentration of 5000 ppm. Subsequently, 4 wt.% of isobutyl trimethoxy silane (IBTMS) was added. The solution was kept under magnetic stirring for 24 h to complete the hydrolysis process of the W$_1$ solution. Second, in another flask, 2 g of PS was dissolved in 10 mL Dichloromethane (DCM), and 0.5 g of CPC was added as a cationic surfactant. Approximately 1 mL of W$_1$ solution was added to the PS mixture under sonication for 2 min, resulting in a W$_1$/O emulsion. Third, in a 200 mL flask, about 30 mL of 1 wt.% of Polyvinyl Alcohol (PVA) aqueous solution was added. Then, under magnetic stirring (300 rpm), W$_1$/O was slowly added to the PVA solution. The last step of W$_1$/O/W$_2$ preparation was the addition of a large amount of 0.3 wt.% PVA aqueous solution. The double emulsion was left for 24 h under magnetic stirring for solvent evaporation. After that time, polystyrene microcapsules (MC) were filtered with a Büchner funnel, and the MC was washed with deionized water several times. Finally, MC appeared as a fine white powder, easily collected in a crucible and kept in a desiccator. The Ce^{+3} ions played the role of corrosion inhibitor and healing agent, whereas Silanol (Si–OH) was used as a corrosion inhibitor and a coupling agent for MC.

2.5. Coating the C-Steel Surface

The C-steel rods have an exposed surface area of 4.75 cm^2, which was calculated using a Mitutoyo gauging tool. These rods went through a polishing process using abrasive papers of various grades 80, 220, 600, and 1000 for the purpose of scraping the surface and removing possible deposits and impurities. Thereafter, they were washed with distilled water, followed by immersion in acetone using an ultrasonic bath for 2 min to ensure the removal of any leftover impurities. Next, the C-steel rods were dried and left aside. There are three types of coating: the first one is PS coatings, the second one is PCN coatings, and third one is PCN (MC). To prepare the PS-coating solution, 2 g of PS was dissolved in 10 mL toluene and was stirred magnetically overnight at room temperature. For casting, the PCN solution or PS solution was applied dropwise on the C-steel as the first layer. A similar application was conducted on the second layer. It is important to note that each layer was left to air-dry for about 4 h, and subsequently dried in a Carbolite furnace for 24 h at 60 °C.

In the case of the new PCN coating impregnated with MCs, the suspended MCs in isopropyl alcohol were distributed on the surface of the C-steel rod and then left under atmospheric air for about 10 min until the alcohol evaporated. Next, the casting of the PCN solution was done following the procedure described previously. Finally, before performing electrochemical measurements, the thickness of the dry-casted coating on the electrode was determined using Elcometer 456 coating thickness gauge, and it was noted that the total thickness was in the range of 100 ± 15 µm. The synthesis of polystyrene–organoclay nanocomposite and the coating of the C-steel is depicted in Scheme 1.

2.6. Fourier Transform Infrared Spectroscopy (FT-IR)

The FT-IR was applied in the Department of Chemistry, College of Science, King Saud University. This technique was used to verify the success of the modification process done on RC by sodium chloride and the organic modifier CPC, as well as PCNs coatings and the MCs. The analysis was performed for all samples using the standard KBr disk method at frequencies in the range of 400–4400 cm^{-1}.

2.7. X-ray Diffraction Analysis (XRD)

This analysis was performed in the Department of Chemistry, College of Science, King Saud University. The aim is to study the characteristic diffraction peak shifts of RC powder as a result of its modification by Na ions and CPC after being ground and dried at 50 °C for 48 h. Additionally, the PCNs were analyzed as films. The anode was CuKα radiation, and the wavelength was equivalent to 1.54060 Å with a scanning rate of 3°/min. Furthermore, the scanning range was (3–30 2θ), with a divergence slit of 0.4, while the current and voltage generator was 40.0 mA and 40.0 kV, respectively, at 25 °C.

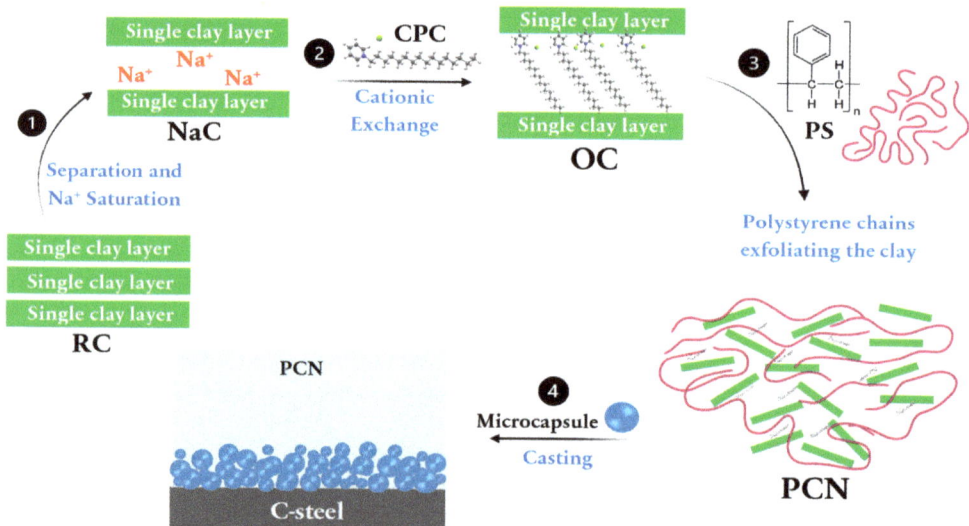

Scheme 1. Polystyrene–organoclay nanocomposite synthesis and the coating of C-steel.

2.8. Transmission Electron Microscopy (TEM)

Transmission Electron Microscopy imaging for the prepared PCNs films was performed in the central laboratory at King Saud University to identify their structure. A small piece of each PCN film was immersed in a suitable resin, then, hardened under high temperature. The diamond knife was then used to cut thin slices of each sample of a thickness of about 70 nm. The slices were then separated by chloroform vapor. Finally, using a thin coated-carbon 200-mesh copper grid held by forceps, the samples were loaded onto the machine. The samples were analyzed under certain conditions with accelerating voltage of 100 kV, and the magnifications were ×100,000 and ×200,000.

2.9. Electrochemical Methods

A Potentiostat/Galvanostat ZRA from Gamry Interface 1000 instrument was used. Three methods were performed to evaluate the protection efficiency of C-steel using the prepared coatings according to the following sequence. First, the open circuit potential (OCP) was conducted for 60 min to stabilize the system and reach the steady-state potential (E_{SS}) before applying the electrochemical methods. After reaching steady-state potential, electrochemical impedance spectroscopy (EIS) was operated, the sweep range was 10^5 to 10^{-1} Hz, with an AC amplitude of 10 mV, and 10 points per decade. Second, electrochemical frequency modulation (EFM) was carried out using the frequencies of 2 and 5 Hz, with a base frequency equal to 0.1 Hz; the perturbation amplitude was 10 mV. For EFM, two sine waves at different frequencies were applied to the electrochemical cell simultaneously. The EFM obtained spectrum, the so-called intermodulation spectrum, is a current response as a function of frequency, with the x- and y-axis representing (frequency) and (current), respectively. Lastly, Potentiodynamic Polarization was performed where the parameters have been set in the Gamry software to align the Tafel conditions. The Tafel plots have been obtained by scanning a potential of ±250 mV according to the potential corrosion value (E_{Corr}). All the electrochemical measurements were performed using a three-electrode electrochemical cell, where the working electrode was the C-steel, the reference electrode was the standard calomel electrode (SCE), and the auxiliary/counter electrode was a rigid platinum foil with a surface area of 100 mm^2 surrounded in a glass body. All the electrodes have been immersed in an aqueous solution of 3.5 wt.% NaCl under room temperature of

(20 ± 0.5 °C), open to atmospheric air. Each test was repeated three-to-four times to ensure the validity of the data.

3. Results and Discussion
3.1. FT-IR Results
3.1.1. FT-IR of the Modified Clay and PCNs

As presented in Figure 1A, the FT-IR spectrum of RC, NaC and OC clearly shows the same characteristic bands. However, the OC shows the characteristic bands of CPC, which indicates the success of the modification process and the transformation of the inorganic NaC into an organic moiety [7–11,13].

The FT-IR spectra of RC, NaC, and OC in Figure 1A show the same characteristic bands of silicates with slight shifts in frequencies. As shown in the clay spectra of RC, NaC and OC, there are absorption bands at the frequencies 3626, 3629, and 3625 cm^{-1}, which refer to the stretching vibrations of the (O–H) bonds in Al–OHs, respectively, and conjugated to them the bending vibrations at 915, 916, and 913 cm^{-1}, respectively. Medium absorption bands can be seen at frequencies of 1636, 1644 and 1638 cm^{-1}, which are the bending vibrations of water molecules in RC, NaC, and OC, respectively. Regarding the absorption bands shown at 1034, 1031, and 1032 cm^{-1}, they refer to the stretching vibrations of the (Si–O) bonds in the tetrahedron layers of SiO_4 for RC, NaC, and OC, respectively.

Moreover, two bands at frequencies of 2922 and 2852 cm^{-1} refer to the asymmetric and symmetric stretching vibrations of (C–H) bonds of methylene groups (–CH2) in the alkyl chain of CPC. In addition, a new weak band in the OC spectrum at 1195 cm^{-1} may refer to the (C–N) bond stretching from CPC, with a little shift to a lower frequency than the present one in CPC spectrum [7–11,13]. The FTIR spectrum of OC differs from the RC and NaC clay spectra with the presence of additional absorption bands, which proves the success of CPC cations intercalation between the silicate layers (d_{001} spacing), since the absorption band at 3065 cm^{-1} is related to the stretching vibrations of sp^2 (C–H) bonds from CPC with its bending at 1493 cm^{-1}.

In Figure 1B, with regards to the PS spectrum, the two absorption bands appearing at 3061 and 3028 cm^{-1} refer to the aromatic stretching vibration of sp^2 (C–H). In addition, two strong absorption bands were observed at about 2920 and 2852 cm^{-1}, which are related to the asymmetric/symmetric vibrations of the aliphatic (C–H) stretching of(–CH_2), with its bending vibration at 1449 cm^{-1}. It was also noted that the aromatic ring-stretching vibrations of (C=C–C) have absorptions at 1601, 1543, and 1493 cm^{-1}, and their several combination or overtone bands can be seen in the range between (1671–1946 cm^{-1}). Moreover, three absorption bands at 1071, 1028, and 966 cm^{-1} are related to (in-plane) bending, while the other three bands at 907, 758, and 699 cm^{-1} referred to (out-of-plane) bending of (C–H). A weak absorption band at 4042 cm^{-1} was also noticed; it may suggest the first overtone or combination of stretching vibrations of (C–H) and (C=C–C) bonds in polystyrene [8–11]. By comparing the infrared spectra of both PS and OC, it is easy to notice the presence of the characteristic absorption bands of PS in the prepared polystyrene clay nanocomposites of 1 wt.% and 3 wt.% PCN, which supports the successful preparation of these nanocomposites by inserting polystyrene chains between the organically modified clay layers (interlayer). With respect to the absorption bands within the spectra of 1% PCN and 3% PCN, the stretching vibrations of the (–OH) of Al–OHs at 3651 and 3650 cm^{-1}, respectively appeared, but their bending vibrations did not appear clearly, probably due to overlapping and (out-of-plane) bending vibrations of (C–H) from PS at the frequency of 907 and 908 cm^{-1} respectively [7–11].

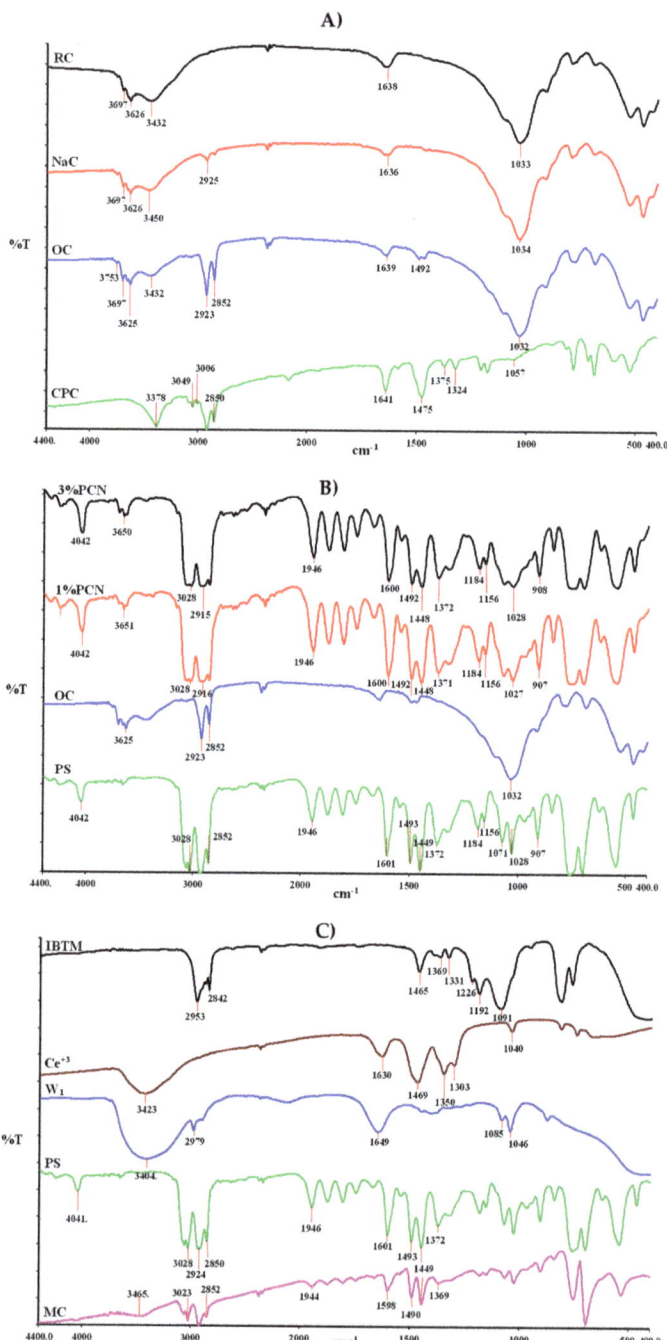

Figure 1. FT-IR of (**A**) RC, NaC, OC and CPC, (**B**) OC, PS, 1% PCN and 3% PCN, and (**C**) PS, Ce^{+3}, IBTMS, and MC.

3.1.2. FT-IR of IBTMS, Ce (NO$_3$)$_3$, W$_1$, PS and MCs

As shown in Figure 1C, the spectrum of IBTMS has an absorption band of 2842 cm^{-1}, which is attributed to the (C–H) bonds stretching vibrations of methoxy groups (–OCH$_3$) in the silane [12]. Also, two sharp absorption bands appeared at 2953 and 2874 cm^{-1} that were assigned to the asymmetric and symmetric stretching vibrations for (C–H) bonds in the methylene group (–CH$_2$–). Its bending can be clearly seen as a medium absorption band at 1465 cm^{-1} [12]. The low, intense peaks between 1400 and 1369 cm^{-1} may be due to the deformation vibration of (–CH$_2$–) in the methylene groups. A wide, high, and intense absorption band present at 1091 cm^{-1} might suggest the stretching vibration of (Si–O–C) and (Si–O–Si) bonds present in the silane, while the two peaks around 816 and 766 cm^{-1} may be due to (Si–C) and (Si–O) bending modes, respectively [12,14,15]. Regarding the cerium ions spectrum, a strong, broad peak with a frequency of 3423 cm^{-1} is related to the (O–H) stretching vibration of the water molecules, with a bending vibration at about 1630 cm^{-1} [16]. The intense absorption bands at 1469, 1350, 1303, 1040, and 816 cm^{-1} are the typical vibrations associated with the nitrate ions (NO$_3^{-1}$), so the 1469 and 1350 cm^{-1} bands are related to the asymmetric and symmetric stretching vibrations of (N–O) bonds, respectively, while the absorption band at 1040 cm^{-1} can be attributed to the stretching vibration of free NO$_3^{-1}$ ions [11,17]. The observed weak peak at a frequency of 550 cm^{-1} corresponds to the stretching vibrations of (Ce–O) [18,19]. When IBTMS and Ce^{+3} spectra are compared, we can easily figure out the characteristic peaks of each of them in the FT-IR spectrum of W$_1$. The stretching and bending vibrations of water molecules from cerium ions appeared at 3405 and 1649 cm^{-1}, with a slight increase in their broadening and intensities, which may be due to strong chemical bonds. The typical vibrations of nitrates were observed but with fewer intensities in the range of (1455–1376 and 1046) cm^{-1}. In contrast, the stretching vibrations of (C–H) bonds related to the methoxy groups from the silanes disappeared, proving that the methoxy groups were completely hydrolyzed during W$_1$ preparation. Additionally, a new weak intense band was observed at approximately 878 cm^{-1}. This new band is likely associated with (Si–O–Ce) stretching vibrations resulting from the bond formation between the cerium and silicon atoms from the silanols [19,20]. Furthermore, all the absorption bands of PS and W$_1$ were observed with little shift toward fewer frequencies, probably due to the conjugation effect, except for the (O–H) stretching vibrations of water molecules, where it was observed at a higher frequency at about 3466 cm^{-1}. A weak absorption band at 4042 cm^{-1} was noticed, suggesting the first overtone or combination of stretching vibrations of (C–H) and (C=C–C) bonds in polystyrene [7,13]. The discussed FT-IR spectra of W$_1$, PS, and the FT-IR spectrum of the prepared microcapsules (MCs), clearly reveals the success of the encapsulation of W$_1$ via PS.

3.2. X-ray Diffraction (XRD) Analysis

The XRD analysis is a powerful technique for assessing the intercalations of organic compounds and their arrangements between the layers of smectites and bentonite subgroups. The intercalated organic cations have various configurations and can be identified based on the d$_{001}$ spacing value, as previously reported by Chen et al. [21] and Zhu et al. [22]. The bentonite clay samples, as well as the PS/OC nanocomposites, were analyzed by XRD, and the results are discussed in detail below.

3.2.1. XRD Analysis of RC, NaC and OC

The X-ray diffraction analysis was applied in order to confirm the intercalation of the CPC cations into the interlayer according to the (d$_{001}$ spacing) value of the clay's characteristic peak. After the inorganic modification with NaCl and the organic modification with the surfactant, NaC and OC show the same characteristic peaks as the RC with minor shifts, which proves that the crystalline structure of bentonite clay did not change. The XRD patterns of the RC, NaC, and OC are shown in Figure 2A. The 2θ angle of the (001) reflection corresponding to the basal spacing of the RC is 7.16°, while the d$_{001}$ spacing is 12.33 Å.

After the first stage of RC modification of sodium chloride, a decrease in d_{001} spacing was observed to be 11.95 Å in the NaC pattern, and the characteristic diffraction peak shifted slightly to a higher angle of about 7.40°. This shifting may be due to the replacement of the common cations present in the RC interlayer, such as K^{+1} and Ca^{+2}, which are known to have a higher radius compared to Na^{+1}. In contrast, the diffraction pattern of the OC shows a remarkable increase in the interlayer (d_{001} spacing) up to 23.079 Å, at 2θ equal to 3.825°. This is a confirmation of the CPC cation intercalations into the clay layers and pushing them into a wider distance [8,20,23]. Numerous studies have reported that the basal spacing up to (14 Å) is a monolayer; a bilayer arrangement occurs when the spacing ranged between (14 to 17.7 Å), whereas the spacing value is approximately in the range of (17–22 Å) is a pseudo-trilayer. Finally, a paraffin-type-monolayer or -bilayer usually has a spacing (>22 Å) [24–27]. From the foregoing discussion, we suggest that the modified organoclay (OC) in our work has two possible configurations: pseudo-tri-layer and paraffin type.

3.2.2. XRD Analysis of PS, 1% PCN and 3% PCN

Further analysis using XRD was applied to OC, PS, 1% and 3% PCN to study the effect of PS chains on OC interlayers using 1 wt.% and 3 wt.% of OC. From the XRD patterns shown in Figure 2B, the PS pattern does not show any diffraction at 2θ from 3° to approximately 14°, but it shows a wide, hill-like diffraction at 2θ in the range of (14°–23°) with low intensity [28,29]. By comparing the previously discussed patterns of OC and PS, we can deduce the effect of PS chain insertion on the layers of OC. In the 1% PCN and 3% PCN patterns, the diffraction peak of polystyrene appeared in the same position. Low crystallinity is the factor behind broad diffraction peaks in PS and PCNs [10]. Importantly, the diffraction peak of the OC at 2θ equal to 3.825° disappeared in the prepared nanocomposites 1% PCN and 3% PCN patterns, which indicates the amorphous PCN structure. This shows the possibility of an exfoliated structure of the prepared PCNs, as found by other researchers [28–30]. To support these XRD results, an additional analysis was performed using transmission electron microscopy (TEM).

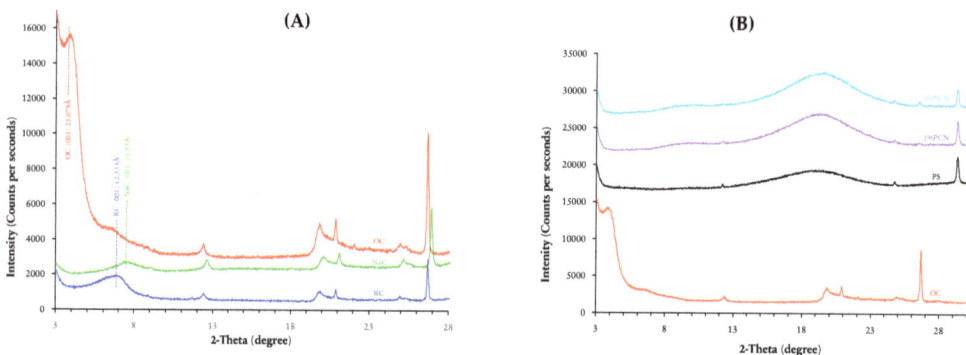

Figure 2. XRD images of (**A**) RC, NaC, OC, and (**B**) OC, PS, 1% PCN, 3% PCN.

3.3. *Scanning Electron Microscopy (SEM)*

Based on the preliminary FT-IR results for the prepared polystyrene microcapsules MCs, we can confirm that the MCs were successfully prepared. To verify, SEM analysis was performed to study the other morphological properties. It is worth saying that the MC's preparation was repeated and when the weight of the starting materials is doubled, the percentage yield increases by about 70%, producing 1.94 g of white MCs with a homogeneous size, which in turn is consistent with the previous studies [31,32]. As shown in the SEM image Figure 3, the perfectly spherical shape of the prepared MCs was obtained with a radius that varies in the range (10–30 µm). In addition, a mononuclear structure was observed with a smooth, nonporous surface, as mentioned by Cotting et al. [5].

Figure 3. SEM images of the obtained polystyrene microcapsules (MCs).

3.4. Transmission Electron Microscopy (TEM)

This technique (TEM) is powerful due to its effectiveness in determining the internal structure of the nanocomposites in the nanoscale range. Thus, the degree of the clay nanolayer exfoliation in polystyrene can be verified. To support the results revealed by FT-IR and XRD, transmission electron microscopy (TEM) analysis was performed on the prepared 1 wt.% and 3 wt.% PCN. The light areas correspond to the polystyrene, while the hair-like dark areas are related to the nanolayers of the clay [28,30]. Both 1% PCN and 3% PCN were imaged at ascending magnifications (100,000 and 200,000). In Figure 4A, the TEM images of 1% PCN are presented. As shown, the clay layers are lined with each other in most areas, maintaining a generally parallel arrangement, indicating the presence of an intercalated structure, while the hair-like areas in the samples provide an exfoliated structure [33,34]. As shown in Figure 4B, free 3% PCN clearly shows that the nanolayers have been totally dispersed throughout the polystyrene matrix. Therefore, combining the XRD and TEM results, it can be observed that the free 3% PCN has an exfoliated structure [35,36]. This result can be strongly observed in the 200,000 magnifications. It is worth noting that the black areas belong to the clay layers that were not dispersed in the polystyrene matrix.

3.5. Electrochemical Measurements

In our previous study, [6] The adhesion using the crosscut (cross-hatch) testing for PS and 1, 3, 5, and 10% PCN were determined. The results showed that for the PS, 1% and 5% PCN coatings, the area removed was 35–65%. However, for 3%, none of the coating areas was removed. Thus, 3% PCN provided the best coating adhesion. This improvement in adhesion indicates that the prepared PCN filled the voids and crevices on the C-steel surface. Therefore, in this study, only the low OC loadings were employed. To evaluate the efficiency of the prepared protective coatings, three sequential electrochemical methods were performed as follows: Electrochemical Impedance Spectroscopy (EIS), Electrochemical Frequency Modulation (EFM), then Potentiodynamic polarization. All the measurements were taken after applying Open Circuit Potential (OCP) technique for 1 h, which is the approximate time for systems to attain the steady-state cell potential (E_{SS}) as indicated in several studies [37–39]. The working electrode (WE) systems are bare and coated C-steel.

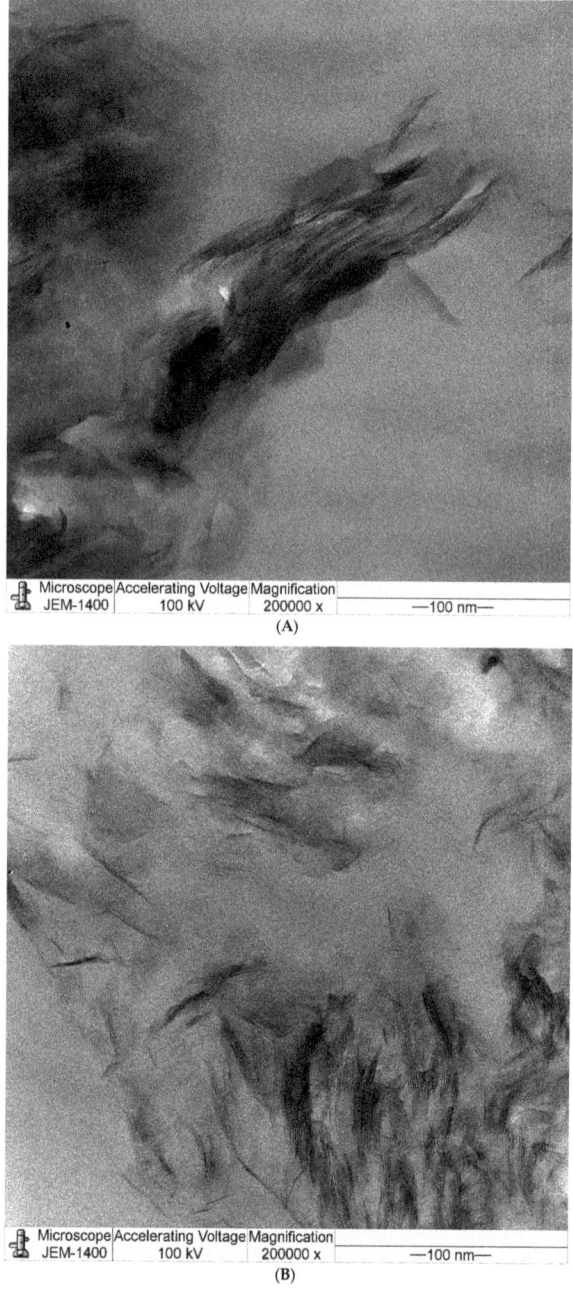

Figure 4. TEM images of (**A**) free 1% PCN at high magnification and (**B**) Free 3% PCN at high magnification.

3.5.1. Electrochemical Impedance Spectroscopy (EIS)

During the coating preparation processes, many defects such as micropores, cavities, as well as free volumes are generated in the coating, resulting in penetration of the coating by corrosive electrolytes, causing a reduction in the barrier performance of coating with time [38–40]. For the organic coating of metals, the usual interpretation of the impedance diagrams in the high-frequency (HF) part is related to the organic coating, while the low-frequency (LF) part corresponds to the reactions occurring on the metal through defects and pores in the coating [38,39]. In this study, we focus on the Nyquist plots to build our interpretation of EIS data based on their fitted (calculated) data via a suitable equivalent circuit. Two equivalent circuits were used, depending on the corroding system. For the bare C-steel electrode, the simplest circuit known as (Randles) was applied. However, the coated electrode has additional components to compensate for the mechanisms or processes when the coating becomes deformed or penetrated. One of the most effective applications of the EIS has been on the evaluation of the anticorrosive coating's efficiency and its changes during exposure to corrosive environments. From the Nyquist plot in Figure 5A, we notice only a single semi-circle for the bare C-steel, whereas the Nyquist plots related to the PS-coated samples in Figure 5B and for PCNs free samples in Figure 5C consist of almost two semi-circles. The PCN (MC) in Figure 5B,C show a semi-circle with a larger diameter than the PCN coating, which reflects higher R_{corr} values and higher protection.

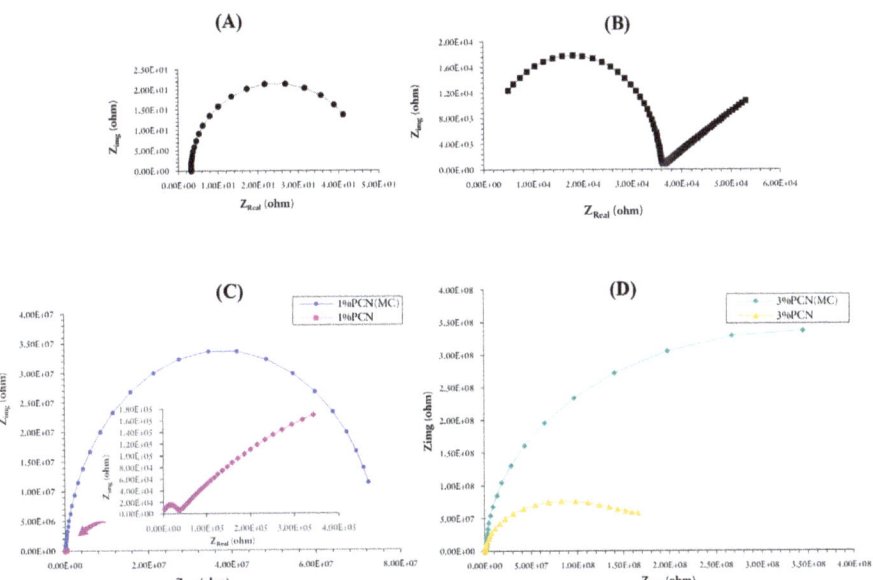

Figure 5. EIS Nyquist plots of (**A**) Bare C-steel, (**B**) PS, (**C**) 1 wt.% PCNs free and impregnated, and (**D**) 3 wt.% PCNs free and impregnated after 1 h of immersion in 3.5 wt.% NaCl at 20 ± 0.5 °C.

The electrical circuit in Figure 6 comprises the electrolyte resistance (R_{Sol}), a constant phase element representing the coating capacitance (C_c), the coating resistance to the passage of electrolytes (R_{Po}), the constant phase elements representing the double layer capacitance between the metal surface/electrolyte solution (C_{Corr}), and the charge transfer resistance across the metal surface (R_{Corr}). As reported in the literature for the coated substrates, the first semi-circle in the high-frequency region was related to the resistance and capacitance of the protective coating and its properties [40–42]. The second semicircle in the low-frequency region was attributed to the electrochemical reactions on the C-steel surface. Illustrations of the bare and coated C-steel equivalent circuits are shown in Figure 6. These electrical circuits were used for the purpose of fitting the experimental results via the Gamry software. The fitted EIS data of bare C-steel, PS, 1% PCN, 3% PCN, 1% PCN (MC) and 3% PCN (MC) are presented in Table 3. It is observed that Ess increased after adding PCN, compared to the bare C-steel and PS-coated C-steel. Thus, it can be considered that the PCN formulation tend to become more noble (electropositive potential), improving the decrease of the anodic current density compared with the bare metal substrate. It can be noted that the pure polystyrene coating played an important role in the improvement of the protection efficiency of the C-steel compared to bare C-steel. This significant improvement is proven by a comparison between the corrosion resistance R_{Corr} values of bare C-steel, which equals 4.295×10^1 Ω, and that of the C-steel coated with pure PS, which was equal to 2.347×10^5 Ω. This increase in the value of corrosion resistance is coupled with a decrease in the corrosion capacitance C_{Corr} value from 1.332×10^{-3} to 5.337×10^{-5} F/cm². Regarding the coated C-steel samples with PCNs, which are 1% PCN, 3% PCN, 1% PCN (MC) and 3% PCN (MC), the enhancement of the anticorrosive coating efficiency has also been observed based on the values shown in Table 3, as they were increased almost a thousand time compared to the pure PS. Besides, the effect of raising the weight percentage of the added organoclay from 1 to 3 wt.%, promoted the corrosion resistance as well as reduced the corrosion capacitance related to the double-layer region.

Table 3. The EIS fitted data of bare C-steel, PS, 1% PCN, 3% PCN, 1% PCN (MC) and 3% PCN (MC) after 1 h of immersion in 3.5 wt.% NaCl at room temperature 20 ± 0.5 °C.

Sample	E_{SS} (mv)	R_{Corr} (MΩ)	R_{Po} (MΩ)	C_{Corr} (nF/cm²)	CC (nF/cm²)	% PE
Bare (C-steel)	−601	4.30×10^{-5}	-	1.332×10^6	-	-
PS	−549	0.235	0.0555	5.337×10^4	0.1108	-
1% PCN	−455	1.181	0.311	2.589×10^3	0.1857	80.127
1% PCN (MC)	−215	1.556	74.10	1.022	0.3885	84.916
3% PCN	−480	27.52	6.16	0.298	0.1430	99.146
3% PCN (MC)	−297	667.8	8.191	0.1142	0.1988	99.964

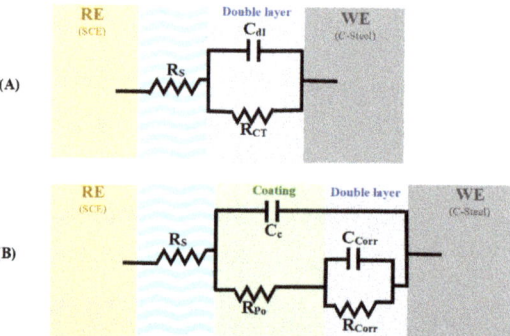

Figure 6. Schematic diagram of the equivalent circuits (**A**) Randles, and (**B**) for free and impregnated PCNs.

Also, the percentage protection efficiency (%PE) for the coated substrate relative to the pure PS has an average value of approximately 99%, which was estimated using the following equation [43]:

$$PE\% = \left[1 - \left(\frac{R_{Corr}\ (PS)}{R_{Corr}\ (PCN)}\right)\right] \times 100 \quad (1)$$

Moreover, the capacitance values of either the corrosion or the coating C_{Corr} and C_C, respectively, gave a slight change in their values. Overall, this significant improvement in the nanocomposite coating efficiency compared to pure polystyrene is due to the addition of organically modified clay. In more detail, the exfoliated clay layers in PS or the clay nanolayers in PCN, play a significant role in blocking or lengthening the diffusion pathway of the corrosive ions such as chloride and moisture whenever the coating gets damaged or penetrated. This is due to the increased tortuosity of the diffusion path in PCN compared to pure PS coating. Altogether, the coated C-steel by 3% PCN (MC) gave the best protection properties after 1 h of immersion in 3.5 wt.% NaCl, with the highest value of R_{Corr} 6.678×10^8 Ω, with a %PE equivalent to 99.996% compared to the pure PS. This superiority in the protection efficiency of the 3% PCN (MC) sample is likely due to several reasons; the most important is the partially exfoliated structure of the prepared 3 wt.% polystyrene clay nanocomposite that has been verified by TEM results. Also, the new PCN formulation through the addition of polystyrene microcapsules impregnated with Ce^{+3} may have contributed to enhancing the protection efficiency of the coating. The trend in the protection efficiency of the prepared formulations based on the EIS results are in the following order:

$$3\% \text{ PCN (MC)} > 3\% \text{ PCN}$$

$$1\% \text{ PCN (MC)} > 1\% \text{ PCN}$$

3.5.2. Electrochemical Frequency Modulation (EFM)

As in the EIS technique, electrochemical frequency modulation (EFM) is a non-destructive corrosion measurement technique that can directly give values of the electrochemical corrosion parameters like corrosion current I_{Corr} (expressed in μA) and corrosion rate CR (expressed in mpy) without a prior knowledge of Tafel constants βa and βc. In general, the intermodulation spectrum of bare C-steel includes something like peaks going up towards higher current values. The intermodulation spectrum of the coated C-steel showed peaks moving down towards less corrosion current. The electrochemical parameters obtained from EFM measurements of bare C-steel, PS, 1% PCN, 3% PCN, 1% PCN (MC) and 3% PCN (MC) are presented in Table 4. It is noticed that PCNs demonstrate a reduction in the corrosion rate, ending with high efficiency in corrosion protection in comparison with bare substrate. In addition, 3% PCN have provided superior corrosion protection for both free and those impregnated with MCs, 3% PCN (MC). Addition of organically modified clay results in the lengthening of the diffusion pathway of corrosive ions and reducing corrosion current density. According to the obtained corrosion parameters, specifically the corrosion rate (CR) and corrosion current density (I_{Corr}) shown in Table 4, the results of the EFM technique coincide with what was obtained from EIS in the reduction of the corrosion rate as well. The measured CR for bare C-steel was 5.365×10^1 mpy, while the CR of other coated C-steel was in the range between 1.583×10^{-1} to 1.040×10^{-5} mpy. Importantly, the CF(2) and CF(3) are related to the causality factor (2) and causality factor (3), respectively. Based on their experimental values shown in Table 4, they are close to their theoretical standard values, which are equal to (2.00) and (3.00), which in turn, reflects the validity of the calculated CR and I_{Corr} data. The deviation of causality factors from their standard values may be due to the perturbation amplitude being too small or the resolution of the frequency spectrum not being high enough, as mentioned in other studies [16,17]. In conclusion, 3% PCN and 3% PCN (MC) samples gave the highest protection efficiency relative to the pure PS with 99.993% and 99.990% PE, respectively. It is worth mentioning that the 3% PCN (MC) is slightly more than 3% PCN, unlike the results of EIS which may be related to method sensitivity.

Table 4. The EFM results of bare C-steel, PS, 1% PCN, 3% PCN, 1% PCN (MC) and 3% PCN (MC) after (1 h) of immersion in 3.5 wt.% NaCl at room temperature 20 ± 0.5 °C.

Sample	I_{Corr} (μA)	CR (mpy)	CF(2)	CF(3)	PE%
Bare (C-steel)	5.577×10^2	5.365×10^1	2.132	2.694	-
PS	1.645	1.583×10^{-1}	1.095	2.416	-
1% PCN	3.697×10^{-4}	3.556×10^{-5}	1.469	1.455	99.977
1% PCN (MC)	5.553×10^{-4}	5.341×10^{-5}	2.101	3.331	99.966
3% PCN	1.081×10^{-4}	1.040×10^{-5}	1.627	1.043	99.993
3% PCN (MC)	1.579×10^{-4}	1.519×10^{-5}	1.743	2.689	99.990

3.5.3. Potentiodynamic Polarization

Potentiodynamic polarization is a destructive technique of applying DC that provides quantitative information as anodic Tafel constant (βa), cathodic Tafel constant (βc), corrosion current (I_{Corr}), corrosion potential (E_{Corr}) and corrosion rate (CR) [1,4,44]. The corrosion current density was calculated for each experiment by the intersection of extrapolating the Tafel anodic and cathodic lines at the corrosion potential (E_{Corr}). The polarization curves using Tafel plot were collected in the range of (±250 mV vs. E_{Corr}). A statistical test was used to estimate how the quality of the obtained measurement data fitted with the expected theoretical results according to the used equivalent circuit, based on its value. In our work, the fitting quality was estimated for each sample by calculating the Chi-squared value. The results reveal a good fitting quality for all the performed measurements, with values ranging between 10^{-1} and 10^{-4}. For further clarification, a decrease in the Chi-squared value resulted in an increase in the fitting quality, as reported by Lim et al. [45].

Initially, we noticed that the corrosion potential (E_{Corr}) values of the coated C-steel shifted slightly towards more positive values compared to the corrosion potential of bare C-steel, which has a value of −634 mV. This can also be seen when C- steel coated with polystyrene is compared to that coated with PCNs formulation, whether free or impregnated, as shown in Figure 7. This behavior has been discussed in previous studies, proving the efficient protection of the prepared PCN coatings [46,47]. The test specimen 3% PCN provided the best corrosion resistance with a corrosion rate equal to 4.128×10^{-6} mpy, and corrosion current (I_{Corr}) equivalent to 4.290×10^{-5} μA after 1 h of immersion. The calculated protection efficiency percentage of 3% PCN relative to the pure PS was the highest, with percentages of about 99.998%. The obtained Potentiodynamic polarization parameters of bare C-steel, PS and coated PCNs are presented in Table 5. The obtained results from these three electrochemical methods after 1 h of immersion are not sufficient to evaluate the enhanced efficiency of the impregnated coatings over time; we therefore studied their efficiency over time using the EIS nondestructive method.

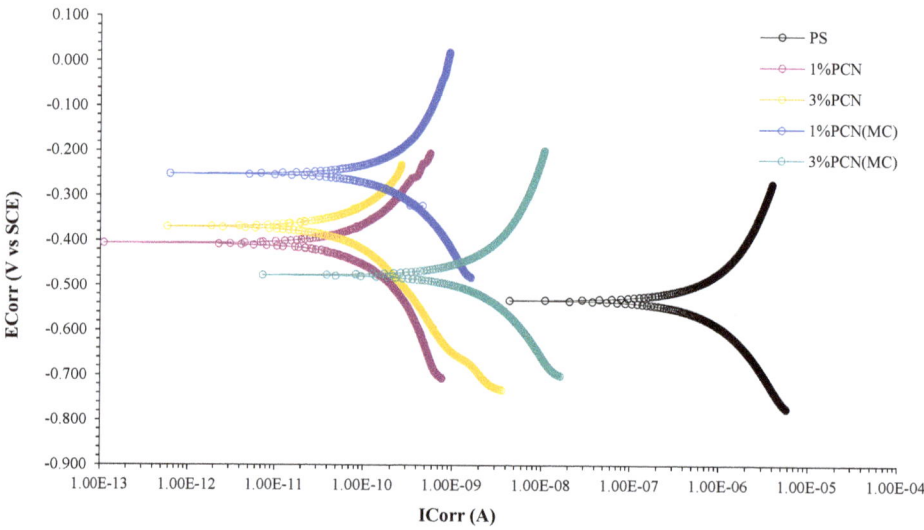

Figure 7. Tafel plots of PS, 1% PCN, 3% PCN, 1% PCN (MC) and 3% PCN (MC) after 1 h of immersion in 3.5 wt.% NaCl at room temperature 20 ± 0.5 °C.

Table 5. Tafel parameters of bare C-steel, PS, 1% PCN, 3% PCN, 1% PCN (MC) and 3% PCN (MC) after (1 h) of immersion in 3.5 wt.% NaCl at room temperature 20 ± 0.5 °C.

Sample	E_{Corr} (mV)	I_{Corr} (µA)	CR (mpy)	Chi.Sq.	PE%
Bare (C-steel)	−634	5.61×10^{1}	5.397	6.359	-
PS	−532	2.660	2.557×10^{-1}	1.8×10^{-1}	-
1% PCN	−404	2.670×10^{-4}	2.569×10^{-5}	6.61×10^{-4}	99.989
1% PCN (MC)	−251	1.820×10^{-3}	1.754×10^{-4}	1.1×10^{-3}	99.931
3% PCN	−369	4.290×10^{-5}	4.128×10^{-6}	2.48×10^{-2}	99.998
3% PCN (MC)	−478	5.820×10^{-3}	5.602×10^{-4}	1.52×10^{-3}	99.781

3.5.4. Evaluation of the Mechanism of Enhanced Coating Protection

Evaluating the healing performance of smart coatings has no standards. There is no guidance on the types of experimental methods that will provide consistent information for understanding the exact process or mechanism of self-healing. Numerous previous studies used the electrochemical impedance spectroscopy (EIS) technique to study the self-healing behavior of protective coatings to confirm the role of MCs on the prepared nanocomposite coating [48–52]. Thus, EIS measurements were set up for both 1 wt.% and 3 wt.% PCNs with free and impregnated PCN coatings. The measurements were performed during exposure to the corrosive electrolyte 3.5 wt.% NaCl at room temperature 20 ± 0.5 °C, after 1, 4, 24, and 48 h. Once the electrolyte diffuses through the coating, an electrochemical corrosion reaction takes place on the non-adherent area of the C-steel. This leads to C-steel dissolution accompanied by a reduction reaction of the dissolved oxygen in the solution. The corrosion product is ferric hydroxide with a formula of $Fe(OH)_3$. As a result of coating penetration, healing agents stored inside MCs will diffuse [5,49]. Thus, the self-healing mechanism may consist of two processes. The silanol functionality relies on its coupling agent property as well as forming a non-soluble adsorbed film on the surface of the C-steel electrode. Condensation of silanol may occur, followed by hydrogen bonding formation with the hydroxyl groups present on the surface of the C-steel. This mechanism finishes with a decrease in the corrosion reaction via adsorption as well as increasing the coating adhesion. The trivalent cerium ions may possibly undergo a series of chemical reactions, depending on the present system components. This is depicted in Scheme 2. Danaee et al. [49] suggested the oxidation reaction of the Ce^{+3} ions to form cerium hydroxide $Ce(OH)_3$. Thereafter, further reactions, including precipitation, might result in the formation of cerium oxide CeO_2. Suggested chemical reactions of trivalent cerium ions are shown below [49]:

$$2H_2O + O_2 + 4e \longrightarrow 4OH^- \quad (2)$$

$$Ce^{+3}(aq) + 3(OH^-) \longrightarrow Ce(OH)_3 \quad (3)$$

$$2Ce(OH)_3 + 2(OH^-) \longrightarrow 2CeO_2(s) + 2H_2O + 2e \quad (4)$$

The deposition of cerium oxides and hydroxides on the ferric hydroxide layer blocks the cracks and causes the "self-healing" ability. Consequently, the coating is healed and adherent to the C-steel substrate by the efforts of both the adsorbed silanol and CeO_2. As a result, the protection efficiency will be enhanced to provide not only a barrier property but long-term protection. From our study, the best evidence for this mechanism is the evaluation of the electrochemical parameters from the EIS measurements after 48 h of immersion. The obtained EIS fitted data for both 1 wt.% and 3 wt.% PCNs free and impregnated are summarized in Table 6 and shown in Figure 8. During the time of exposure, it was observed that the steady-state potential (E_{SS}) shifted towards more positive values, indicating a strengthening in the protective film. One of the most important parameters for assessing the coating integrity is the (C_C) values. Unlike the expected, the impregnated coating capacitance of 1% PCN (MC) decreased significantly from its initial value 3.885×10^{-10} F.cm^{-2} towards lesser values as follows: 1.628×10^{-10}, 1.467×10^{-10}

and 1.562×10^{-10} F.cm^{-2} during immersion times of 4–48 h. These changes in decreasing coating capacitance values can be attributed to the diffusion of the electrolyte through the coating as well as the MCs shell, leading to the release and diffusion of corrosion inhibitors. Pore resistance (R_{Po}) of 1% PCN (MC) after 4 h of immersion was reduced from 74.1 MΩ to 8.839 MΩ which was associated with C_C reduction as well. Moreover, after 24 h, a noticeable increment in the pore resistance to a value of 11.53 MΩ was noticed, reflecting that the coating may have undergone a healing process by blocking the electrolyte diffusion pathways with the formation of CeO_2. In addition, the increase in corrosion resistance (R_{Corr}) was noted, in addition to ascending behavior during the time of exposure. It is shown in Figure 8 that the effectiveness of 1% PCN (MC) continued even after 48 h of exposure to the electrolyte with the inhibitor's releasing process. The same argument applies to 3% PCN (MC) but with higher corrosion parameters, which may be due to the exfoliated structure, as shown from TEM results in Figure 4. It is worth mentioning that although the 3% PCN (MC) has a higher R_{corr} at each time interval, it reached its highest R_{corr} after 4 h then gradually decreased with time. These results are consistent with the results of previous studies on the self-healing performance in protective coatings [5,49–53]. In conclusion, the enhanced properties of the developed PCN (MC) coatings make them attractive for their potential application in the oil industry and many industrial applications.

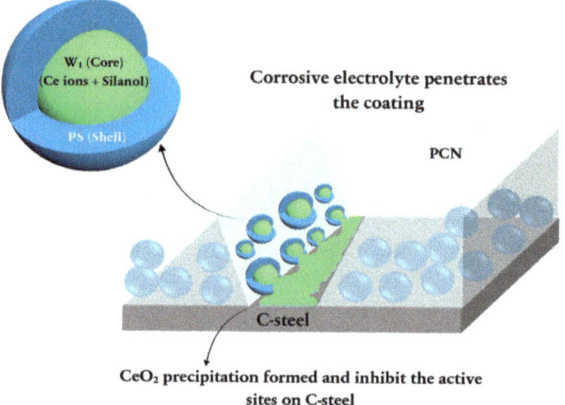

Scheme 2. The role of microcapsules in enhancing the efficiency Coating.

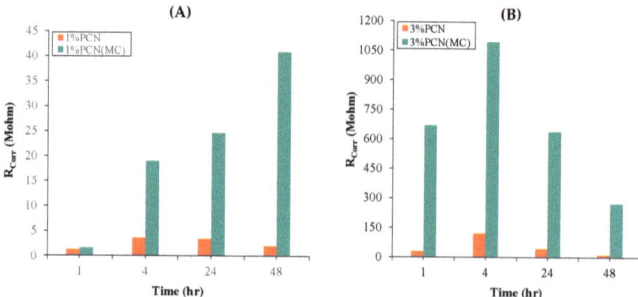

Figure 8. Corrosion resistance R_{Corr} (MΩ) vs. time at 1, 4, 24 and 48 h for (**A**) Free and impregnated 1%PCNs, and (**B**) Free and impregnated 3%PCNs immersed in 3.5 wt.% NaCl at room temperature 20 ± 0.5 °C.

Table 6. EIS fitted data of 1% PCN, 3% PCN, 1% PCN (MC) and 3% PCN (MC) after immersion in 3.5 wt.% NaCl for 1, 4, 24, and 48 h at room temperature 20 ± 0.5 °C.

Sample	Time (Hour)	R_{Corr} (MΩ)	R_{Po} (MΩ)	C_{Corr} (nF.cm^{-2})	C_C (nF.cm^{-2})
1% PCN	1	1.181	0.3108	2.584×10^3	0.1857
	4	3.536	0.5156	3.725×10^3	0.1399
	24	3.354	3.122	172.2	0.1493
	48	1.936	3.295	285.1	0.1624
1% PCN(MC)	1	1.556	74.10	1.022	0.3885
	4	18.97	8.839	0.4099	0.1623
	24	24.64	11.53	0.3899	0.1467
	48	40.91	7.013	0.1520	0.1562
3% PCN	1	27.52	6.16	0.298	0.1430
	4	118.6	42.65	0.1067	0.1152
	24	42.06	18.58	0.1791	0.1526
	48	11.41	3.514	24.05	0.1497
3% PCN(MC)	1	667.8	8.191	0.1142	0.1988
	4	1092	17.03	0.0113	0.3412
	24	636.1	8.951	0.1682	0.2586
	48	270.5	7.109	0.1635	0.1221

4. Conclusions

Extending our previous findings that low organic clay loading gave the best coating protection, only 1% and 3% PCN formulations were investigated in this study. Moreover, new protective coatings were prepared from polystyrene/organoclay nanocomposites (PCNs) impregnated with polystyrene microcapsules (MCs) loaded with inhibitors (Ce^{+3} ions and silanol), labeled as 1% PCN (MC) and 3% PCN (MC). The Structural and morphological characterization techniques confirmed the success of preparing the new coating formulations. The electrochemical measurements confirmed the enhancement in the protection efficiency of PCN (MC) compared with the PS-coated samples and the unimpregnated samples. The prepared, 3% PCN(MC clearly exhibited more efficient protection properties reaching 270.5 MΩ after 48 h of immersion, higher than all the other formulations. The diffusion mechanism of the inhibitors released from the MCs has been suggested through CeO_2 precipitation on the C-steel surface after being released and by the silanol functionality as a coupling agent. The enhanced properties of the developed PCN (MC) coatings make them attractive for their potential application in the oil industry.

Author Contributions: This paper contains the data from a Master's thesis of A.M.A. under the chief supervision of L.A.A.J. and assisted by W.K.M. Conceptualization: L.A.A.J. and W.K.M.; Funding acquisition, L.A.A.J.; Methodology, A.M.A., L.A.A.J. and W.K.M.; Data Validation, A.M.A.; L.A.A.J. and W.K.M.; Writing—original draft, A.M.A.; Writing—review and editing: all authors; Investigation A.M.A., L.A.A.J. and W.K.M. All authors have read and agreed to the published version of the manuscript.

Funding: The authors extend their appreciation to the Deanship of Research and Innovation, King Saud University, Saudi Arabia for funding this research work through project no. (IFKSURG-2-480).

Institutional Review Board Statement: Not applicable.

Informed Consent Statement: Not applicable.

Data Availability Statement: Some of the raw/processed data required to reproduce these findings cannot be shared at this time as the data also forms part of an ongoing study.

Acknowledgments: We want to thank Asmaa Al Angari for operating the FT-IR instrument and Sharefa Al Ahmarey for operating the XRD instrument. We extend our gratitude to Leonel S.J. Bautista for performing the SEM.

Conflicts of Interest: We declare that the reported work is original, and it has not been submitted elsewhere for publication. The authors declare no conflict of interest. The funders had no role in the design of the study; in the collection, analyses, or interpretation of data; in the writing of the manuscript; or in the decision to publish the results.

References

1. Jones, D.A. *Principles and Prevention of Corrosion*; Macmillan Publishing Company: New York, NY, USA, 1991.
2. Olad, A. Polymer/clay nanocomposites. In *Advances in Diverse Industrial Applications of Nanocomposites*; InTech: Nordersteds, Germany, 2011.
3. Bergaya, F.; Lagaly, G. *Handbook of Clay Science*; Newnes: London, UK, 2013; Volume 5.
4. Makhlouf, A.S.H. *Handbook of Smart Coatings for Materials Protection*; Elsevier: Amsterdam, The Netherlands, 2014.
5. Cotting, F.; Aoki, I.V. Smart protection provided by epoxy clear coating doped with polystyrene microcapsules containing silanol and Ce (III) ions as corrosion inhibitors. *Surf. Coat. Technol.* **2016**, *303*, 310–318. [CrossRef]
6. Al Juhaiman, L.; Al-Enezi, D.; Mekhamer, W. Polystyrene/Organoclay Nanocomposites as Anticorrosive Coatings of C-Steel. *Int. J. Electrochem. Sci.* **2016**, *11*, 5618–5630. [CrossRef]
7. Workman, J., Jr.; Weyer, L. *Practical Guide and Spectral Atlas for Interpretive Near-Infrared Spectroscopy*; CRC Press: Boca Raton, FL, USA, 2012.
8. Ben-Yahia, A.; El Kazzouli, S.; Essassi, E.M.; Bousmina, M.M. Synthesis and characterization of new organophilic clay. Preparation of polystyrene/clay nanocomposite. *Sci. Study Res. Chem. Chem. Eng. Biotechnol. Food Ind.* **2018**, *19*, 193–202.
9. Yu, C.; Ke, Y.; Deng, Q.; Lu, S.; Ji, J.; Hu, X.; Zhao, Y. Synthesis and Characterization of Polystyrene-Montmorillonite Nanocomposite Particles Using an Anionic-Surfactant-Modified Clay and Their Friction Performance. *Appl. Sci.* **2018**, *8*, 964. [CrossRef]
10. Paul, P.K.; Hussain, S.A.; Bhattacharjee, D.; Pal, M. Preparation of polystyrene–clay nanocomposite by solution intercalation technique. *Bull. Mater. Sci.* **2013**, *36*, 361–366. [CrossRef]
11. Coates, J. Interpretation of infrared spectra, a practical approach. In *Encyclopedia of Analytical Chemistry: Applications, Theory and Instrumentation*; John Wiley & Sons Ltd.: Chichester, UK, 2006.
12. Violeta, P.; Raluca, I.; Valentin, R.; Andi, N.C.; Ilie, S.C. Preparation and characterization of acrylic hybrid materials. *Int. Multidiscip. Sci. GeoConf. SGEM Surv. Geol. Min. Ecol. Manag.* **2017**, *17*, 293–300.
13. Al Juhaiman, L.; Al-Enezi, D.; Mekhamer, W. Preparation and characterization of polystyrene/organoclay nanocomposites from raw clay. *Dig. J. Nanomater. Biostruct.* **2016**, *11*, 105–114.
14. Zhang, Y.; Li, S.; Zhang, W.; Chen, X.; Hou, D.; Zhao, T.; Li, X. Preparation and mechanism of graphene oxide/isobutyltriethoxysilane composite emulsion and its effects on waterproof performance of concrete. *Constr. Build. Mater.* **2019**, *208*, 343–349. [CrossRef]
15. Yong, W.Y.D.; Zhang, Z.; Cristobal, G.; Chin, W.S. One-pot synthesis of surface functionalized spherical silica particles. *Colloids Surf. A Physicochem. Eng. Asp.* **2014**, *460*, 151–157. [CrossRef]
16. Zheludkevich, M.L.; Tedim, J.; Freire, C.S.R.; Fernandes, S.C.; Kallip, S.; Lisenkov, A.; Gandini, A.; Ferreira, M.G.S. Self-healing protective coatings with "green" chitosan based pre-layer reservoir of corrosion inhibitor. *J. Mater. Chem.* **2011**, *21*, 4805–4812. [CrossRef]
17. Ozkazanc, E.; Zor, S.; Ozkazanc, H.; Guney, H.Y.; Abaci, U. Synthesis, characterization and dielectric behavior of (ES)-form polyaniline/cerium(III)-nitrate-hexahydrate composites. *Mater. Chem. Phys.* **2012**, *133*, 356–362. [CrossRef]
18. Pop, O.L.; Diaconeasa, Z.; Mesaroş, A.; Vodnar, D.C.; Cuibus, L.; Ciontea, L.; Socaciu, C. FT-IR Studies of Cerium Oxide Nanoparticles and Natural Zeolite Materials. *Bull. UASVM Food Sci. Technol.* **2015**, *72*, 50–55. [CrossRef]
19. Thakur, S.; Patil, P. Rapid synthesis of cerium oxide nanoparticles with superior humidity-sensing performance. *Sens. Actuators B Chem.* **2014**, *194*, 260–268. [CrossRef]
20. Song, X.; Jiang, N.; Li, Y.; Xu, D.; Qiu, G. Synthesis of CeO_2-coated SiO_2 nanoparticle and dispersion stability of its suspension. *Mater. Chem. Phys.* **2008**, *110*, 128–135. [CrossRef]
21. Chen, B.; Zhu, L.; Zhu, J.; Xing, B. Configurations of the Bentonite-Sorbed Myristylpyridinium Cation and Their Influences on the Uptake of Organic Compounds. *Environ. Sci. Technol.* **2005**, *39*, 6093–6100. [CrossRef]
22. Zhu, R.; Zhou, Q.; Zhu, J.; Xi, Y.; He, H. Organo-Clays As Sorbents of Hydrophobic Organic Contaminants: Sorptive Characteristics and Approaches to Enhancing Sorption Capacity. *Clays Clay Miner.* **2015**, *63*, 199–221. [CrossRef]
23. Selvaraj, M.; Kim, B.H.; Lee, T.G. FT-IR studies on selected mesoporous metallosilicate molecular sieves. *Chem. Lett.* **2005**, *34*, 1290–1291. [CrossRef]
24. Li, J.; Zhu, L.; Cai, W. Characteristics of organobentonite prepared by microwave as a sorbent to organic contaminants in water. *Colloids Surf. A Physicochem. Eng. Asp.* **2006**, *281*, 177–183. [CrossRef]
25. Sreedharan, V.; Sivapullaiah, P.V. Effect of Organic Modification on Adsorption Behaviour of Bentonite. *Indian Geotech. J.* **2012**, *42*, 161–168. [CrossRef]
26. Li, Z.; Wang, C.-J.; Jiang, W.-T. Intercalation of Methylene Blue in a High-Charge Calcium Montmorillonite—An Indication of Surface Charge Determination. *Adsorpt. Sci. Technol.* **2010**, *28*, 297–312. [CrossRef]
27. Bonczek, J.L.; Harris, W.G.; Nkedi-Kizza, P. Monolayer to Bilayer Transitional Arrangements of Hexadecyltrimethylammonium Cations on Na-montmorillonite. *Clays Clay Miner.* **2002**, *50*, 11–17. [CrossRef]

28. Yeh, J.-M.; Liou, S.-J.; Lin, C.-G.; Chang, Y.-P.; Yu, Y.-H.; Cheng, C.-F. Effective enhancement of anticorrosive properties of polystyrene by polystyrene-clay nanocomposite materials. *J. Appl. Polym. Sci.* **2004**, *92*, 1970–1976. [CrossRef]
29. Sasaki, A.; White, J.L. Polymer nanocomposites formation by the use of sodium montmorillonite dispersion in alcohol and a cationic surfactant. *J. Appl. Polym. Sci.* **2004**, *91*, 1951–1957. [CrossRef]
30. Fu, X.; Qutubuddin, S. Polymer–clay nanocomposites: Exfoliation of organophilic montmorillonite nanolayers in polystyrene. *Polymer* **2001**, *42*, 807–813. [CrossRef]
31. Blaiszik, B.; Caruso, M.; McIlroy, D.; Moore, J.; White, S.; Sottos, N. Microcapsules filled with reactive solutions for self-healing materials. *Polymer* **2009**, *50*, 990–997. [CrossRef]
32. Valero-Gómez, A.; Molina, J.; Pradas, S.; López-Tendero, M.J.; Bosch, F. Microencapsulation of cerium and its application in sol-gel coatings for the corrosion protection of aluminum alloy AA2024. *J. Sol-Gel Sci. Technol.* **2019**, *93*, 36–51. [CrossRef]
33. Dike, A.S.; Yilmazer, U. Mechanical, thermal and rheological characterization of polystyrene/organoclay nanocomposites containing aliphatic elastomer modifiers. *Mater. Res. Express* **2020**, *7*, 015055. [CrossRef]
34. Alshabanat, M.; Al-Arrash, A.; Mekhamer, W. Polystyrene/Montmorillonite Nanocomposites: Study of the Morphology and Effects of Sonication Time on Thermal Stability. *J. Nanomater.* **2013**, *2013*, 650725. [CrossRef]
35. Hwu, J.M.; Ko, T.H.; Yang, W.-T.; Lin, J.C.; Jiang, G.J.; Xie, W.; Pan, W.P. Synthesis and properties of polystyrene-montmorillonite nanocomposites by suspension polymerization. *J. Appl. Polym. Sci.* **2004**, *91*, 101–109. [CrossRef]
36. Fu, X.; Qutubuddin, S. Synthesis of polystyrene–clay nanocomposites. *Mater. Lett.* **2000**, *42*, 12–15. [CrossRef]
37. Baldissera, A.F.; Freitas, D.B.; Ferreira, C.A. Electrochemical impedance spectroscopy investigation of chlorinated rubber-based coatings containing polyaniline as anticorrosion agent. *Mater. Corros.* **2010**, *61*, 790–801. [CrossRef]
38. Abdel-Rehim, S.; Khaled, K.; Abd-Elshafi, N. Electrochemical frequency modulation as a new technique for monitoring corrosion inhibition of iron in acid media by new thiourea derivative. *Electrochim. Acta* **2006**, *51*, 3269–3277. [CrossRef]
39. Fedrizzi, L.; Rodriguez, F.; Rossi, S.; Deflorian, F.; Di Maggio, R. The use of electrochemical techniques to study the corrosion behaviour of organic coatings on steel pretreated with sol–gel zirconia films. *Electrochim. Acta* **2001**, *46*, 3715–3724. [CrossRef]
40. Kuriyama, N.; Sakai, T.; Miyamura, H.; Uehara, I.; Ishikawa, H.; Iwasaki, T. Electrochemical impedance and deterioration behavior of metal hydride electrodes. *J. Alloys Compd.* **1993**, *202*, 183–197. [CrossRef]
41. Es-Saheb, M.; Sherif, E.S.M.; El-Zatahry, A.; El Rayes, M.M.; Khalil, A.K. Corrosion passivation in aerated 3.5% NaCl solutions of brass by nanofiber coatings of polyvinyl chloride and polystyrene. *Int. J. Electrochem. Sci.* **2012**, *7*, 10442–10455.
42. Behzadnasab, M.; Mirabedini, S.; Kabiri, K.; Jamali, S. Corrosion performance of epoxy coatings containing silane treated ZrO_2 nanoparticles on mild steel in 3.5% NaCl solution. *Corros. Sci.* **2011**, *53*, 89–98. [CrossRef]
43. Chang, K.-C.; Chen, S.-T.; Lin, H.-F.; Lin, C.-Y.; Huang, H.-H.; Yeh, J.-M.; Yu, Y.-H. Effect of clay on the corrosion protection efficiency of PMMA/Na^+-MMT clay nanocomposite coatings evaluated by electrochemical measurements. *Eur. Polym. J.* **2008**, *44*, 13–23. [CrossRef]
44. Banik, N.; Jahan, S.; Mostofa, S.; Kabir, H.; Sharmin, N.; Rahman, M.; Ahmed, S. Synthesis and characterization of organoclay modified with cetylpyridinium chloride. *Bangladesh J. Sci. Ind. Res.* **2015**, *50*, 65–70. [CrossRef]
45. Lim, A.B.; Neo, W.J.; Yauw, O.; Chylak, B.; Gan, C.L.; Chen, Z. Evaluation of the corrosion performance of Cu–Al intermetallic compounds and the effect of Pd addition. *Microelectron. Reliab.* **2016**, *56*, 155–161. [CrossRef]
46. ElShami, A.A.; Bonnet, S.; Makhlouf, M.H.; Khelidj, A.; Leklou, N. Novel green plants extract as corrosion inhibiting coating for steel embedded in concrete. *Pigment Resin Technol.* **2020**, *49*, 501–514. [CrossRef]
47. Afsharimani, N.; Talimian, A.; Merino, E.; Durán, A.; Castro, Y.; Galusek, D. Improving corrosion protection of Mg alloys (AZ31B) using graphene-based hybrid coatings. *Int. J. Appl. Glass Sci.* **2021**, *13*, 143–150. [CrossRef]
48. Onofre-Bustamante, E.; Dominguez-Crespo, M.A.; Torres, A.; Olvera-Martínez, A.; Genescá-Llongueras, J.; Rodriguez-Gomez, F. Characterization of cerium-based conversion coatings for corrosion protection of AISI-1010 commercial carbon steel. *J. Solid State Electrochem.* **2009**, *13*, 1785–1799. [CrossRef]
49. Danaee, I.; Darmiani, E.; Rashed, G.R.; Zaarei, D. Self-healing and anticorrosive properties of Ce (III)/Ce (IV) in nanoclay–epoxy coatings. *Iran. Polym. J.* **2014**, *23*, 891–898. [CrossRef]
50. Ubaid, F.; Radwan, A.B.; Naeem, N.; Shakoor, R.; Ahmad, Z.; Montemor, F.; Kahraman, R.; Abdullah, A.M.; Soliman, A. Multifunctional self-healing polymeric nanocomposite coatings for corrosion inhibition of steel. *Surf. Coat. Technol.* **2019**, *372*, 121–133. [CrossRef]
51. Stankiewicz, A.; Szczygieł, I.; Szczygiel, B. Self-healing coatings in anti-corrosion applications. *J. Mater. Sci.* **2013**, *48*, 8041–8051. [CrossRef]
52. Jackson, A.C.; Bartelt, J.A.; Marczewski, K.; Sottos, N.R.; Braun, P.V. Silica-Protected Micron and Sub-Micron Capsules and Particles for Self-Healing at the Microscale. *Macromol. Rapid Commun.* **2011**, *32*, 82–87. [CrossRef]
53. Huang, Y.; Liu, T.; Ma, L.; Wang, J.; Zhang, D.; Li, X. Saline-responsive triple-action self-healing coating for intelligent corrosion control. *Mater. Des.* **2022**, *214*, 110381. [CrossRef]

Disclaimer/Publisher's Note: The statements, opinions and data contained in all publications are solely those of the individual author(s) and contributor(s) and not of MDPI and/or the editor(s). MDPI and/or the editor(s) disclaim responsibility for any injury to people or property resulting from any ideas, methods, instructions or products referred to in the content.

Article

Ascorbic Acid-Modified Silicones: Crosslinking and Antioxidant Delivery

Guanhua Lu, Akop Yepremyen, Khaled Tamim, Yang Chen and Michael A. Brook *

Department of Chemistry and Chemical Biology, McMaster University, 1280 Main St. W., Hamilton, ON L8S 4M1, Canada
* Correspondence: mabrook@mcmaster.ca

Abstract: Vitamin C is widely used as an antioxidant in biological systems. The very high density of functional groups makes it challenging to selectively tether this molecule to other moieties. We report that, following protection of the enediol as benzyl ethers, the introduction of an acrylate ester at C1 is straightforward. Ascorbic acid-modified silicones were synthesized via aza-Michael reactions of aminoalkylsilicones with ascorbic acrylate. Viscous oils formed when the amine/acrylate ratios were <1. However, at higher amine/acrylate ratios with pendent silicones, a double reaction occurred to give robust elastomers whose modulus is readily tuned simply by controlling the ascorbic acid amine ratio that leads to crosslinks. Reduction with H_2/Pd removed the benzyl ethers and led to increased crosslinking, and either liberated the antioxidant small molecule or produced antioxidant elastomers. These pro-antioxidant elastomers show the power of exploiting natural materials as co-constituents of silicone polymers.

Keywords: ascorbic acid; silicone elastomer; antioxidant activity; reductive cleavage; aza-Michael addition

Citation: Lu, G.; Yepremyen, A.; Tamim, K.; Chen, Y.; Brook, M.A. Ascorbic Acid-Modified Silicones: Crosslinking and Antioxidant Delivery. *Polymers* 2022, 14, 5040. https://doi.org/10.3390/polym14225040

Academic Editors: Tomasz Makowski and Sivanjineyulu Veluri

Received: 18 October 2022
Accepted: 18 November 2022
Published: 21 November 2022

Publisher's Note: MDPI stays neutral with regard to jurisdictional claims in published maps and institutional affiliations.

Copyright: © 2022 by the authors. Licensee MDPI, Basel, Switzerland. This article is an open access article distributed under the terms and conditions of the Creative Commons Attribution (CC BY) license (https://creativecommons.org/licenses/by/4.0/).

1. Introduction

Antioxidants are needed by both biological and synthetic materials for protection against the detrimental effects of oxidative radical species [1–7]. Their presence has been demonstrated to preserve the mechanical and other properties of polymers, especially in high oxidative stress environments, including high temperatures or biological environments. Frequently, antioxidants are simply added to a material and their efficacy and longevity depend both on their specific chemistry—their response to oxidative stress [4,7]—and whether they leach from the material to adjacent media [7,8]. Covalently attaching antioxidants to polymer matrices avoids the latter problem [9–15]. Some simple examples of grafted antioxidants include gallic acid or catechin grafted to gelatin [11], and the use of grafted phenolic antioxidants on fuel cells [12] or polyisobutylene [10].

Silicone polymers well known for their biocompatibility, electrical resistance, thermostability, and high oxidative resistance [16]; they are redox insensitive. However, in many of their applications, there would be a benefit if they could convey antioxidant activity to adjacent materials. For example, biomaterial applications ranging from topical contact lenses/cosmetics products to implanted biomaterials such as breast implants and catheters would benefit from the presence of antioxidants [17]. However, release of any bioactive from the silicone polymer could be disadvantageous [18].

Leivo et al. demonstrated the use of ascorbic acid as a linker between amine-modified silicone elastomer surfaces and collagen for cell culture [19]. The enediol was involved in forming one imine with amines from each entity and, eventually, undergoing oxidative cleavage and ceasing to function as an antioxidant. In essence, ascorbic acid was analogous to, but less toxic than, glutaraldehyde because both can react twice with an amine to form imines.

Our objective was to graft ascorbic acid to silicones while maintaining antioxidant activity. Ascorbic acid/vitamin C was chosen as the candidate antioxidant modification to graft to silicones because of its robust antioxidant and antiviral properties [1,5,20,21], which may be due to its redox properties [22]. It is found in a wide variety of fresh vegetables and fruits, and at the highest concentrations in citrus fruits and green leafy vegetables [3]. Due to the hydrophilic nature of ascorbic acid, it is challenging to incorporate it into very low-energy, hydrophobic silicone matrices. We report the formation of more hydrophobic, protected ascorbic acid-modified silicones that can be crosslinked and, when desired, deprotected with concomitant release or generation of antioxidants.

2. Materials and Methods

2.1. Materials

Potassium carbonate, sodium sulfate benzyl bromide, acryloyl chloride, triethylamine, ascorbic acid (vitamin C), deuterated methanol (MeOH-d_4), deuterated chloroform ($CDCl_3$), Pd/C (palladium, 5%wt. % (dry basis) on activated carbon), EtOAc, hexanes, DMF and all other solvents were purchased from Sigma Aldrich (Burlington, MA, USA). H_2 (ultra-high purity 5.0) was taken from a Praxair gas cylinder. Telechelic aminopropylsilicone **T334** (DMS-A31, 0.11–0.12% mol aminopropylmethylsiloxane, molar mass ~25,000 g mol^{-1}); a lower molar mass **P21** (AMS-152, 4–5% mol aminopropylmethylsiloxane, molar mass ~8000 g mol^{-1}) pendent silicone and an analogous higher mass material **P22** (AMS-1203, 20–25% mol aminopropylmethylsiloxane, molar mass ~20,000 g mol^{-1}) were purchased from Gelest (Morrisville, PA, USA).

2.2. Methods

^{1}H NMR spectra were recorded on Bruker NEO 600 MHz or NEO 500 MHz nuclear magnetic resonance spectrometer. A Shore OO durometer (Rex Gauge Company, Inc., Buffalo Grove, IL, USA) was used to characterize the hardness of the elastomer. A centrifuge was used for sedimentation of charcoal during purification of the hydrogenated product.

2.3. Synthesis of Benzyl-Protected Ascorbic Acid and Modification with Acrylate 1

Ascorbic acid (6.0 g, 34 mmol) was dissolved in DMF (20 mL). K_2CO_3 (11.8 g, 85 mmol) was added and the mixture was stirred for 1 h at 50 °C. A solution of benzyl bromide (12 g, 70 mmol) in DMF (15 mL) was added dropwise to the ascorbic acid mixture and stirred for 5 h at room temperature under an N_2 blanket. The reaction solution was filtered through a pad of Celite and washed with ethyl acetate. The combined organic phases were extracted with H_2O (3 × 100 mL). The organic layer was collected, dried over Na_2SO_4, and filtered. Following concentration by rotary evaporation, the crude product was purified by flash column chromatography (hexanes: EtOAc 1:3 to 1:1) to afford benzylated ascorbic acid (4.3 g, 36%) as light-yellow oil (for NMR and mass spectrum, see Supporting Information SI). ^{1}H NMR ($CDCl_3$, 600 MHz): 7.41–7.16 (m, 10H), 5.46–5.05 m, 4H), 4.68 (d, J = 2.0 Hz, 1H), 3.91 (m, 1H), 3.82–3.70 (m, 2H).

A stirred solution of the benzylated product (4.01 g, 11.3 mmol) was added to anhydrous CH_2Cl_2 (50 mL) and stirred over ice for 10 min. Et_3N (1.57 mL, 11.3 mmol) was added to the reaction mixture and let stir for 5 min. Acryloyl chloride (0.91 mL, 11.3 mmol) was first dissolved in 10 mL of anhydrous CH_2Cl_2 and added into the reaction mixture dropwise over 1 h. The reaction was stirred for 5 h at 0 °C and filtered over Celite. The organic layer was washed with brine (3 × 40 mL), dried over (Na_2SO_4), and filtered. Following concentration, the crude product was purified by flash column chromatography (hexanes: EtOAc 2:1) to afford **1** (858 mg, 19%) as a white solid (for NMR, see Supporting Information). Note: the isolated yield was low due to the formation of di-adduct at C2 of ascorbic acid (in addition to C1) that was difficult to separate from the monoadduct; the isolated yield of the mixture of mono- and diadducts was 77%.

^{1}H NMR ($CDCl_3$, 600 MHz): 7.52–7.29 (m, 10H), 6.57 (dd, J = 17.2, 1.5 Hz, 1H), 6.25 (dd, J = 17.2, 10.3 Hz, 1H), 5.99 (dd, J = 10.3, 1.5 Hz, 1H), 5.38-5.18 (m, 4H), 4.83 (d, J = 1.9 Hz, 1H), 4.49 (qd, J = 11.5, 6.2 Hz, 2H), 4.33–4.17 (m, 1H), 3.06 (br, 1H).

2.4. Reactions with Benzylated Acryl Ascorbic Acid 1 by Butylamine (Bn2AA)

Benzylated acryl ascorbic acid **1** (0.026 g, 0.06 mmol) and excess butylamine (0.3 g, 4 mmol) were mixed neat and stirred overnight. The product mixture was concentrated over vacuum and dried under nitrogen for 2 h before NMR was taken (SI).

2.5. Benzylated, Acryl Ascorbic Acid-Modified Silicones

2.5.1. Telechelic Silicone

Benzylated acryl ascorbic acid **1** (0.05 g, 0.12 mmol) and **T334** (1.16, 0.12 mmol amine) were dissolved in IPA (5 mL) and stirred overnight. The reaction solution was concentrated, and a yellow oil was obtained (SI).

2.5.2. Pendant Silicones

Benzylated acryl ascorbic acid **1** (0.4g, 0.96 mmol) was first dissolved in IPA (20 mL) to generate a 0.02 mg/mL stock solution. The stock solution (2.5 mL) and different quantities of **P22** (2% **P22-2**, 5% **P22-5**, 10% **P22-10**, 15% **P22-15**, 20% **P22-20**, 50% **P22-50**, 75% **P22-75**, and 100% **P22-100**) and **P21** (25% **P21-25**) (Table S1), in quantities based on 1:1 amine:acrylate) were added and stirred in additional IPA (5 mL total volume) overnight; the product solution was concentrated by evaporating the solvent in oven overnight at 50 °C. NMR was taken for resulting yellow oil (SI).

2.6. Debenzylation (Hydrogenation) of Benzylated Ascorbic Acid Silicones

Hydrogenation was performed by first dissolving benzylated ascorbic acid silicone **P22-20** (0.049 g, 0.12 mmol) in IPA (50 mL) in a 100 mL round-bottomed flask equipped with stir bar. Based on the benzyl group 15% mole Pd (0.039 g 10% Pd/C, 0.037 mmol) was then added to the solution. The round-bottomed flask was then connected to a dual manifold, after 10 × de-gas/ nitrogen purges, the manifold was then connected to an H_2 balloon; the system was then 5 × de-gas/hydrogen purged before switching to hydrogen overnight. After the reaction was performed, the solution was then centrifuged at 14,000 rpm for 30 min to give a slightly grey solution that was vacuum filtered through a Celite plug followed by concentration using rotary evaporation, the resulting clear oil was then washed with $CDCl_3$ and centrifuged at 14,000 rpm for 5 min. Two phases resulted: a $CDCl_3$ phase (from which NMR was measured) supernatant and a cloudy oil. MeOH-d_4 was added to the oily residue with shaking. After centrifugation at 14,000 rpm for 5 min, the solution phase was collected and the NMR spectrum was recorded, the remaining solid (ascorbic acid) was then dissolved in D_2O and an NMR spectrum was recorded (SI).

For most compounds, however, including **P22-10**, and **P21-25**, the reduction was accompanied by a change in color: **P22-2** went from a pale-yellow oil to a brown oil; **P22-10** and **P21-25** yielded black elastomers.

2.7. Kinetic Study of Benzyl Acryl Ascorbic Acid and Pendant Silicone

A kinetic study was conducted using NMR. Benzylated acryl ascorbic acid (0.05 g, 0.12 mmol) was first dissolved in deuterated chloroform or MeOH-d_4 (0.35 mL); **P22** (0.049 g, 0.12 mmol) was dissolved separately in deuterated chloroform or MeOH-d_4 (0.35 mL). The two components were combined right before the first NMR spectrum was collected at time 0 min, and then at 0.5 h, 1 h, 2 h, 4 h, 8 h, 12 h, and 24 h to monitor the reaction process (SI).

2.8. DPPH Assay for Elastomer Samples

For quantitative analyses, the debenzylated products of **P21-25** (84.6 mM, based on the concentration of ascorbic acid in 50 mg of **P21**) were swelled in IPA (1 mL) in a 1.5 mL centrifuge tube in quantities (Table S2); the sample was allowed to swell for 2 h. The DPPH solution (0.5 mL of 0.2 mM) was then added to the sample and the mixture sat in the dark for 30 min to react. The resulting solution was then filtered, and 200 µL aliquot of the resulting solution was added to a 96-well plate in triplicate. Scans were taken for each well at 520 nm from the plate reader and the results were recorded (Table S3). **P21-25** and **P22-2** samples, after hydrogenation, were similarly treated. **P22-10** and **P21-25** elastomers showed moderate antioxidant activity, whereas **P22-2** showed no significant antioxidant activity.

2.9. DPPH Assay of Ascorbic Acid and Benzylated Ascorbic Acid Control

DPPH assays were performed for ascorbic acid (AA) and benzylated ascorbic acid as controls to the ascorbic acid-modified elastomer samples. Stock solutions were prepared by dissolving ascorbic acid (74.5 mg) in DI water (5 mL) or benzylated ascorbic acid (150.7 mg) in IPA (5 mL), respectively. The stock solution was then diluted 2-fold, 4-fold, 10-fold, or 20-fold. Each concentration (0.5 mL) of the control solution was added to a 1.5 mL centrifuge tube, subsequently (0.5 mL of 0.2 mM DPPH solution) was added to the tube, mixed, and allowed to rest in the dark for 30 min. An aliquot (200 µL) of the resulting solution was added to a 96-well plate in triplicate. Scans were taken for each well at 520 nm from the plate reader and the results were recorded (Table S3).

3. Results

3.1. Synthesis of Benzyl-Protected Ascorbic Acid and Modification with Acrylate

Survey experiments demonstrated that ascorbic acid (AA) was both too polar and too reactive, in particular to oxidation, for many of the desired synthetic processes to succeed. Therefore, the enols in AA were protected as benzyl ethers using a simple Williamson approach (Figure 1A). Acrylic ester formation using acryloyl chloride preferentially occurred at the primary alcohol to give **1** (Figure 1B); no secondary alcohol modification was observed, as shown by NMR (SI).

3.2. Benzylated, Acryl Ascorbic Acid-Modified Silicones and Cleavage of Acylated, Benzylated Ascorbic Acid by Butyl Amine

Under oxidizing conditions, ascorbic acid can be induced to react twice with amines to (putatively) form a 1,2-dimine from dienol [19]. Model studies were undertaken with the protected derivative **1** to understand how the functional differences with the protected compound would manifest when aminosilicones were present. ^{1}H NMR showed that two equivalents of butylamine also reacted with benzylated acryl ascorbic acid **1**: the first performed an aza-Michael addition with the acrylate; and the second led to amidation and cleavage of the aza-Michael acrylate (Figure 1C,D and Supporting Information). Other motifs are also likely involved, including ring-opening cleavage or secondary Michael additions (**2**, **3**). It was, therefore, anticipated that linear silicone oils, modified with ascorbic acid, would arise from aza-Michael reactions between **1** and aminoalkylsilicones provided that the stoichiometry of [H$_2$N]/[acrylate] was kept below 1:1.

The aza-Michael process was both trivial and facile, requiring only stirring in IPA (isopropanol). A library of ascorbic acid-modified silicones could then be prepared from this key functional molecule **1** by the aza-Michael reaction with either pendent (Figure 1E) or telechelic (Figure 1F) aminoalkylsilicones containing different amine densities.

The telechelic sample **T334** (nomenclature: **Tn**, where n is the number of Me$_2$SiO units in the chain, **T334**, n = 334 Figure 1) was modified completely at both termini with **1**. With the pendent silicones both partial **P22-x** (nomenclature: **Pt-x** where t is the % of aminopropyl monomers m in the chain (m/(m + n) × 100, normally t = 22, and x = 2, 5, 10, 15, 20, 50, 75, and 100, Figure 1) and complete modification **P22-100** with **1** was performed. The telechelic products and pendent products made with lower equivalents of AA (**P22-2** → **P22-15**), or 100% **P22-100** were yellow oils that were stable for extended periods of time; so far, over one year. A lower molecular weight analog **P21-25** was also prepared as a yellow oil. The rates of reaction were shown to be faster in more polar methanol (2 h) than in chloroform (12 h, Supporting Information).

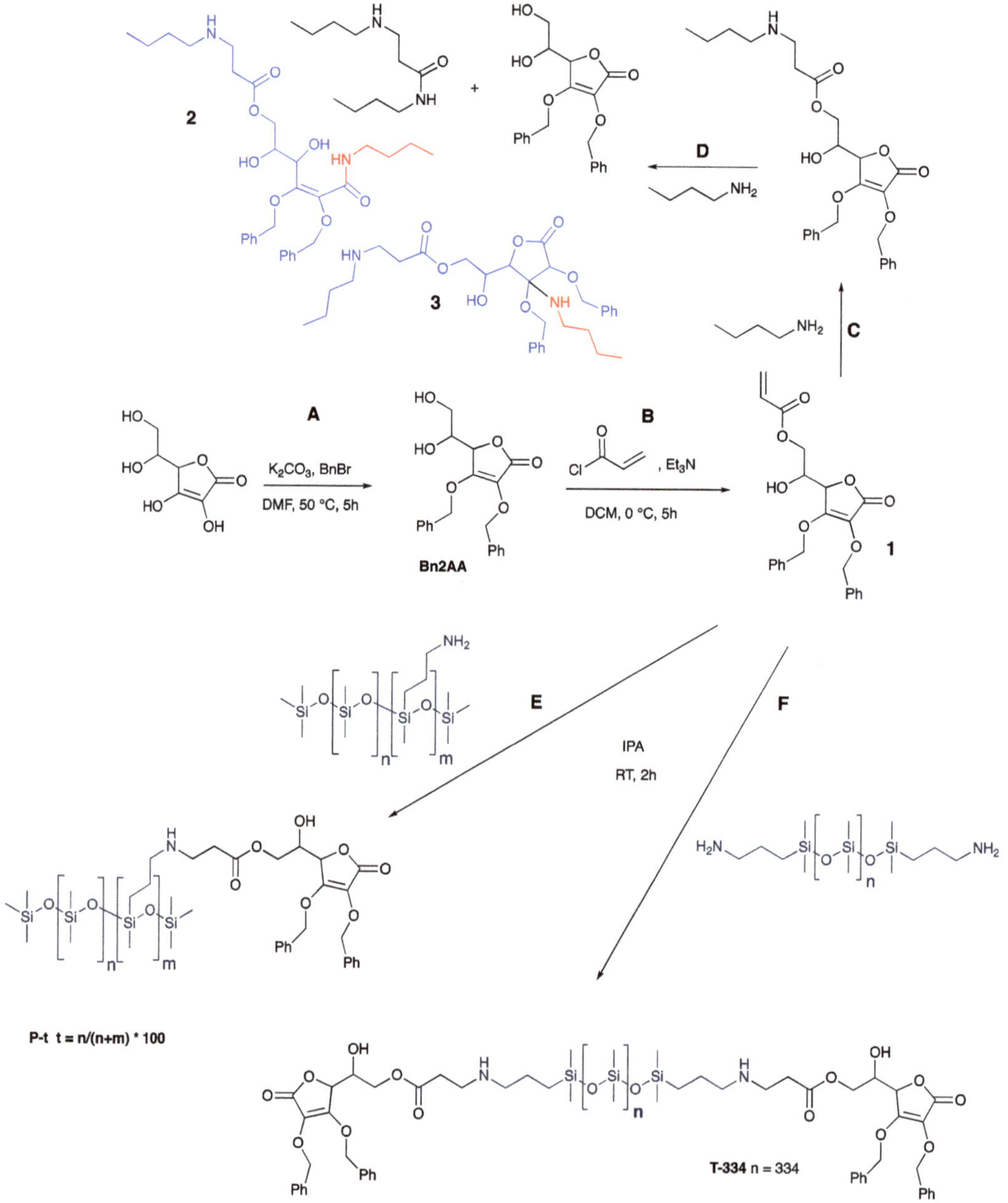

Figure 1. Synthesis of (**A**,**B**): benzyl protected, acrylated ascorbic acid and conversion to (**C**): mono or (**D**): dibutyl amine derivatives, or conversion to (**E**): pendent, or (**F**): telechelic ascorbic acid-modified silicone polymers.

If the higher molar mass pendent polymers based on **P22** were modified with higher quantities of **1** they ceased to be oils and were instead isolated as elastomers (**P22-20** → **P22-75**). There was a direct correlation between the quantity of AA 'crosslinker' **1** added and the Shore hardness of the resulting elastomer, consistent with the formation of typical silicone elastomers [23,24]. The model study with butylamine suggests the origin of the observed crosslinking. When the stoichiometric excess of amines to acylates exceeds 1:1, the initial aza-Michael (similar to Figure 1C) is accompanied by a secondary reaction (similar to Figure 1D) where compound **1** bridges polymer chains leading to crosslinks analogous to **2**, **3**. One is obliged to explain, however, why there is an onset of elastomer formation only at 20% **1**. The silicone polymer has about 1 aminopropyl-containing monomer for each 5 D unit (Me$_2$SiO). At low concentrations of **1**, the secondary reaction process will lead to both chain extension and intramolecular processes giving cycles (Figure 2B,C). In addition, not all the added **1** will undergo both processes. Thus, at lower concentrations of **1**, the aza-Michael reaction will lead to higher molecular weight silicone oils of viscosities that increase with the available fraction of **1**. At higher concentrations, sufficient crosslink arises that elastomers form, with crosslink density and durometer increasing in line with the relative quantity of **1** added. (Figure 3A).

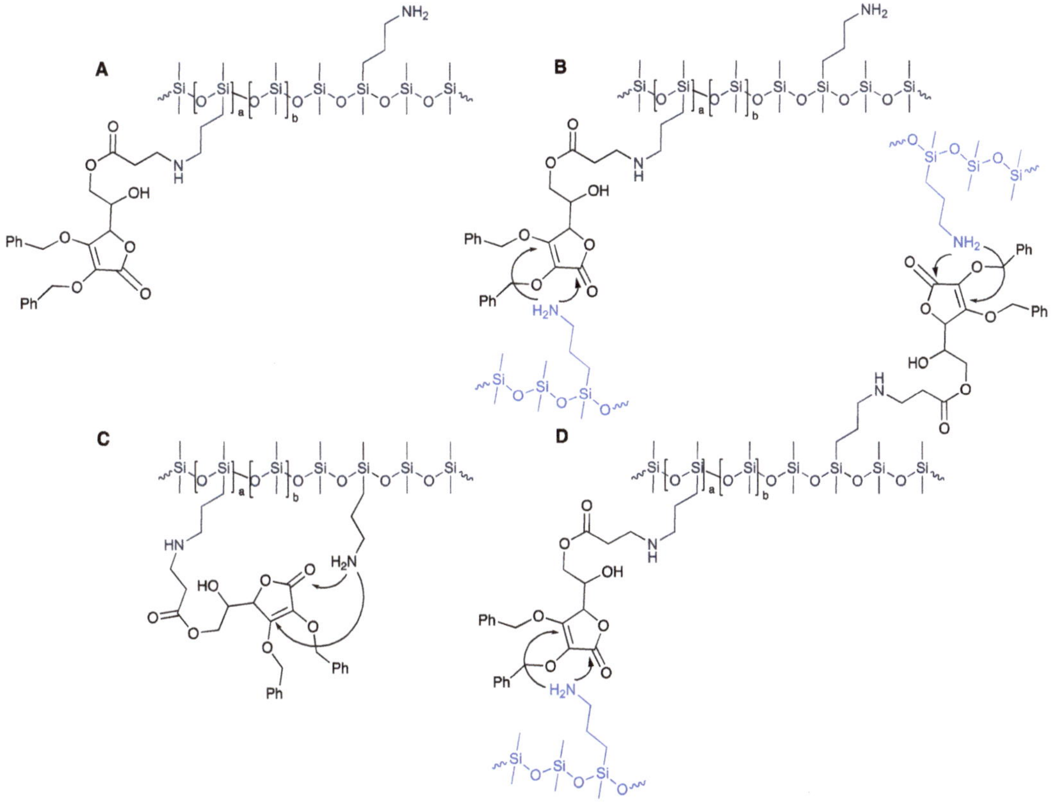

Figure 2. (**A**): monofunctional modifier; (**B**): chain extender; (**C**): loop reagent; or, at higher concentrations, (**D**): crosslinker.

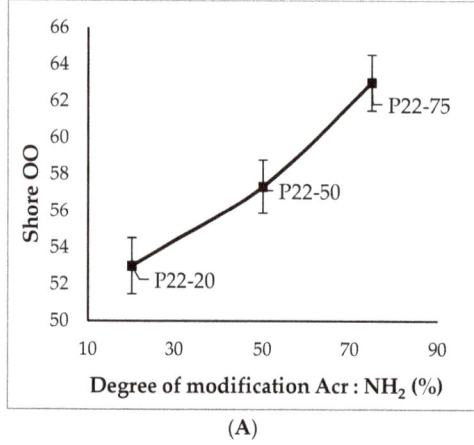 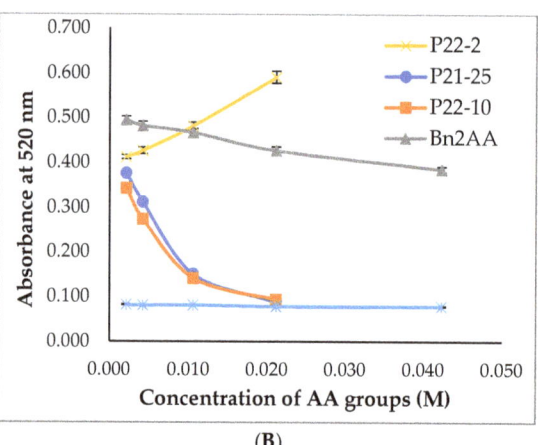

Figure 3. (**A**): Shore hardness data of benzylated ascorbic acid crosslinked aminoalkylsilicones for samples **P22-20**, **P22-50**, and **P22-75** (**B**): DPPH assay results showing antioxidant activity of ascorbic acid and benzylated ascorbic acid control compared with different debenzylated ascorbic acids.

3.3. Antioxidant Activity

The antioxidant activity of vitamin C is associated with the relative ease with which the ene-diol can undergo oxidation [25]. The ene-diols in products **T334**, **P21-25**, and **P22-x** were protected and, therefore, were not expected to have antioxidant activity. DPPH (2,2-diphenyl-1-picrylhydrazyl), a stable radical species, is a particularly convenient reagent for colorimetrically observing qualitatively, or determining quantitatively, antioxidant activity [26]. Neither **T334** nor any of the **P22-x** products exhibited significant antioxidant activity, as shown qualitatively when tested with 0.2 mM DPPH; over a period of 2 h the solution only very slowly turned yellow for oil samples, and 6–12 h for elastomeric samples, whereas ascorbic acid control solutions exhibited high antioxidant activity, immediately turning yellow. In quantitative DPPH assays, the benzylated ascorbic acid control also showed nearly no antioxidant activity (Figure 3B). It was inferred that, in order to reveal antioxidant activity, deprotection of the benzyl ethers to regenerate the ene-diol would first be necessary.

Benzyl ethers are conveniently cleaved by hydrogenation of Pd/C to yield the free alcohol and toluene. In our hands, the reduction process with both oils and elastomers led to the release of ascorbic acid (Figure 4) or its derivatives from the silicone. The reaction could be capricious; in one case, free ascorbic acid was isolated in an aqueous extract. More commonly, upon reduction of oils such as **P22-2**, **P22-10**, **P22-100**, or **P21-25** in IPA, the products took on a darker color and, in the case, of **P22-10** and **P21-25**, yielded black elastomers. That is, deprotection led to further crosslinking/chain extension. However, it also led to the liberation of antioxidant activity, as shown by DPPH assays (Figure 3B, Supporting Information). This suggests free enediols present in the product either as liberated ascorbic acid, or as tethered, crosslinking AA moieties.

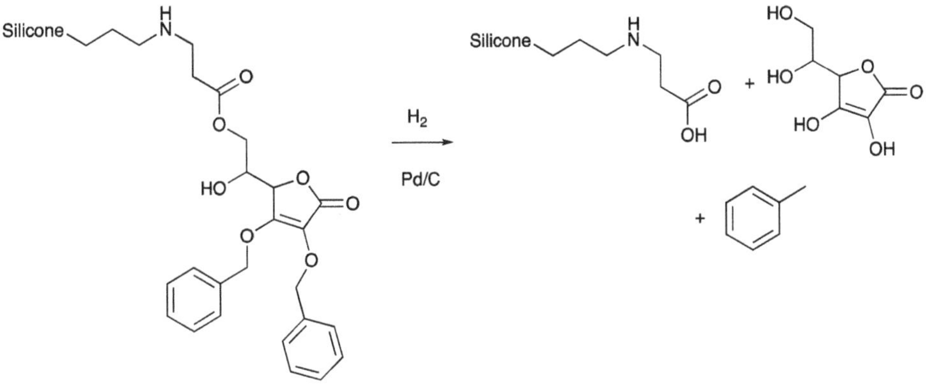

Figure 4. Reductive cleavage of benzyl ethers and the linking ester.

4. Discussion

For the reasons articulated above, there remains much interest in the use of/release of vitamin C because of its powerful biological activities, including as an antioxidant. We do not consider materials in which vitamin C that is simply mixed into a matrix, and focus on materials in which the ascorbic acid is chemically grafted. There are surprisingly few examples of Vitamin C being used in a prodrug format. These include reports of the formation of esters of the ene-diol or at the C1 position to give materials that exhibit biological activity of various types after exposure to a biological environment. Proof of release of the ascorbic acid via ester hydrolysis is typically inferred. A vitamin C–ibuprofen ester was shown, for example, to cross the blood–brain barrier where a response to the ibuprofen was shown [27]; in this case, vitamin C was the carrier. Other examples describe the use of glycosides [28], or a combination of glycosides + aliphatic esters to link to vitamin C. In these cases, the biological release of vitamin C was reported after exposure to the spleen homogenates [29] or esterases [30]; in neither case was proof of the release of free vitamin C shown. In these examples the ester linkages operate, in part, to stabilize the vitamin C from degradation.

Although vitamin C has been shown to be involved in various forms of crosslinking of polymers, particularly biological polymers, its role is generally to mediate the chemistry of the polymers themselves, including the crosslinking of proteins, for example, by inducing the Maillard reaction [31]. An important exception is the work of Leivo et al. who showed that the direct reaction of amine-modified silicone elastomer surfaces and with collagen permitted with ascorbic acid to link the two materials together via imines; the ascorbic acid moiety was then subject to autoxidative decomposition [19].

In the reactions described here, it is clear that—even when protected as benzyl ethers, **1** can undergo at least 2 sequential reactions under mild conditions (Figure 1C,D and Figure 2) leading first to chain extension and then crosslinking to give robust silicone elastomers that do not have antioxidant activity—the enediol is protected. This form of vitamin C is thus a convenient, natural crosslinking agent.

Upon liberation of the enediol by reductive deprotection of the benzyl ethers further crosslinking ensued. The accompanying darkening in color is consistent with a Maillard reaction. Elaborating the subtleties of these processes is a current occupation. Regardless, the products are also efficient antioxidants whether free ascorbic acid is liberated, or the crosslinker retains the enediol.

Simple silicone fluids undergo rather efficient environmental depolymerization [32,33]. While speculation only, it is expected that silicone fluids modified by **1**, and elastomers formed following reductive deprotection, will be subject to ester hydrolysis to regenerate silicon oils (Figure 4) that will also undergo facile depolymerization. The new crosslinks formed by ascorbic acid self-reaction should be analogous to the normal outcomes of ascorbic acid self-condensation and should also be readily degraded. The conditions for reductive cleavage of benzyl ethers are mild, but require the transition metal catalyst for efficient cleavage. This is an aspect that is clearly disadvantageous. However, it may be possible to elicit reductive cleavage without the need for platinum; it is noted that some benzyl ethers are susceptible to both oxidative and reductive cleavage under biological conditions [34].

The Green Chemistry rules call for materials that make better use of natural feedstocks [35]. In the present case, while **1** does dilute the synthetic silicone, it is to a small extent only (and we note that not all aspects of the synthesis are consistent with Green Chemistry, e.g., the (de)protection sequences). However, the use of vitamin C provides both a useful mechanism for crosslinking and delivery of new functionality—natural antioxidant activities—during cleavage. We hope to demonstrate this utility will be accompanied by the more facile decomposition of the silicone component at end of life.

5. Conclusions

Benzyl-protected ascorbic acid-modified silicones were successfully synthesized using an aza-Michael addition; the ascorbic acid ranged from 2% to 100% on both telechelic and aminoalkylsilicones. The ascorbic acid acts as a crosslinker for pendent silicones to give robust silicone elastomers without significant antioxidant behavior. Reductive debenzylation was expected to liberate antioxidant activity but, surprisingly, also lead to cleavage of the crosslink to give silicone oils and vitamin C. Thus, aspects of this work: natural materials, function, and programmed degradation fall within the rules of Green Chemistry.

Supplementary Materials: The following supporting information can be downloaded at: https://www.mdpi.com/article/10.3390/polym14225040/s1, Table S1: quantities for polymer synthesis, and Tables S2 and S3 for DPPH analyses; Figures S1–S10 ^{1}H, ^{13}C NMR spectra and 1 mass spectrum of starting materials and silicone products; and Figure S11 photo of DPPH assay.

Author Contributions: Conceptualization, G.L., A.Y. and M.A.B., methodology, G.L. and A.Y.; experimental optimization, K.T. and Y.C., writing—original draft preparation, G.L.; writing—review and editing, G.L. and M.A.B.; supervision, M.A.B.; project administration, M.A.B.; funding acquisition, M.A.B. All authors have read and agreed to the published version of the manuscript.

Funding: This research was funded by the Natural Sciences and Engineering Research Council of Canada.

Institutional Review Board Statement: Not Applicable.

Informed Consent Statement: Not Applicable.

Data Availability Statement: Spectroscopic data may be found in the Supporting Information.

Acknowledgments: We gratefully acknowledge the financial support of the Natural Sciences and Engineering Research Council of Canada (NSERC). G.L. also expresses thanks for receipt from NSERC of a PGSD PhD scholarship.

Conflicts of Interest: The authors declare no conflict of interest.

References

1. Mangir, N.; Bullock, A.J.; Roman, S.; Osman, N.; Chapple, C.; MacNeil, S. Production of Ascorbic Acid Releasing Biomaterials for Pelvic Floor Repair. *Acta Biomater.* **2016**, *29*, 188–197. [CrossRef] [PubMed]
2. Nagarajan, S.; Nagarajan, R.; Kumar, J.; Salemme, A.; Togna, A.R.; Saso, L.; Bruno, F. Antioxidant Activity of Synthetic Polymers of Phenolic Compounds. *Polymers* **2020**, *12*, 1646. [CrossRef] [PubMed]
3. Rasheed, A.; Azeez, R.F.A.; Rasheed, A.; Azeez, R.F.A. *A Review on Natural Antioxidants*; IntechOpen: London, UK, 2019; ISBN 978-1-78985-377-3.
4. Pospíšil, J. Mechanistic Action of Phenolic Antioxidants in Polymers-A Review. *Polym. Degrad. Stab.* **1988**, *20*, 181–202. [CrossRef]
5. Estevinho, B.N.; Carlan, I.; Blaga, A.; Rocha, F. Soluble Vitamins (Vitamin B12 and Vitamin C) Microencapsulated with Different Biopolymers by a Spray Drying Process. *Powder Technol.* **2016**, *289*, 71–78. [CrossRef]

6. Paradiso, P.; Serro, A.P.; Saramago, B.; Colaço, R.; Chauhan, A. Controlled Release of Antibiotics from Vitamin E-Loaded Silicone-Hydrogel Contact Lenses. *J. Pharm. Sci.* **2016**, *105*, 1164–1172. [CrossRef]
7. Al-Malika, S. *Reactive Modifiers for Polymers*, 1st ed.; Blackie Academic & Professional: London, UK, 1997.
8. Grassie, N. *Developments in Polymer Degradation—7*; Springer Science & Business Media: New York, NY, USA, 2012; ISBN 978-94-009-3425-2.
9. Huang, S.; Kong, X.; Xiong, Y.; Zhang, X.; Chen, H.; Jiang, W.; Niu, Y.; Xu, W.; Ren, C. An Overview of Dynamic Covalent Bonds in Polymer Material and Their Applications. *Eur. Polym. J.* **2020**, *141*, 110094. [CrossRef]
10. Parada, C.M.; Parker, G.L.; Storey, R.F. Polyisobutylene Containing Covalently Bound Antioxidant Moieties. *J. Polym. Sci. Part A Polym. Chem.* **2019**, *57*, 1836–1846. [CrossRef]
11. Spizzirri, U.G.; Iemma, F.; Puoci, F.; Cirillo, G.; Curcio, M.; Parisi, O.I.; Picci, N. Synthesis of Antioxidant Polymers by Grafting of Gallic Acid and Catechin on Gelatin. *Biomacromolecules* **2009**, *10*, 1923–1930. [CrossRef]
12. Buchmüller, Y.; Wokaun, A.; Gubler, L. Polymer-Bound Antioxidants in Grafted Membranes for Fuel Cells. *J. Mater. Chem. A* **2014**, *2*, 5870–5882. [CrossRef]
13. Buchmüller, Y.; Zhang, Z.; Wokaun, A.; Gubler, L. Antioxidants in Non-Perfluorinated Fuel Cell Membranes: Prospects and Limitations. *RSC Adv.* **2014**, *4*, 51911–51915. [CrossRef]
14. Zhang, W.; Li, J.X.; Tang, R.C.; Zhai, A.D. Hydrophilic and Antibacterial Surface Functionalization of Polyamide Fabric by Coating with Polylysine Biomolecule. *Prog. Org. Coat.* **2020**, *142*, 105571. [CrossRef]
15. Wang, H.; Zhu, D.; Paul, A.; Cai, L.; Enejder, A.; Yang, F.; Heilshorn, S.C. Covalently Adaptable Elastin-Like Protein–Hyaluronic Acid (ELP–HA) Hybrid Hydrogels with Secondary Thermoresponsive Crosslinking for Injectable Stem Cell Delivery. *Adv. Funct. Mater.* **2017**, *27*, 1605609. [CrossRef] [PubMed]
16. Brook, M.A. *Silicon in Organic, Organometallic, and Polymer Chemistry*; Wiley: Hoboken, NJ, USA, 1999.
17. Kaliyathan, A.V.; Mathew, A.; Rane, A.V.; Kanny, K.; Thomas, S. *Natural Rubber and Silicone Rubber-Based Biomaterials*; Elsevier: Amsterdam, The Netherlands, 2018; ISBN 9780081021958.
18. Ibrahim, M.; Bond, J.; Medina, M.A.; Chen, L.; Quiles, C.; Kokosis, G.; Bashirov, L.; Klitzman, B.; Levinson, H. Characterization of the Foreign Body Response to Common Surgical Biomaterials in a Murine Model. *Eur. J. Plast. Surg.* **2017**, *40*, 383–392. [CrossRef] [PubMed]
19. Leivo, J.; Virjula, S.; Vanhatupa, S.; Kartasalo, K.; Kreutzer, J.; Miettinen, J.; Kallio, P. A Durable and Biocompatible Ascorbic Acid-Based Covalent Coating Method of Polydimethylsiloxane for Dynamic Cell Culture. *J. R. Soc. Interface* **2017**, *14*, 20170318. [CrossRef] [PubMed]
20. Marik, P.E.; Khangoora, V.; Rivera, R.; Hooper, M.H.; Catravas, J. Hydrocortisone, Vitamin C, and Thiamine for the Treatment of Severe Sepsis and Septic Shock: A Retrospective Before-After Study. *Chest* **2017**, *151*, 1229–1238. [CrossRef]
21. Suraeva, O.; Champanhac, C.; Mailänder, V.; Wurm, F.R.; Weiss, H.; Berger, R.; Mezger, M.; Landfester, K.; Lieberwirth, I. Vitamin C Loaded Polyethylene: Synthesis and Properties of Precise Polyethylene with Vitamin C Defects via Acyclic Diene Metathesis Polycondensation. *Macromolecules* **2020**, *53*, 2932–2941. [CrossRef]
22. Sivakanthan, S.; Rajendran, S.; Gamage, A.; Madhujith, T.; Mani, S. Antioxidant and Antimicrobial Applications of Biopolymers: A Review. *Food Res. Int.* **2020**, *136*, 109327. [CrossRef]
23. Meththananda, I.M.; Parker, S.; Patel, M.P.; Braden, M. The Relationship between Shore Hardness of Elastomeric Dental Materials and Young's Modulus. *Dent. Mater.* **2009**, *25*, 956–959. [CrossRef]
24. Bui, R.; Brook, M.A. Catalyst Free Silicone Sealants That Cure Underwater. *Adv. Funct. Mater.* **2020**, *30*, 2000737. [CrossRef]
25. Lü, J.-M.; Lin, P.H.; Yao, Q.; Chen, C. Chemical and Molecular Mechanisms of Antioxidants: Experimental Approaches and Model Systems. *J. Cell Mol. Med.* **2010**, *14*, 840–860. [CrossRef]
26. Ogliani, E.; Skov, A.L.; Brook, M.A. Purple to Yellow Silicone Elastomers: Design of a Versatile Sensor for Screening Antioxidant Activity. *Adv. Mater. Technol.* **2019**, *4*, 1900569. [CrossRef]
27. Wu, X.-Y.; Li, X.-C.; Mi, J.; You, J.; Hai, L. Design, Synthesis and Preliminary Biological Evaluation of Brain Targeting l-Ascorbic Acid Prodrugs of Ibuprofen. *Chin. Chem. Lett.* **2013**, *24*, 117–119. [CrossRef]
28. Jacques, C.; Genies, C.; Bacqueville, D.; Borotra, N.; Noizet, M.; Tourette, A.; Bessou-Touya, S.; Duplan, H. 776 Optimized Vitamin C Prodrug for Controlled Release and Antioxidant Activity. *J. Investig. Dermatol.* **2020**, *140*, S103. [CrossRef]
29. Tai, A.; Goto, S.; Ishiguro, Y.; Suzuki, K.; Nitoda, T.; Yamamoto, I. Permeation and Metabolism of a Series of Novel Lipophilic Ascorbic Acid Derivatives, 6-O-Acyl-2-O-α-d-Glucopyranosyl-l-Ascorbic Acids with a Branched-Acyl Chain, in a Human Living Skin Equivalent Model. *Bioorg. Med. Chem. Lett.* **2004**, *14*, 623–627. [CrossRef] [PubMed]
30. Shibayama, H.; Hisama, M.; Matsuda, S.; Ohtsuki, M. Permeation and Metabolism of a Novel Ascorbic Acid Derivative, Disodium Isostearyl 2-O-L-Ascorbyl Phosphate, in Human Living Skin Equivalent Models. *Skin Pharmacol. Physiol.* **2008**, *21*, 235–243. [CrossRef] [PubMed]
31. Mohammadi Nafchi, A.; Tabatabaei, R.H.; Pashania, B.; Rajabi, H.Z.; Karim, A.A. Effects of Ascorbic Acid and Sugars on Solubility, Thermal, and Mechanical Properties of Egg White Protein Gels. *Int. J. Biol. Macromol.* **2013**, *62*, 397–404. [CrossRef] [PubMed]
32. Lehmann, R.G.; Varaprath, S.; Annelin, R.B.; Arndt, J.L. Degradation of Silicone Polymer in a Variety of Soils. *Environ. Toxicol. Chem.* **1995**, *14*, 1299–1305. [CrossRef]
33. Lehmann, R.G.; Varaprath, S.; Frye, C.L. Degradation of Silicone Polymers in Soil. *Environ. Toxicol. Chem.* **1994**, *13*, 1061–1064. [CrossRef]

34. Kinne, M.; Poraj-Kobielska, M.; Ralph, S.A.; Ullrich, R.; Hofrichter, M.; Hammel, K.E. Oxidative Cleavage of Diverse Ethers by an Extracellular Fungal Peroxygenase. *J. Biol. Chem.* **2009**, *284*, 29343–29349. [CrossRef]
35. Anastas, P.; Warner, J. *Green Chemistry: Theory and Practice*; Oxford University Press: Oxford, UK, 2000; ISBN 978-0-19-850698-0.

Article

Preparation of Biocidal Nanocomposites in X-ray Irradiated Interpolyelectolyte Complexes of Polyacrylic Acid and Polyethylenimine with Ag-Ions

Kristina V. Mkrtchyan [1], Vladislava A. Pigareva [2], Elena A. Zezina [2], Oksana A. Kuznetsova [3], Anastasia A. Semenova [3], Yuliya K. Yushina [3], Etery R. Tolordava [3], Maria A. Grudistova [3], Andrey V. Sybachin [2], Dmitry I. Klimov [1,2], Sergey S. Abramchuk [4], Alexander A. Yaroslavov [2] and Alexey A. Zezin [1,2,*]

1. Enikolopov Institute of Synthetic Polymeric Materials, Russian Academy of Sciences, Profsoyuznaya St. 70, 117393 Moscow, Russia
2. Department of Chemistry, M.V. Lomonosov Moscow State University, Leninskie Gory 1-3, 119991 Moscow, Russia
3. Gorbatov Federal Research Centre for Food Systems, Talalikhina St. 26, 109316 Moscow, Russia
4. Nesmeyanov Institute of Organoelement Compounds, Russian Academy of Sciences, Vavilova St. 28, 119334 Moscow, Russia
* Correspondence: aazezin@yandex.ru

Citation: Mkrtchyan, K.V.; Pigareva, V.A.; Zezina, E.A.; Kuznetsova, O.A.; Semenova, A.A.; Yushina, Y.K.; Tolordava, E.R.; Grudistova, M.A.; Sybachin, A.V.; Klimov, D.I.; et al. Preparation of Biocidal Nanocomposites in X-ray Irradiated Interpolyelectolyte Complexes of Polyacrylic Acid and Polyethylenimine with Ag-Ions. *Polymers* 2022, *14*, 4417. https://doi.org/10.3390/polym14204417

Academic Editors: Tomasz Makowski and Javier González-Benito

Received: 1 September 2022
Accepted: 15 October 2022
Published: 19 October 2022

Publisher's Note: MDPI stays neutral with regard to jurisdictional claims in published maps and institutional affiliations.

Copyright: © 2022 by the authors. Licensee MDPI, Basel, Switzerland. This article is an open access article distributed under the terms and conditions of the Creative Commons Attribution (CC BY) license (https://creativecommons.org/licenses/by/4.0/).

Abstract: Due to the presence of cationic units interpolyelectrolyte complexes (IPECs) can be used as a universal basis for preparation of biocidal coatings on different surfaces. Metallopolymer nanocomposites were successfully synthesized in irradiated solutions of polyacrylic acid (PAA) and polyethylenimine (PEI), and dispersions of non-stoichiometric IPECs of PAA–PEI containing silver ions. The data from turbidimetric titration and dynamic light scattering showed that pH 6 is the optimal value for obtaining IPECs. Metal polymer complexes based on IPEC with a PAA/PEI ratio equal to 3/1 and 1/3 were selected for synthesis of nanocomposites due to their aggregative stability. Studies using methods of UV-VIS spectroscopy and TEM have demonstrated that the size and spatial organization of silver nanoparticles depend on the composition of polymer systems. The average sizes of nanoparticles are 5 nm and 20 nm for complexes with a molar ratio of PAA/PEI units equal to 3/1 and 1/3, respectively. The synthesized nanocomposites were applied to the glass surface and exhibited high antibacterial activity against both gram-positive (*Staphylococcus aureus*) and gram-negative bacteria (*Salmonella*). It is shown that IPEC-Ag coatings demonstrate significantly more pronounced biocidal activity not only in comparison with macromolecular complexes of PAA–PEI, but also coatings of PEI and PEI based nanocomposites.

Keywords: polyacrylic acid; polyethyleneimine; interpolyelectrolyte complexes; metal–polymer nanocomposites; radiation-induced reduction; silver nanoparticles; biocidal activity

1. Introduction

The composites containing silver nanoparticles are a prospective base for design of optical, catalytic, diagnostic or sensor devices [1–7]. However, the greatest efforts were focused on the production of biocidal materials with prolonged effect, because application of metal–polymer nanocomposites did not lead to bacterial adaptation effects, in contrast to antibiotics [6–13]. Metal polymer nanocomposites are able to release biologically active substances into the environment in a controlled and gradual manner [11–19]. Reduction of metal ions in aqueous solutions and polymer dispersions is a general method for producing nanocomposites. Functional groups of polyelectrolytes can effectively bind metal ions, and in addition, they are powerful stabilizers of metal nanostructures. Interpolymer and interpolyelectrolyte complexes have been widely used in recent decades as scaffolds for the synthesis of nanocomposites with controlled sizes and different spatial organization of nanoparticles [2,20–25]. The use of macromolecular complexes provides more efficient synthesis of nanocomposites, better stabilization conditions and effective size control due to the wide possibilities of tuning the interactions of diverse units with metal ions/nanoparticles.

Nowadays much attention is paid to combat against microbiological contamination of surfaces, since bacterial and fungal infections are an important problem for medicine, the food industry and other areas of management. One of the most promising approaches to the production of biocidal films/coatings is the synthesis of the polymers with cationic groups [17,26–28]. Cationic polymers bind to the negatively charged surface of the cell and initiate the processes [29] that lead to its death. Dispersions of non-stoichiometric interpolyelectrolyte complexes (IPECs) are a flexible basis for obtaining coatings. Soluble IPECs may have an excessive content of positively and negatively charged groups, which ensures interaction with charged surfaces of various types. The mutual neutralization of the charges of the cationic and anionic polymer in the IPEC results in the formation of hydrophobic regions, which also leads to its binding to hydrophobic surfaces. In this way, macromolecular complexes can be fine-tuned to the properties of various surfaces. The use of water as a solvent has obvious advantages in terms of accessibility and environmental safety.

IPECs of polyacrylic acid (PAA) and polyethylenimine (PEI) capable of binding large amounts of metal ions (up to 50 wt.%) [20,21], due to the formation of strong triple complexes with carboxylate and nitrogen-containing units. The films and coatings of triple metal–polymer complexes were used for efficient preparation of biocidal composites with silver nanoparticles [2,30,31]. The radiation induced approach is especially promising due to the possibility of effective preparation of metal polymer nanocomposites without contaminants [1,15,32,33]. Among various potential applications of nanocomposites, we may highlight the preparation of antimicrobial films and coatings based on the insoluble films and coatings of stoichiometric complexes PAA–PEI loaded by silver nanoparticles [2,30–32]. It is important to note that the matrix of PAA–PEI complexes is expected to be stable under irradiation to moderate absorbed doses used for both sterilization and synthesis of nanocomposites [34].

The aim of the work was to obtain biocidal nanocomposites in soluble non–stoichiometric IPECs with PAA/PEI ratios equal to 1/3 and 3/1 for further obtaining coatings due to their aggregative stability. Complexes with both an excess of anionic and cationic groups have been used because they promise the potential adjustment of adhesive interactions to surfaces of various types and have different effects on the generation of metal nanostructures. This paper discusses the features of radiation-initiated formation of silver nanoparticles and the biocidal properties of applied coatings from IPECs and nanocomposites. Silver nanoparticles were obtained for the first time in irradiated dispersions of PAA–PEI complexes. The use of the radiation–chemical approach made it possible to correctly compare the formation processes and sizes of nanoparticles in complexes of various compositions and their components.

2. Materials and Methods

The following reagents were used to prepare the samples: polyacrylic acid (Mw = 100,000) from Sigma-Aldrich (St. Louis, MO, USA), polyethylenimine (Mw = 60,000) from Serva (Heidelberg, Germany) and silver nitrate of analytical grade from Reachim (Moscow, Russia). To obtain metal polymer complexes, a silver nitrate solution was added to polyelectrolyte solutions or IPEC dispersions in low light conditions, after which the values were adjusted to pH 6 using 0.1 M KOH or H_2SO_4 solutions-both from Reachim (Moscow, Russia). To study the formation of metal polymer nanocomposites, 0.1 wt.% nonstoichiometric complexes of the PAA and PEI units with a molar ratio 3/1 and 1/3 with concentration of 0.026 wt.% of $AgNO_3$ moles of silver ions were used.

Samples of metal polymer complexes were irradiated on an X-ray machine with a 5-BKhV-6W tube from Svetlana X-ray (St. Petersburg, Russia) with tungsten anode (applied voltage 45 kV, anode current 80 mA) in polymer test tubes of 5 mm diameter. The above conditions provided a uniform generation of radiolysis products. Irradiation was carried out in a water–alcohol mixture with an ethanol content of 10 vol.%. To prevent oxidation with oxygen dissolved in water, the irradiated solution was bubbled with argon of a special purity grade during the irradiation process. The dose rate was determined using a ferrosulfate dosimeter irradiated in the same geometry as the test sample. The dose rate was calculated taking into account the mass absorption coefficients and the effective energy of X-ray quanta [35], the value of which was 17 Gy/s.

The structure of the nanocomposite material was studied using a transmission microscope "Leo-912 AB OMEGA" with a resolution of 0.3 nm. Data on the size and spatial distribution of nanoparticles were obtained from analysis of about 150–300 objects in the TEM images. The UV-VIS spectra and turbidimetric data were measured by a spectrometer from Perkin Elmer Lambda 9 instruments with the optical range was 200–900 nm (Überlingen, Germany). For measuring turbidity, the portions of the guest polyelectrolyte (GPE) solutions were successively added to a solution of the host polyelectrolyte (HPE) at 1-min intervals between titrant additions. Measurements were performed under constant stirring at room temperature directly in a quartz cuvette.

The objects of biocidal research were glass slides (76 mm × 26 mm) with polymer coatings applied from polyelectrolyte solutions or IPEC dispersions with concentration 0.1 wt.% using an airbrush. Daily suspensions were prepared from daily cultures of microorganisms of the genus *Salmonella* (amount was 2×10^6 CFU) and *St. aureus* (amount was 2.4×10^6 CFU). 100 µL of suspensions were applied on the slides with polymer coatings and samples were kept at 37 °C for 2 h. After the exposure, the glass was washed off with a swab soaked in saline solution. Washout screening was carried out on the surface of meat-peptone agar. The number of CFU was determined using a standard protocol [36]. Slide glasses without applied coatings were used as control samples.

3. Results

The interaction of PEI with PAA was studied by means of turbidimetric titration. The addition of PEI to a 0.1 wt.% solution of PAA or PAA 0.1 wt.% solution of PEI leads to a gradual turbidity of the mixtures.

The results are presented in Figure 1 as dependence of relative turbidity upon the ratio of monomer units of polyacid and polybase. The addition of the solution of PAA to the solution of PEI in buffer with pH = 6, resulted in a progressive increase in the turbidity of the system reflecting formation of the IPEC and phase separation (see curve 1). This behavior is typical for mixtures of oppositely charged polyelectrolytes and indicates the formation of IPEC, which is insoluble in aqueous media, but can exist as colloidal particles with some excess of positively or negatively charged ionogenic groups. The polyelectrolyte that provides a molar excess of charged groups is named the "host" polyelectrolyte while the second component of IPEC is named the "guest". The structure of the IPEC could be presented as a particle with several macromolecules included, so that areas of contact between guest and host polyelectrolytes form hydrophobic regions, excess units of host polyelectrolyte form (Scheme 1) charged loops and tails [37,38]. In turn, hydrophobic regions could contain areas of non-compensated charge of guest macromolecules due to non-complimentary location of cationic and anionic units and the presence of defects in polyelectrolytes. The structures of IPECs that could be formed upon mixing of the solutions of oppositely charged polyelectrolytes strongly depends on the mixing ratio, the consequence of adding the components to the mixture, degrees of polymerization of polymers and their ratio, ionic strength of the solution and so on [39–41]. As a result, structures with non-uniform distribution of charges, hydrophobic patches, and defect areas with non-compensated charged groups could be formed. As a result, these structures could serve as a good scaffold for the preparation of polymer–nanoparticle composites [42].

Maximum turbidity was reached at an equimolar ratio of monomer units of the polyelectrolytes. This proves that only electrostatic interactions take place between PEI and PAA at pH 6 i.e., all PEI units are protonated while all PAA units are dissociated, and no hydrogen bonds are formed. The effect of the additional charging of weak polyelectrolytes in the presence of the oppositely charged polyelectrolytes due to formation of the IPEC was described earlier [43]. In an excess of PEI, no water-soluble IPEC was obtained (see curve 2). The system with overall negative charge demonstrated phase separation at any polycation-to-polyanion ratio. The formation of non-soluble IPECs could be explained by the relatively high hydrophobicity of PEI, relatively high molecular weight dispersion of polymers and branched structure of PEI that could restrict distribution of polymer chains in IPECs [44].

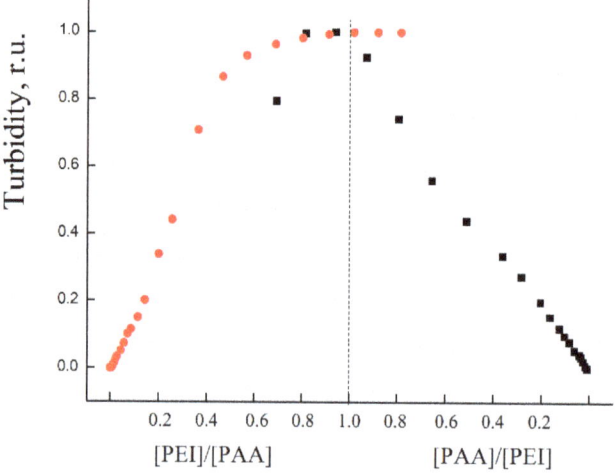

Figure 1. The dependence of relative turbidity at pH = 6 upon [PEI]/[PAA] ratio (curve 1 red) and [PAA]/[PEI] ratio (curve 2 black).

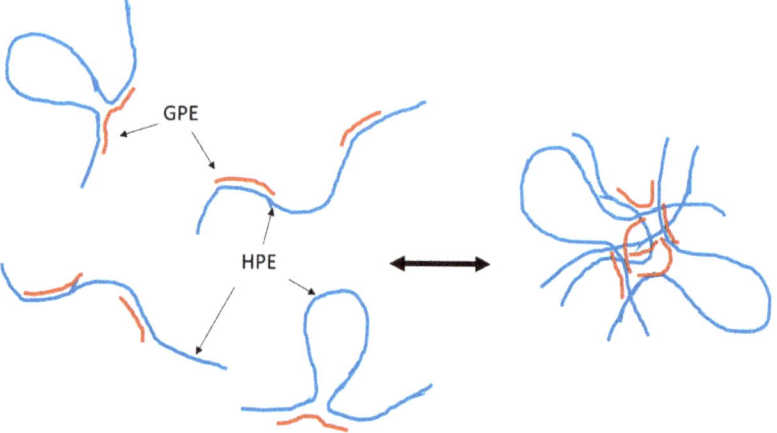

Scheme 1. Non-stoichiometric interpolyelectrolyte complexes.

Two IPECs with polyelectrolyte ratios [PAA]/[PEI] 3/1 and 1/3 were chosen to prepare ternary polyanion–(metal ion)–polycation complexes. The binary IPECs were characterized with DLS and laser microelectrophoresis with the following results: the IPEC with excess of polycation had a mean diameter of 200 nm with PDI 0.192, and an EPM value of 1.30 ± 0.11; the IPEC with an excess of polyanion had a mean diameter of 175 nm with PDI of 0.291 and an EPM value of -82 ± 0.17.

Studies of complexation in IPEC PAA–PEI have shown [21] that triple "sandwich" complexes PAA–(metal ions)–PEI are much stronger (Scheme 2) than the complexes of metal ions with functional groups of macromolecules PAA or PEI. The total content of polymer units in the IPEC dispersions was 1.8×10^{-2} M. To obtain metal polymer complexes, dispersions with a concentration of 1.5×10^{-3} M silver ions were used so that content of metal ions did not exceed the sorption capacity of the IPEC parts of non-stoichiometric complexes.

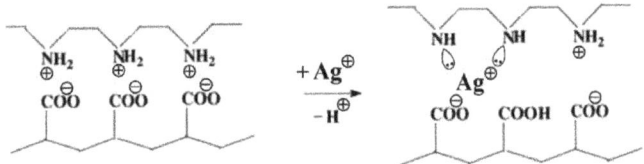

Scheme 2. Formation of triple metal polymer complexes PAA–PEI-Ag$^+$.

The preparation of metal polymer complexes as precursors for the synthesis of metal polymer nanocomposites was carried out at pH 6, because in acidic media the efficiency of the complexation of silver ions with functional groups is suppressed due to the competition of metal ion binding processes and protonation of functional groups of IPECs (Scheme 2). In alkaline media, precipitation of silver oxide occurs. The use of a pH value of 6 also ensures the effective formation of IPECs of PAA–PEI, so the protonation of nitrogen-containing units of PEI is 0.41 [21] and the degree of dissociation of carboxylate groups in this case is 0.44 [45].

Irradiation of the samples with silver ions led to their coloring to yellow–gray. The formation of silver nanoparticles is proven by the fact that absorption bands with maxima from 400 nm to 414 nm (Figures 2–5) are present on the optical spectra of all irradiated samples. An increase in the radiation dose leads to growth in the intensity of the absorption bands, which cease to increase after irradiation in a dose range of 15–30 kGy. To characterize nanoparticles on an electron microscope, samples irradiated to a dose of 15 or 30 kGy were used, which ensured the completion of the formation of nanoparticles in all irradiated systems. The reflections corresponding to interplane distances are observed on micro diffractograms: 2.35; 2.04; 1.45; 1.23 Å (Figures 2–5), which corresponds to the values for the silver crystal lattice [46]. Thus, the analysis of diffraction data confirms that radiation-induced reduction of silver ions leads to the formation of metal nanoparticles.

With a minimum size, these particles are formed in irradiated solutions of polyethylenimine and complexes with an excess of polyethylenimine. The sizes of nanoparticles of particles are in the range 1–14 nm with a maximum distribution at 3 nm and in the range 1–24 nm with a maximum distribution at 5 nm, obtained, respectively, in irradiated PEI solutions and complexes with a molar ratio of PAA/PEI units equal to 1/3 (Figures 2 and 3). Average sizes of nanoparticles synthesized in PAA solutions and complexes with a molar ratio of PAA/PEI units equal to 3/1 were 12 and 20 nm, respectively (Figures 4 and 5), and the size distribution becomes broader.

Studies of the biocidal properties of coatings were carried out using both gram-negative and gram-positive bacteria. The results of the study of antibacterial action of the polymer-based coatings on glass are presented in Figure 6. For the species of *Salmonella* the coating of PEI demonstrated an efficiency to suppress bacterial growth—only 1/5 of the number of CFUs were detected on the coating in comparison to controls. Formation of the coatings from both negatively charged and positively charged IPECs resulted in a partial loss of the biocidal activity of PEI, however, the pronounced suppression of the bacterial growth was detected. Formation of coatings from the PEI/Ag nanocomposite resulted in a slight increase in biocidal activity in relation to PEI. The most effective antibacterial action was observed for the ternary nanocomposites, IPEC/Ag. Coatings from both negatively charged and positively charged nanocomposites allowed suppression of bacterial growth almost completely.

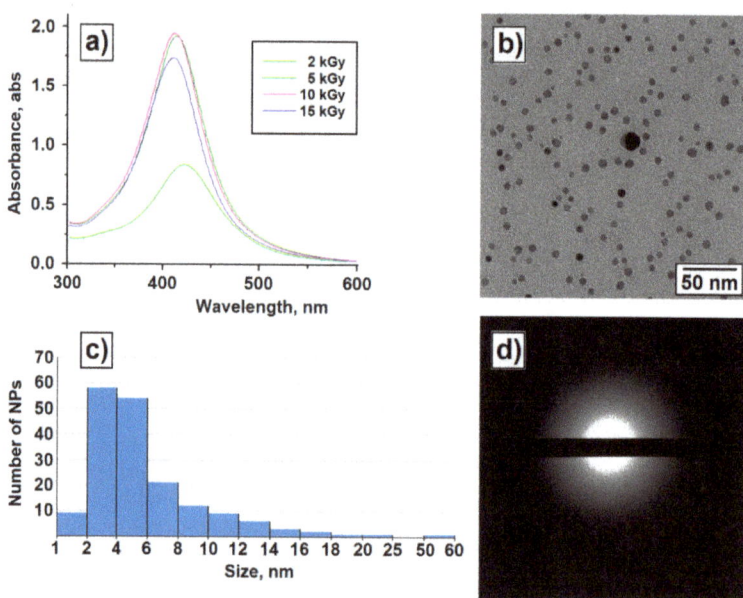

Figure 2. UV–visible spectra (**a**), micrography (**b**), histogram of the size distribution of nanoparticles (**c**), microdiffraction image (**d**) of irradiated PEI-Ag$^+$ complexes.

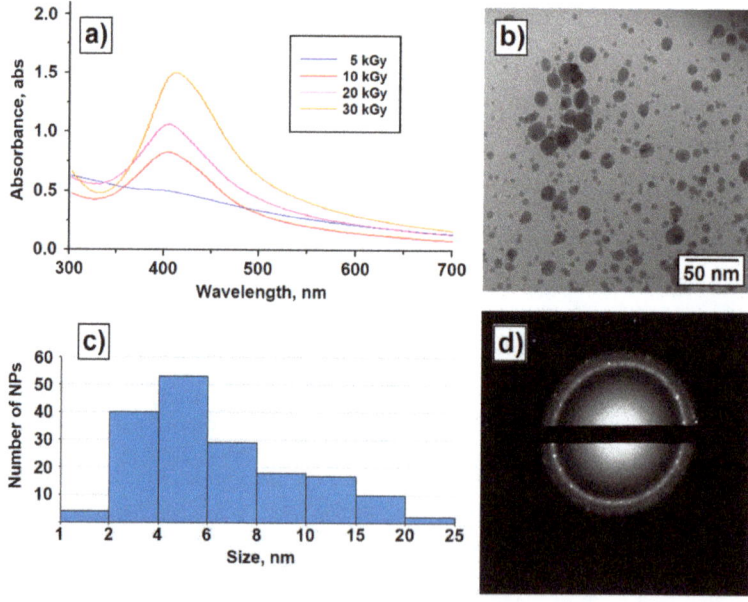

Figure 3. UV–visible spectra (**a**), micrography (**b**), histogram of the size distribution of nanoparticles (**c**), microdiffraction image (**d**) of irradiated PAA–PEI-Ag$^+$ complexes, the molar ratio of PAA/PEI units equal to 1/3.

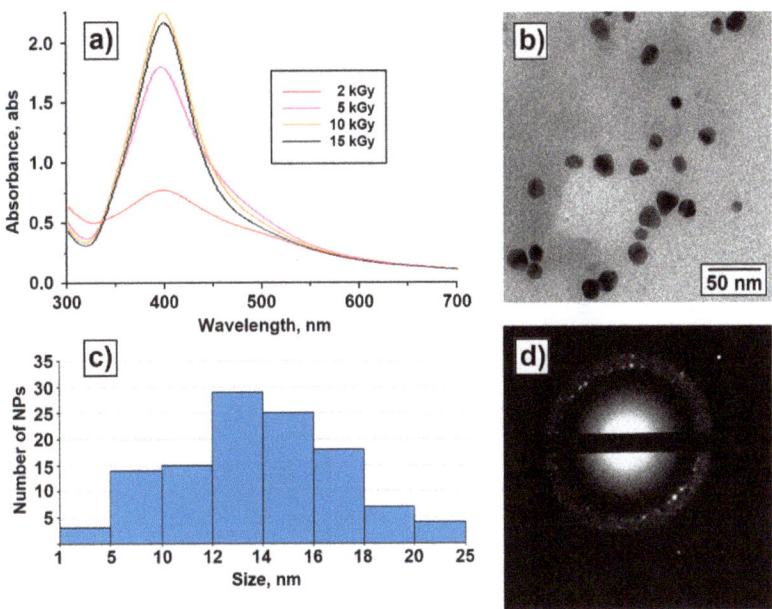

Figure 4. UV–visible spectra (**a**), micrography (**b**), histogram of the size distribution of nanoparticles (**c**), microdiffraction image (**d**) of irradiated PAA–Ag$^+$ complexes.

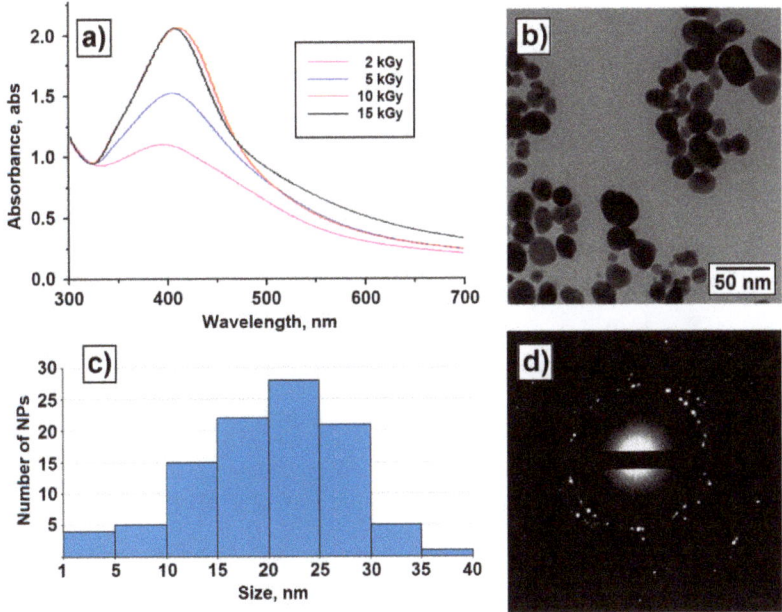

Figure 5. UV–visible spectra (**a**), micrography (**b**), histogram of the size distribution of nanoparticles (**c**), microdiffraction image (**d**) of irradiated PAA–PEI–Ag$^+$ complexes with the molar ratio of PAA/PEI units equal to 3/1.

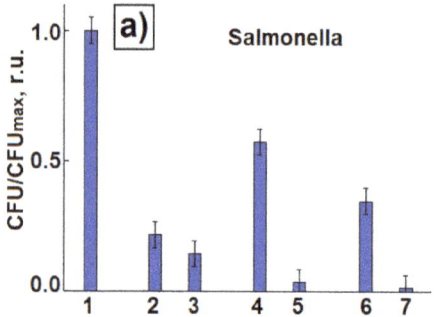 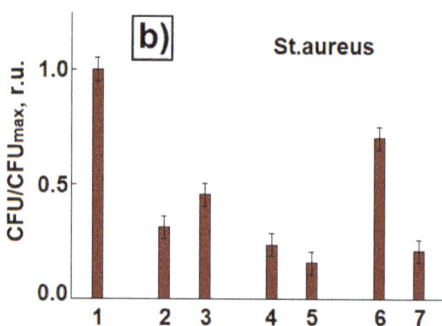

Figure 6. Biocidal activity of polymer coatings and nanocomposite coatings towards *Salmonella* (**a**) and *St. Aureus* (**b**): control (1); PEI (2); PEI/Ag (3); PAA/PEI 1/3 (4); PAA/PEI 1/3 + Ag (5); PAA/PEI 3/1 (6); PAA/PEI 3/1 +Ag (7).

For the species of *St. Aureus*, the coating of PEI also demonstrated efficiency to suppress bacterial growth—only 1/3 of CFUs were detected on the coating in comparison to control. The coating from IPEC with PAA in excess demonstrated lower efficiency towards *St. Aureus* but the antibacterial properties of this coating were preserved. In contrast to *Salmonella* species, the coating from the PEI/Ag complex demonstrated a slight decrease in biocidal activity in comparison to the PEI coating. The coatings from the ternary IPEC/Ag nanocomposites demonstrated the most effective biocide activity towards *St. Aureus*—the suppression of the bacterial growth down to 15–20% was detected.

4. Discussion

For radiation-induced synthesis of nanoparticles in solutions and dispersions with 0.1 wt.% of polymers, the proportion of water is the overwhelming content. Due to this, the main role in the processes of reduction of metal ions and the formation of nanoparticles is played by processes involving products formed during radiolysis of water:

$$H_2O \rightarrow e_{aq}^-, \cdot OH, H_3O^+, H\cdot, H_2, H_2O_2, HO_2\cdot \qquad (1)$$

The main products of radiolysis of water are hydrated electrons, which have extremely high reduction potentials of -2.9 V [47] and such strong oxidizers as OH-radicals. To neutralize OH radicals and create favorable conditions for the reduction of metal ions, an additive of ethyl alcohol was used, while the radical $CH_3\cdot CHOH$ (reaction (2)) is formed, which has reducing properties ($E_0(CH_3\cdot CHOH) = -1.4$ V) [48]:

$$CH_3CH_2OH + \cdot OH \rightarrow CH_3\cdot CHOH + H_2O \qquad (2)$$

Oxidation of $CH_3\cdot CHOH$ leads to the formation of a weak reducing agent-acetaldehyde [32]. Thus, when ethyl alcohol is used as a scavenger of OH radicals, only reducing particles are generated in water–organic mixtures.

The mechanisms of radiation-chemical formation of silver nanoparticles have been studied in detail [49–51]. In the first stage, silver ions are reduced to isolated atoms and clusters are formed:

$$(Ag)^+ + e^-_{aq} \rightarrow (Ag)^0 \qquad (3)$$

$$(Ag)^0 + (Ag)^+ \rightarrow (Ag_2)^+ \rightarrow (Ag_m)^{n+} \qquad (4)$$

The growth of clusters by the reduction of silver ions on their surface (reaction (5)) and coalescence processes (reaction (6)) lead to the formation and enlargement of nanoparticles.

$$(Ag_m)^{n+} + (Ag)^+ \rightarrow (Ag_{m+1})^{n+1} \rightarrow (Ag_{m+1})^n \qquad (5)$$

$$(Ag_m)^{n+} + (Ag_k)^{g+} \rightarrow (Ag_{m+k})^{(n+g)+} \qquad (6)$$

The process of reducing silver ions to isolated atoms (reaction (3), $E^0_{(Ag+/Ag0)}$ is more negative than -1.8 V [52]), which limits nucleation, and can provide such strong reducing agents as hydrated electrons. Reduction of silver ions on the surface of nanoparticles, which ensure the growth of nanoparticles (reaction (5)), may occur due to reactions of reducing agents with lower reduction potentials, including acetaldehyde [32].

Analysis of the UV–visible spectra and microdiffraction data (Figures 2–5) shows that irradiation leads to the formation of silver nanoparticles with metal lattice in all the studied samples. Electron microscopy data also show the spatial ordering of nanoparticles. Previously, similar hierarchical structures were observed in IPEC PAA–polyalliamine and PAA–polyvinylimidazole [25,53] due to the formation of micro- and nano-particles of IPEC complexes filled by metal nanoparticles.

Experimental data show that the composition of polymer solutions and dispersions affects the size of the resulting nanoparticles and their spatial organization. In polymer systems based on polyelectrolytes, the formation of nanoparticles occurs under conditions of electrostatic interactions of charged functional groups and hydrophobic interactions of the main chain with the surface of nanoparticles. From the point of view of electrostatic interactions, polycations should provide the least favorable conditions for the stabilization of positively charged nanoparticles. At pH 6, the degree of protonation of PEI is about 0.41. Taking into account the fact that every second nitrogen-containing group is charged in fully protonated polyethylenimine, the fraction of charged groups in experimental conditions is about 20%. Nevertheless, polyethylenimine solutions have demonstrated the highest stabilizing ability. In this case, nanoparticles with an average size of 3 nm and the narrowest size distribution occur (Figure 2). In this situation, the stabilization of nanoparticles can be provided by hydrophobic interactions of hydrocarbon groups with the surface of nanoparticles. Nitrogen-containing groups are strong ligands for silver ions, therefore, another powerful stabilizing factor we can assume due to the interaction of a significant proportion of uncharged amino groups and silver ions adsorbed on the surface of nanoparticles due to incomplete reduction processes. Micrography analysis also reveals specifically elongated structures of 3–5 isolated nanoparticles. The observed phenomenon is caused by mutual repulsion of both positively charged nanoparticles immobilized in macromolecules and positively charged PEI units in which metal nanostructures are immobilized.

The results obtained, show that the electrostatic interactions of PAA polyanions with the oppositely charged surface of nanoparticles provide significantly less favorable conditions for their stabilization than hydrophobic interactions in the case of PEI. The average size of nanoparticles in this case is 12 nm (Figure 3). Micrographs of irradiated solutions also show a specific spatial organization of nanoparticles. The balance of repulsion of positively charged nanoparticles and negatively charged links and attraction of opposite charges leads to their close location in aggregates.

The formation of nanoparticles of nanometer scale also occurs in the irradiated dispersions of interpolyelectrolyte complexes with an excess of polyanionic or polycationic component (Figures 3 and 5). However, in both cases, their sizes are larger than the sizes of nanoparticles formed in solutions of polyelectrolytes (Figures 2 and 4). On TEM images of irradiated IPEC with an excess of polyethylenimine both relatively small nanoparticles (up to 6 nm in size) and large nanoparticles with a tendency to aggregation are clearly visible (Figure 3). The obtained results clearly show that in this case, the immobilization of nanostructures occurs mainly separately in PEI and PAA units. However, experimental results do not exhibit such separation of nanostructures in IPEC with PAA excess. In this case, aggregates of relatively large nanoparticles are observed also as for individual PAA macromolecules. Thus, analysis of micrographs does not reveal an effect of the polycationic component on sizes and spatial distribution of nanoparticles, unlike IPEC with an excess of PEI. Consequently, the presence of an excess of PAA units prevents the hydrophobic interactions of PEI directly with nanoparticles. Moreover, nanoparticles with an average size exceeding the average size of nanoparticles formed in PAA solutions are formed (Figures 4 and 5). Thus, in this case, the presence of polycations also worsens the electrostatic stabilization conditions, possibly due to partial blocking of negatively charged carboxylate groups.

Study of antibacterial activity have demonstrated that PEI coatings provide a multiple increase in biocidal activity for both *Salmonella* and *Staphylococcus aureus* compared to the control sample. Recent publications have demonstrated antimicrobial activity of branched polyelectrolytes like PEI and silver nanoparticles [54–59]. We focus on possible cumulative action of polycation and nanoparticles. In this case the presence of silver nanoparticles either slightly decreases the antibacterial activity during tests of *Staphylococcus aureus*, or in the case of gram-negative bacteria (*Salmonella*) there is a slight increase in antibacterial effect. A probable explanation for this effect may be the strong interaction of PEI with the surface of nanoparticles, which not only leads to the formation of relatively small nanoparticles, but also ensures mutual passivation of both biocidal components. With a decrease in the content of PEI in the IPEC samples, as a rule, a significant decrease in biocidal activity occurs. Such phenomenon is logically interpreted as a diminution in the content of the polycationic biocidal component. The presence of silver nanoparticles in IPEC coatings leads to a significant increase in their antibacterial activity. The effect is more pronounced in the case of biocidal activity for *Salmonella*. Therefore, in this case, the role of silver nanoparticles for the destruction of *Salmonella* is many times higher than for the effect against *Staphylococcus aureus* which correlates with the fact that gram-positive microorganisms have less sensitivity to silver colloids [60]. Deposition of IPEC on the surfaces, either hydrophilic or hydrophobic, results in formation of a coating that has high resistance to wash-off in comparison to cationic polyelectrolyte [61]. Thus, preparation of nanocomposites of IPECs with silver nanoparticles has great potential for the creation of effective biocidal coatings.

5. Conclusions

The results of the work reveal the possibility of effective radiation chemical production of silver nanoparticles, both in the studied solutions of polyelectrolytes and in non-stoichiometric IPEC PAA–PEI of various compositions. The PAA–PEI–silver coatings applied to glass slides retained their inherent yellow–gray color for 14 days before antibacterial tests, which shows the high stability of the obtained nanocomposites. PEI solutions demonstrated the highest ability to stabilize nanoparticles among studied polymers. However, the reverse side of the coin in this case, is the passivation of the antibacterial activity of silver nanoparticles. A decrease in the content of the cationic component in IPEC, as a rule, leads to a significant decrease in the biocidal activity of macromolecular complexes compared with coatings of individual PEI. However, IPEC coatings containing silver nanoparticles show multiple higher biocidal activity, not only in comparison with macromolecular complexes of PAA–PEI, but also PEI coatings. Thus, the results obtained show the prospects of using IPEC–Ag nanocomposites for the further development of biocidal coatings with adjustable adhesion to surfaces of various types.

Author Contributions: Conceptualization, A.A.Z.; methodology, K.V.M., E.A.Z., D.I.K., O.A.K., A.A.S., Y.K.Y. and E.R.T.; validation, A.A.Z. and A.A.Y.; formal analysis, A.V.S., D.I.K. and A.A.Z.; investigation, K.V.M., V.A.P., E.A.Z., O.A.K., M.A.G. and S.S.A.; data curation, A.V.S., D.I.K. and A.A.Z.; writing—original draft preparation, A.A.Z. and A.V.S.; writing—review and editing, D.I.K. and A.A.Y.; visualization, K.V.M. and D.I.K.; supervision, A.A.Z.; project administration, A.A.Z.; funding acquisition, A.A.Z. All authors have read and agreed to the published version of the manuscript.

Funding: This research was funded by Ministry of Science and Higher Education of the Russian Federation, project no. 075-15-2020-775.

Institutional Review Board Statement: Not applicable.

Informed Consent Statement: Not applicable.

Data Availability Statement: Not applicable.

Conflicts of Interest: The authors declare no conflict of interest.

References

1. Belloni, J. Nucleation, growth and properties of nanoclusters studied by radiation chemistry. *Catal. Today* **2006**, *113*, 141–156. [CrossRef]
2. Dai, J.H.; Bruening, M.L. Catalytic nanoparticles formed by reduction of metal ions in multilayered polyelectrolyte films. *Nano Lett.* **2002**, *2*, 497–501. [CrossRef]
3. Pomogailo, A.D.; Kestelman, V.N. *Metallopolymer nanocomposites*; Springer Science & Business Media: Berlin/Heidelberg, Germany, 2006; Volume 81.
4. Rosi, N.L.; Mirkin, C.A. Nanostructures in biodiagnostics. *Chem. Rev.* **2005**, *105*, 1547–1562. [CrossRef] [PubMed]
5. Zhang, Z.; Shen, W.; Xue, J.; Liu, Y.; Liu, Y.; Yan, P.; Liu, J.; Tang, J. Recent advances in synthetic methods and applications of silver nanostructures. *Nanoscale Res. Lett.* **2018**, *13*, 54. [CrossRef]
6. Dawadi, S.; Katuwal, S.; Gupta, A.; Lamichhane, U.; Thapa, R.; Jaisi, S.; Lamichhane, G.; Bhattarai, D.P.; Parajuli, N. Current research on silver nanoparticles: Synthesis, characterization, and applications. *J. Nanomater.* **2021**, *2021*, 6687290. [CrossRef]
7. Calderón-Jiménez, B.; Johnson, M.E.; Bustos, A.R.M.; Murphy, K.E. Winchester; J.R., Vega Baudrit. Silver Nanoparticles: Technological Advances, Societal Impacts, and Metrological Challenges. *Front. Chem.* **2017**, *5*, 6. [CrossRef]
8. Volker, C.; Oetken, M.; Oehlmann, J. The Biological Effects and Possible Modes of Action of Nanosilver. *Rev. Environ. Contam. Toxicol.* **2013**, *223*, 81–106. [CrossRef]
9. Sharma, V.K.; Yngard, R.A.; Lin, Y. Silver nanoparticles: Green synthesis and their antimicrobial activities. *Adv. Colloid Interface Sci.* **2009**, *145*, 83–96. [CrossRef] [PubMed]
10. Malaekeh-Nikouei, B.; Bazzaz, B.S.F.; Mirhadi, E.; Tajani, A.S.; Khameneh, B. The role of nanotechnology in combating biofilm-based antibiotic resistance. *J. Drug Deliv. Sci. Technol.* **2020**, *60*, 15. [CrossRef]
11. Kim, J.S.; Kuk, E.; Yu, K.N.; Kim, J.H.; Park, S.J.; Lee, H.J.; Kim, S.H.; Park, Y.K.; Park, Y.H.; Hwang, C.Y.; et al. Antimicrobial effects of silver nanoparticles. *Nanomed. Nanotechnol. Biol. Med.* **2007**, *3*, 95–101. [CrossRef]
12. Lee, S.H.; Jun, B.-H. Silver nanoparticles: Synthesis and application for nanomedicine. *Int. J. Mol. Sci.* **2019**, *20*, 865. [CrossRef]
13. Das, G.; Patra, J.K.; Debnath, T.; Ansari, A.; Shin, H.-S. Investigation of antioxidant, antibacterial, antidiabetic, and cytotoxicity potential of silver nanoparticles synthesized using the outer peel extract of *Ananas comosus* (L.). *PLoS ONE* **2019**, *14*, e0220950. [CrossRef]
14. Kvitek, L.; Panacek, A.; Soukupova, J.; Kolar, M.; Vecerova, R.; Prucek, R.; Holecova, M.; Zboril, R. Effect of surfactants and polymers on stability and antibacterial activity of silver nanoparticles (NPs). *J. Phys. Chem. C* **2008**, *112*, 5825–5834. [CrossRef]
15. Yoksan, R.; Chirachanchai, S. Silver nanoparticles dispersing in chitosan solution: Preparation by gamma-ray irradiation and their antimicrobial activities. *Mater. Chem. Phys.* **2009**, *115*, 296–302. [CrossRef]
16. Manikandan, A.; Sathiyabama, M. Green synthesis of copper-chitosan nanoparticles and study of its antibacterial activity. *J. Nanomed. Nanotechnol.* **2015**, *6*, 1.
17. Misin, V.M.; Zezin, A.A.; Klimov, D.I.; Sybachin, A.V.; Yaroslavov, A.A. Biocidal Polymer Formulations and Coatings. *Polym. Sci. Ser. B* **2021**, *63*, 459–469. [CrossRef]
18. Zhang, X.; Qu, Q.; Cheng, W.; Zhou, A.; Deng, Y.; Ma, W.; Zhu, M.; Xiong, R.; Huang, C. A Prussian blue alginate microparticles platform based on gas-shearing strategy for antitumor and antibacterial therapy. *Int. J. Biol. Macromol.* **2022**, *209*, 794–800. [CrossRef]
19. Rurarz, B.P.; Gibka, N.; Bukowczyk, M.; Kadłubowski, S.; Ulański, P. Radiation synthesis of poly (acrylic acid) nanogels for drug delivery applications–post-synthesis product colloidal stability. *Nukleonika* **2021**, *66*, 179–186. [CrossRef]
20. Zezin, A.A. Synthesis of metal-polymer complexes and functional nanostructures in films and coatings of interpolyelectrolyte complexes. *Polym. Sci. Ser. A* **2019**, *61*, 754–764. [CrossRef]
21. Pergushov, D.V.; Zezin, A.A.; Zezin, A.B.; Müller, A.H.E. Advanced functional structures based on interpolyelectrolyte complexes. In *Polyelectrolyte Complexes in the Dispersed and Solid State I*; Springer: Berlin/Heidelberg, Germany, 2013; pp. 173–225.
22. Demchenko, V.; Riabov, S.; Sinelnikov, S.; Radchenko, O.; Kobylinskyi, S.; Rybalchenko, N. Novel approach to synthesis of silver nanoparticles in interpolyelectrolyte complexes based on pectin, chitosan, starch and their derivatives. *Carbohydr. Polym.* **2020**, *242*, 116431. [CrossRef]
23. Demchenko, V.; Riabov, S.; Kobylinskyi, S.; Goncharenko, L.; Rybalchenko, N.; Kruk, A.; Moskalenko, O.; Shut, M. Effect of the type of reducing agents of silver ions in interpolyelectrolyte-metal complexes on the structure, morphology and properties of silver-containing nanocomposites. *Sci. Rep.* **2020**, *10*, 7126. [CrossRef]
24. Schacher, F.H.; Rudolph, T.; Drechsler, M.; Müller, A.H.E. Core-crosslinked compartmentalized cylinders. *Nanoscale* **2011**, *3*, 288–297. [CrossRef]
25. Dağaş, D.E.; Danelyan, G.V.; Ghaffarlou, M.; Zezina, E.A.; Abramchuk, S.S.; Feldman, V.I.; Güven, O.; Zezin, A.A. Generation of spatially ordered metal–polymer nanostructures in the irradiated dispersions of poly (acrylic acid)–poly (vinylimidazole)–Cu^{2+} complexes. *Colloid Polym. Sci.* **2020**, *298*, 193–202. [CrossRef]
26. Periyasamy, T.; Asrafali, S.; Shanmugam, M.; Kim, S.-C. Development of sustainable and antimicrobial film based on polybenzoxazine and cellulose. *Int. J. Biol. Macromol.* **2021**, *170*, 664–673. [CrossRef]
27. Koufakis, E.; Manouras, T.; Anastasiadis, S.H.; Vamvakaki, M. Film properties and antimicrobial efficacy of quaternized PDMAEMA brushes: Short vs long alkyl chain length. *Langmuir* **2020**, *36*, 3482–3493. [CrossRef]

28. Gao, J.; White, E.M.; Liu, Q.; Locklin, J. Evidence for the phospholipid sponge effect as the biocidal mechanism in surface-bound polyquaternary ammonium coatings with variable cross-linking density. *ACS Appl. Mater. Interfaces* **2017**, *9*, 7745–7751. [CrossRef]
29. Parhamifar, L.; Andersen, H.; Wu, L.P.; Hall, A.; Hudzech, D.; Moghimi, S.M. Polycation-Mediated Integrated Cell Death Processes. *Adv. Genet.* **2014**, *88*, 353–398. [CrossRef]
30. Klimov, D.I.; Zezina, E.A.; Lipik, V.C.; Abramchuk, S.S.; Yaroslavov, A.A.; Feldman, V.I.; Sybachin, A.V.; Spiridonov, V.V.; Zezin, A.A. Radiation-induced preparation of metal nanostructures in coatings of interpolyelectrolyte complexes. *Radiat. Phys. Chem.* **2019**, *162*, 23–30. [CrossRef]
31. Mkrtchyan, K.V.; Zezin, A.A.; Zezina, E.A.; Abramchuk, S.S.; Baranova, I.A. Formation of metal nanostructures under X-ray radiation in films of interpolyelectrolyte complexes with different silver ion content. *Russ. Chem. Bull.* **2020**, *69*, 1731–1739. [CrossRef]
32. Zezin, A.A.; Klimov, D.I.; Zezina, E.A.; Mkrtchyan, K.V.; Feldman, V.I. Controlled radiation-chemical synthesis of metal polymer nanocomposites in the films of interpolyelectrolyte complexes: Principles, prospects and implications. *Radiat. Phys. Chem.* **2020**, *169*, 108076. [CrossRef]
33. Long, D.; Wu, G.; Chen, S. Preparation of oligochitosan stabilized silver nanoparticles by gamma irradiation. *Radiat. Phys. Chem.* **2007**, *76*, 1126–1131. [CrossRef]
34. Sosulin, I.S.; Zezin, A.A.; Feldman, V.I. Effect of irradiation on poly(acrylic acid)-polyethyleneimine interpolyelectrolyte complexes: An electron paramagnetic resonance study. *Rad. Phys. Chem.* **2022**, *167*, 110198. [CrossRef]
35. Hubbell, J.H.; Seltzer, S.M. *Tables of X-ray Mass Attenuation Coefficients and Mass Energy-Absorption Coefficients 1 keV to 20 MeV for Elements Z = 1 to 92 and 48 Additional Substances of Dosimetric Interest*; National Inst. of Standards and Technology-PL: Gaithersburg, MD, USA, 1995.
36. Nastulyavichus, A.; Tolordava, E.; Rudenko, A.; Zazymkina, D.; Shakhov, P.; Busleev, N.; Romanova, Y.; Ionin, A.; Kudryashov, S. In Vitro Destruction of Pathogenic Bacterial Biofilms by Bactericidal Metallic Nanoparticles via Laser-Induced Forward Transfer. *Nanomaterials* **2020**, *10*, 2259. [CrossRef]
37. Zezin, A.B.; Mikheikin, S.V.; Rogacheva, V.B.; Zansokhova, M.F.; Sybachin, A.V.; Yaroslavov, A.A. Polymeric stabilizers for protection of soil and ground against wind and water erosion. *Adv. Colloid Interface Sci.* **2015**, *226*, 17–23. [CrossRef]
38. Pergushov, D.V.; Müller, A.H.E.; Schacher, F.H. Micellar interpolyelectrolyte complexes. *Chem. Soc. Rev.* **2012**, *41*, 6888–6901. [CrossRef]
39. Synatschke, C.V.; Löbling, T.I.; Förtsch, M.; Hanisch, A.; Schacher, F.H.; Müller, A.H.E. Micellar Interpolyelectrolyte Complexes with a Compartmentalized Shell. *Macromolecules* **2013**, *46*, 6466–6474. [CrossRef]
40. Müller, M.; Keßler, B.; Fröhlich, J.; Poeschla, S.; Torger, B. Polyelectrolyte Complex Nanoparticles of Poly(ethyleneimine) and Poly(acrylic acid): Preparation and Applications. *Polymers* **2011**, *3*, 762–778. [CrossRef]
41. Izumrudov, V.A.; Sybachin, A.V. Phase separation in solutions of polyelectrolyte complexes: The decisive effect of a host polyion. *Polym. Sci. Ser. A* **2006**, *48*, 1098–1104. [CrossRef]
42. Zezin, A.B.; Rogacheva, V.B.; Feldman, V.I.; Afanasiev, P.; Zezin, A.A. From triple interpolyelectrolyte-metal complexes to polymer-metal nanocomposites. *Adv. Colloid Interface Sci.* **2010**, *158*, 84–93. [CrossRef]
43. Kabanov, V.A. Polyelectrolyte complexes in solution and in bulk. *Russ. Chem. Rev.* **2005**, *74*, 3. [CrossRef]
44. Curtis, K.A.; Miller, D.; Millard, P.; Basu, S.; Horkay, F.; Chandran, P.L. Unusual salt and pH induced changes in polyethylenimine solutions. *PLoS ONE* **2016**, *11*, e0158147. [CrossRef] [PubMed]
45. Zezin, A.A.; Feldman, V.I.; Abramchuk, S.S.; Danelyan, G.V.; Dyo, V.V.; Plamper, F.A.; Müller, A.H.E.; Pergushov, D.V. Efficient size control of copper nanoparticles generated in irradiated aqueous solutions of star-shaped polyelectrolyte containers. *Phys. Chem. Chem. Phys.* **2015**, *17*, 11490–11498. [CrossRef] [PubMed]
46. Hanawalt, J.D.; Rinn, H.W.; Frevel, L.K. Chemical analysis by X-ray diffraction. *Ind. Eng. Chem. Anal. Ed.* **1938**, *10*, 457–512. [CrossRef]
47. Wardman, P. Reduction potentials of one-electron couples involving free radicals in aqueous solution. *J. Phys. Chem. Ref. Data* **1989**, *18*, 1637–1755. [CrossRef]
48. Ershov, B.G. Colloidal copper in aqueous solutions: Radiation-chemical reduction, mechanism of formation, and properties. *Russ. Chem. Bull.* **1994**, *43*, 16–21. [CrossRef]
49. Ershov, B.G.; Janata, E.; Henglein, A. Growth of silver particles in aqueous solution: Long-lived" magic" clusters and ionic strength effects. *J. Phys. Chem.* **1993**, *97*, 339–343. [CrossRef]
50. Mostafavi, M.; Keghouche, N.; Delcourt, M.-O.; Belloni, J. Ultra-slow aggregation process for silver clusters of a few atoms in solution. *Chem. Phys. Lett.* **1990**, *167*, 193–197. [CrossRef]
51. Lampre, I.; Pernot, P.; Mostafavi, M. Spectral properties and redox potentials of silver atoms complexed by chloride ions in aqueous solution. *J. Phys. Chem. B* **2000**, *104*, 6233–6239. [CrossRef]
52. Henglein, A. The reactivity of silver atoms in aqueous solutions (A γ-radiolysis study). *Ber. Der Bunsenges. Für Phys. Chem.* **1977**, *81*, 556–561. [CrossRef]
53. Bakar, A.; De, V.V.; Zezin, A.A.; Abramchuk, S.S.; Güven, O.; Feldman, V.I. Spatial organization of a metal–polymer nanocomposite obtained by the radiation-induced reduction of copper ions in the poly (allylamine)–poly (acrylic acid)–Cu2+ system. *Mendeleev Commun.* **2012**, *22*, 211–212. [CrossRef]

54. Xu, D.; Wang, Q.; Yang, T.; Cao, J.; Lin, Q.; Yuan, Z.; Li, L. Polyethyleneimine Capped Silver Nanoclusters as Efficient Antibacterial Agents. *Int. J. Environ. Res. Public Health* **2016**, *13*, 334. [CrossRef] [PubMed]
55. Chrószcz, M.; Barszczewska-Rybarek, I. Nanoparticles of Quaternary Ammonium Polyethylenimine Derivatives for Application in Dental Materials. *Polymers* **2020**, *12*, 2551. [CrossRef]
56. Pigareva, V.A.; Stepanova, D.A.; Bolshakova, A.V.; Marina, V.I.; Osterman, I.A.; Sybachin, A.V. Hyperbranched kaustamin as an antibacterial for surface treatment. *Mendel. Commun.* **2022**, *32*, 561–563. [CrossRef]
57. Raza, M.A.; Kanwal, Z.; Rauf, A.; Sabri, A.N.; Riaz, S.; Naseem, S. Size- and Shape-Dependent Antibacterial Studies of Silver Nanoparticles Synthesized by Wet Chemical Routes. *Nanomaterials* **2016**, *6*, 74. [CrossRef]
58. Bruna, T.; Maldonado-Bravo, F.; Jara, P.; Caro, N. Silver Nanoparticles and Their Antibacterial Applications. *Int. J. Mol. Sci.* **2021**, *22*, 7202. [CrossRef] [PubMed]
59. Kaur, A.; Kumar, R. Enhanced bactericidal efficacy of polymer stabilized silver nanoparticles in conjugation with different classes of antibiotics. *RSC Adv.* **2019**, *9*, 1095–1105. [CrossRef]
60. Afonina, I.A.; Kraeva, L.A.; Gia, T. Bactericidal activity of colloidal silver against grampositive and gramnegative bacteria. *Antibiot. i Khimioterapiia = Antibiot. Chemoterapy [Sic]* **2010**, *55*, 11–13.
61. Pigareva, V.A.; Senchikhin, I.N.; Bolshakova, A.V.; Sybachin, A.V. Modification of Polydiallyldimethylammonium Chloride with Sodium Polystyrenesulfonate Dramatically Changes the Resistance of Polymer-Based Coatings towards Wash-Off from Both Hydrophilic and Hydrophobic Surfaces. *Polymers* **2022**, *14*, 1247. [CrossRef] [PubMed]

Article

Surface Modification of Carbon Fiber for Enhancing the Mechanical Strength of Composites

Ryoma Tokonami, Katsuhito Aoki, Teruya Goto and Tatsuhiro Takahashi *

Department of Organic Materials Science, Graduated School of Organic Materials Science, Yamagata University, 4-3-16 Jonan, Yonezawa 992-8510, Yamagata, Japan
* Correspondence: effort@yz.yamagata-u.ac.jp

Abstract: The surface of carbon fibers (CFs) is often modified by multi-walled carbon nanotubes (MWCNTs), and the effect of the interface on the mechanical properties has been reported mostly for epoxy matrices. We achieved effective surface modification of CFs by a simple two-step process to graft a large amount of MWCNTs using a highly reactive polymer to enhance the bonding between CFs and MWCNTs. The first step was the reactive mono-molecular coating of a reactive polymer (poly-2-isopropenyl-2-oxazoline; Pipozo) that has high reactivity with COOH from CFs and MWCNTs. The high reactivity between the oxazoline group and COOH or phenol OH was confirmed for low-molecular-weight reactions. The second step was the coating of MWCNTs from a dispersion in a solvent. This simple process resulted in a substantial amount of MWCNTs strongly bonded to CF, even after washing. The MWCNTs grafted onto CFs remained even after melt-mixing. The effect on the interface, i.e., physical anchoring, led to an improvement of the mechanical properties. The novelty of the present study is that Pipozo acted as a molecular bonding layer between CFs and MWCNTs as a physical anchoring structure formed by a simple process, and the interface caused a 20% improvement in the tensile strength and modulus. This concept of a composite having a physical anchoring structure of MWCNTs on CFs has potential applications for lightweight thermoplastics, such as in the automotive industry.

Citation: Tokonami, R.; Aoki, K.; Goto, T.; Takahashi, T. Surface Modification of Carbon Fiber for Enhancing the Mechanical Strength of Composites. *Polymers* **2022**, *14*, 3999. https://doi.org/10.3390/polym14193999

Academic Editors: Tomasz Makowski and Sivanjineyulu Veluri

Received: 20 August 2022
Accepted: 20 September 2022
Published: 24 September 2022

Publisher's Note: MDPI stays neutral with regard to jurisdictional claims in published maps and institutional affiliations.

Copyright: © 2022 by the authors. Licensee MDPI, Basel, Switzerland. This article is an open access article distributed under the terms and conditions of the Creative Commons Attribution (CC BY) license (https://creativecommons.org/licenses/by/4.0/).

Keywords: composite; carbon fiber; interface; carbon nanotubes; reactive polymer; layer-by-layer

1. Introduction

Carbon fiber-reinforced plastics (CFRP) are attractive with respect to their strong mechanical properties, lightweight, long continuous, or short carbon fibers (CFs), and are fabricated based on either thermosetting or a thermoplastic matrix. Research on composites including multi-walled carbon nanotubes (MWCNTs) in a matrix together with CF has been conducted with the aim of a synergistic effect of MWCNTs and CFs to enhance the mechanical strength [1–4]. On the other hand, to improve the mechanical properties of CFRP, many approaches focused on the surface treatment of CFs have been reported, the treatments of which can be categorized as wet, dry, nano, oxidation, and non-oxidation, as described in a recent review article [5]. Related with the surface technology, surface analysis methods, surface control, and surface modification were described in another review article [6]. There have been reports on the use of amines to achieve a chemical reaction between CFs and a matrix polymer [7–9]. However, it is considered that there would not be very effective chemical bonding between amines and carboxylic acid because of the fundamental characteristics of the reversible reaction at equilibrium, even when using a catalyst.

As a typical method for the surface modification of CFs with nanomaterials, there is the surface modification of CFs using MWCNTs. To coat carbon nanotubes (CNTs) onto CF surfaces, various methods such as chemical vapor deposition (CVD) [10–13], electrophoretic deposition (EPD) [14], and chemical functionalization [15,16] were reported in the review article by Zakaria et al. [17]. Surfactants with either high or low molecular weight have been utilized for the chemical functionalization method [17]. Another review article reported CVD, a spray-coating method, and dip coating [18,19] methods using MWCNTs [20]. Pezegic et al. grew MWCNTs on long continuous CF surfaces by the CVD method and to prepare a composite using these CFs with infusion molding for the evaluation of electric and thermal conductivities [10]. Boroujeni et al. grew MWCNTs by the CVD method and investigated the effect of the MWCNT length and distribution on the surface properties [11].

Singh et al. produced an effective electromagnetic wave absorber using MWCNTs coated onto CF surfaces by the electrophoretic deposition method [14]. As an approach to achieve chemical bonding between a MWCNT coating and CFs, Zhao et al. used melamine to realize chemical bonding between MWCNTs and CFs, and evaluated the interfacial properties using a microdroplet method [15]. Wu et al. prepared MWCNT-grafted CFs using a reactive, bi-functional, low-molecular-weight chemical, 3-aminopropyltriethoxysilane (APS) [16]. These two approaches utilize the chemical reaction between carboxylic acid on the surfaces of MWCNTs and CF, and amine with a catalyst. The use of amines to achieve chemical bonding between MWCNTs and CFs is of interest; however, it should be noted that the amount of MWCNTs on CFs is small, and not sufficient for scanning electron microscopy (SEM) observation [15,16]. Gamage et al. reported the enhancement of an electromagnetic wave absorber with CF fabric by MWCNT coating [18]. Despite many former challenges, there have been difficulties with the complex process of the CVD of MWCNTs on CFs and the electrodeposition of MWCNTs on CFs. In addition, there has been no strong chemical bonding reported, especially with methods such as the spray coating or dip-coating of MWCNTs onto CF. Table 1 shows the surface modifications to the CF surface.

Table 1. Surface modification using MWCNTs on CFs.

	Without MWCNTs		With MWCNTs		
Method	Chemical bonding	Chemical vapor deposition (CVD)	Electrophoretic deposition (EPD)	Chemical bond	Dip coating
Reference	[7–9]	[10–13]	[14]	[15,16]	[18,19]

There is a functional group that has high reactivity and an irreversible reaction with carboxylic acid and without the need for a catalyst, which is the oxazoline group, and there are low- and high-molecular-weight chemicals that include oxazoline groups. With regard to reactive polymers that include oxazoline, a fundamental polymerization study using 2-substituted-2-oxazoline was reported [21]. There are two typical reactive polymers that include oxazoline, poly-2-vinyl-2-oxazoline [22] from the polymerization of the 2-vinyl-2-oxazoline monomer, and poly-2-isopropenyl-2-oxazoline [23,24] from the polymerization of the 2-isopropenyl-2-oxazoline (Pipozo) monomer. There is a commercially available copolymer of 2-isopropenyl-2-oxazoline that has been utilized as a water soluble cross-linking agent [25]. However, reactive polymers including oxazoline have not typically been utilized for surface treatment and surface modification to date.

We conducted research on the surface modification of carbon materials using reactive polymers including oxazoline, which has high reactivity and an irreversible reaction with carboxylic acid present on carbon materials, with the aim to enhance the mechanical properties of composites. We previously conducted a quantitative investigation of the uniform formation of a monolayer reactive polymer, and its reacted and unreacted oxazoline on diamond particle surfaces using poly-2-vinyl-2-oxazoline [22]. We also found that a uniform coating of MWCNTs on the diamond particle surface through uniform formation of a monolayer reactive polymer of poly-2-vinyl-2-oxazoline with carboxylic acid on the MWCNT surfaces and unreacted oxazoline. However, there have been no reports about the effective modification of CFs with a large amount of MWCNTs through chemical bonding and its effect on the mechanical properties of the composite, which is melt-mixed based on the thermoplastics.

In the present study on the effective modification of CFs with a large amount of MWCNTs, i.e., physical anchoring, we attempted a first layer fabrication of a copolymer of Pipozo on CF (after elimination of sizing agents with an organic solvent) and a second layer fabrication of a uniform MWCNT coating on the first layer. To realize stronger bonding between MWCNTs and CFs through Pipozo, we focused on using not only carboxylic acid on MWCNTs and CFs, but also phenolic OH on MWCNTs and CFs to achieve more reaction sites with oxazoline, which could result in the formation of amide ester and amide ether groups, respectively. We conducted a quantitative analysis of the MWCNTs on CFs using UV-visible spectroscopy and SEM observation. The effect of the interface structure on the mechanical properties of the composite was investigated by melt-mixing of CFs and a polystyrene (PS) matrix. PS was used as a non-reactive matrix with high solubility in organic solvents for the surface analysis of CFs by elimination of the matrix after melt-mixing. PS composites with MWCNTs chemically bonded to CFs were prepared by melt-mixing and the tensile strength and tensile modulus were evaluated to identify the interfacial effect of MWCNTs bonded on the CF structure.

2. Materials and Methods

2.1. Materials

CF (T-700-12K, Toray Industries, Inc., Tokyo, Japan) was used after washing with acetone and ethanol to eliminate sizing agents. MWCNTs (NC-7000, Nanocyl SA, Sambreville, Belgium) [26] were used after the milling treatment with beads (RMB, Aimex Co., Ltd., Tokyo, Japan; zirconia beads, diameter = 0.5 mm, 1000 rpm) for 1.5 h to break up agglomerates and achieve dispersion of individual MWCNTs in N-methyl-2-pyrrolidone (NMP; Kanto Chemical Co., Inc., Tokyo, Japan) solvent. A copolymer of 2-isopropenyl-2-oxazoline (EPOCROS® WS-300 10 wt% in water, Pipozo, 86 mol% oxazoline units in a copolymer Nippon Shokubai Co., Ltd., Osaka, Japan) [25] was used as the reactive polymer that includes oxazoline functional groups. PS (G100C, Toyo Styrene Co., Ltd., Tokyo, Japan) was used as a matrix for the composites because it is unreactive with MWCNTs and CFs, and is soluble in organic solvents, which facilitates elimination of the matrix [27].

2.2. Model Reaction of Phenol-OH and Oxazoline Using Low-Molecular-Weight Compound

Although CFs and MWCNTs are inert carbon materials, the surface of commercially available CFs is intentionally oxidized to enhance the adhesive interface and that of MWCNT also includes oxygen by the presence of oxygen impurities [7,9,15,16,28–32]. The most typical reaction-available acid functional groups by oxidation are COOH and phenol OH. There have been reports on the presence of acid functional groups such as COOH and phenol OH, and quantitative analyses have also been frequently reported [33–36]. The utilization of these acid functional groups for bonding and reaction has also been previously reported [9,15,16]. In this study, we focused on the use of the oxazoline reactive functional group, which has highly irreversible reactivity with COOH and phenol OH.

Evidence of the reaction between COOH or phenol OH, from the surface of CFs or MWCNTs, and oxazoline should be obtained using attenuated total reflectance-Fourier transform infrared (ATR-FTIR) spectroscopy. However, because of the absorption by carbon sp^2 hybrid orbitals and the limited surface area, evidence of the bonding could not be obtained. Therefore, we used diamond particles with carbon sp^3 hybrid orbitals having COOH and phenol OH acid functional groups as a model carbon material instead of CF and MWCNT. When the particle size was less than 1 µm, evidence that bonding was successfully detected using FTIR to confirm the reaction between oxazoline and COOH [22]. Therefore, based on the evidence that there are COOH and phenol OH on the surface of CFs and MWCNTs, and the evidence of FTIR using the diamond particles, we consider that this indicates that oxazoline reacts with COOH or phenol OH.

In the present experiments, we attempted to achieve a chemical reaction between oxazoline (from WS300, Nippon Shokubai Co., Ltd., Osaka, Japan) and phenol-OH, in addition to the reaction between oxazoline and carboxylic acid, which has already been confirmed to be 100% reaction at 100 °C after 3 h using low-molecular-weight oxazoline and nonanoic acid [22]. To examine the chemical reaction between oxazoline and phenol-OH in the present study, phenyl oxazoline (Tokyo Chemical Industry Co., Ltd., Tokyo, Japan) and 3-methoxy phenol (Tokyo Chemical Industry Co., Ltd., Tokyo, Japan) were used as low-molecular-weight model chemicals. The reactivity was evaluated using ^{1}H nuclear magnetic resonance spectroscopy (^{1}H-NMR; JNM-EC500, JEOL, Tokyo, Japan, 500 MHz) after 0, 1, 3, 6, and 24 h at 200 °C. Table 2 summarizes the model reaction compounds.

Table 2. Model reaction of oxazoline and acidic functional groups (COOH and phenol OH groups) and its application to surface modification, i.e., reaction between poly(2-isopropenyl-2-oxazoline) and COOH and phenol OH groups on the CF surface.

	Oxazoline	Acid	Evaluation of Chemical Reaction
(1) Model reaction with low-molecular-weight compounds.	2-Ethyl-2-oxazoline	Nonanoic acid	^{1}H-NMR (evidence of quantitative reaction amount)
(2) Model reaction with low-molecular-weight compounds.	Phenyl-oxazoline	3-Methoxyphenol	^{1}H-NMR (evidence of quantitative reaction amount)
(3) High reactivity oxazoline and acid functionality onto CF.	Polyisopropenyloxazoline-copolymer(Pipozo)	Acid functional groups (such as COOH) on the surface of CF, MWCNT interface	Layer-by-Layer

2.3. Analysis of Acid Groups on CF by the Neutralization Titration Method

Neutralization titration using sodium hydroxide and sodium hydrogen carbonate can provide a quantitatively accurate number of the carboxylic acid and phenol-OH groups on MWCNTs [31–33]. We already established the accuracy of this method by comparison with previously reported results [22]. Firstly, 10 g of CFs was washed with acetone and ethanol, and then treated with 0.01 M NaOH solution (500 mL) to determine the total acid functional groups (COOH and phenol-OH). The CFs were removed by filtration. 400 mL of 0.01 M HCl aqueous solution was added to 400 mL of the NaOH aqueous solution after CF filtration. The neutralization titration was performed using 0.002 M NaOH aqueous solution. In the same way, sodium hydrogen carbonate was used instead of sodium hydroxide to determine the COOH functional groups. The titration was carefully performed using a blank standard water sample to eliminate the effect of carbon dioxide dissolved in the water.

2.4. Layer-By-Layer Grafting of MWCNTs onto CFs

Uniform dense coating of MWCNTs on CFs with chemical bonding using Pipozo was conducted using a layer-by-layer method to obtain 2 layers. The first layer was a layer of the reactive polymer, Pipozo, on CFs through chemical bonding using a Pipozo aqueous solution at 80 °C for 1 h. This process gave the chemical reaction between the oxazoline of Pipozo and COOH present on the CF surfaces. The unreacted Pipozo was then completely removed by washing with methanol, followed by drying for 30 min at 100 °C. This provided a uniform monolayer layer of the reactive Pipozo, including unreacted oxazoline groups, which is described as CF/Pipozo. Due to the small surface area of CFs, quantitative evaluation of Pipozo using thermogravimetric analysis (TGA; TG-DTA8122, Rigaku Cooperation, Tokyo, Japan) is not possible.

A MWCNT dispersion was then prepared using 1-methoxy-2-propanol (PGME; Kanto Chemical Co., Inc., Tokyo, Japan) as a solvent with polyvinylpyrrolidone (PVP; Mw = 40,000, Tokyo Chemical Industry Co., Ltd., Tokyo, Japan) as a dispersant with the composition PGME 99.4 wt%, PVP 0.4 wt%, and MWCNT 0.2 wt%. CF/Pipozo was added to the dispersion and treated at 100 °C for 1 h to allow for the chemical reaction between oxazoline (from unreacted oxazoline of the Pipozo layer on CFs) and COOH (from the MWCNT surfaces). Unreacted MWCNTs were completely removed by washing with methanol and then drying at 100 °C for 30 min, which resulted in the formation of the second layer of uniform dense MWCNTs.

Under the present conditions, the chemical bonding for the two interfaces between CF and Pipozo, and between Pipozo and MWCNT, was only between COOH and oxazoline, i.e., without chemical reaction between phenol-OH and oxazoline, which requires a higher temperature for the reaction. To realize stronger bonding between MWCNTs and CFs through Pipozo, additional heat treatment at 200 °C for 6 h was performed to promote the reaction between phenol OH and oxazoline, between CFs and Pipozo, and between MWCNT and Pipozo. The MWCNTs were chemically grafted onto CFs through Pipozo, which is represented as CF/Pipozo/MWCNT, as illustrated in Figure 1.

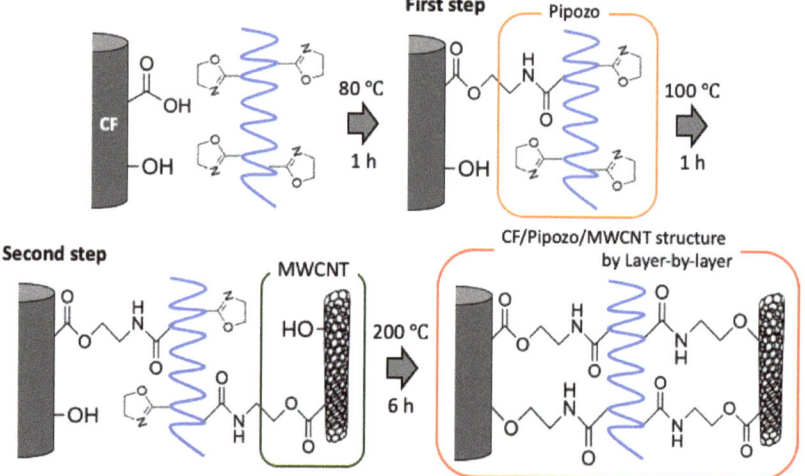

Figure 1. Schematic illustration of layer-by-layer coating Pipozo and MWCNTs onto CF surfaces by chemical reaction.

2.5. Observation of CF/Pipozo/MWCNT Surface and Quantitative Evaluation of MWCNTs

SEM (JSM-7401F, JEOL Tokyo, Japan) was used for observation of the uniformly dense MWCNT-coated CFs. Quantitative evaluation of the amount of MWCNTs on the CFs was conducted using UV-vis spectroscopy measurements at 800 nm (U-4100, Hitachi High-Tech Corporation, Tokyo, Japan) of the MWCNT dispersion after removal of the MWCNTs from CFs by the decomposition of Pipozo and using a calibration curve of absorption with known MWCNT concentrations, as shown in Figure 2. Figure 3 shows the calibration curve for MWCNT concentrations of 0.001, 0.002, 0.003, 0.004, and 0.005 wt% with absorptions of 0.357, 0.709, 1.072, 1.445, and 1.743, respectively (calibration line equation: y = 352.63x + 0.0061).

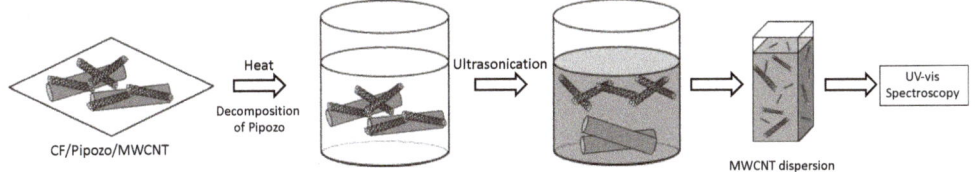

Figure 2. Process to remove MWCNTs (the second layer) from CF surfaces by Pipozo (the first layer) through the decomposition of Pipozo and the complete elimination of MWCNTs from the surfaces of the CFs by ultrasonication in a solvent for quantitative analysis of the amount of MWCNTs using UV-vis spectroscopy.

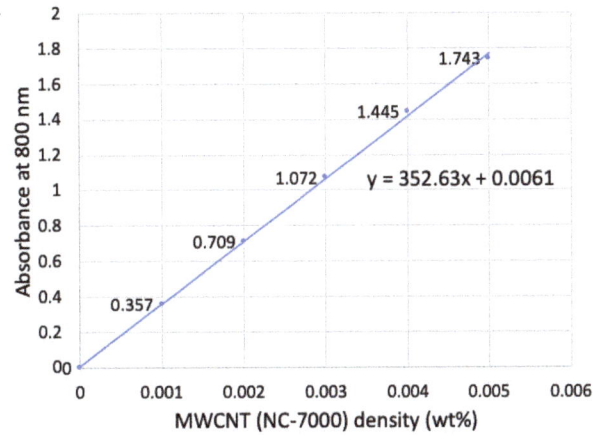

Figure 3. Calibration curve for MWCNT dispersions in NMP solvent.

The MWCNT dispersion for measurement was prepared from the MWCNTs eliminated from MWCNT-grafted CFs by the decomposition of Pipozo at 400 °C for 10 min using CF/Pipozo/MWCNT prepared by the layer-by-layer method. After the heat treatment, ultrasonic treatment (2 h) was conducted for complete removal of the MWCNTs from CFs and preparation of a stable dispersion in NMP solvent with PVP as a dispersant. The UV-vis absorption spectrum was measured twice using the prepared dispersion.

2.6. Fabrication and Physical Properties of PS Composites

Table 3 summarizes the three PS/CF composite samples prepared to evaluate the interfacial physical anchoring effect of the MWCNT-grafted structure, PS, PS-CF, and PS-CF/Pipozo/MWCNT. The PS composites were produced with the CF content (Pipozo layer/MWCNT layer) at 10 wt% using a batch type melt-mixer (Labo Plastomill Micro, Toyo Seiki Co., Ltd., Hyogo, Japan) at 200 °C and 30 rpm for 1 min. The mixed samples were crushed into small particles for hot compression molding to prepare mini dumbbell test pieces (45 mm × 5 mm × 0.5 mm) by pre-heat treatment at 200 °C for 15 min followed by treatment at 200 °C under 5 MPa pressure for 3 min, and then rapidly cooled at 5 MPa for 3 min using cold compression molding (Mini test press, Toyo Seiki Co., Ltd., Hyogo, Japan).

Tensile tests were conducted using a vertical tensile test machine (MCT-1150, A&D Company Limited, Tokyo, Japan) with a crosshead speed of 10 mm/min (distance between chucks: 12 mm, width: 2 mm). The actual thickness was carefully determined from the average of three measurements. In addition, the fractured cross sections after freeze fracture were observed using SEM to evaluate the adhesion of the matrix resin onto the CF surface between the PS-CF and PS-Pipozo/MWCNT/CF composites. Figure 4 shows a process flowchart from the fabrication of the PS/CF composite to the evaluation of interface adhesion (SEM) and tensile testing.

Table 3. Prepared samples and their abbreviations.

	Abbreviation	PS (wt%)	CF (wt%)	CF/Pipozo/MWCNT (wt%)
Samples	PS	100	0	0
	PS-CF	90	10	0
	PS-CF/Pipozo/MWCNT	90	0	10 (Pipozo *: n.d, MWCNT **: 0.04 wt%)

Remarks * Not detectable by TGA; ** Determined from Figure 3.

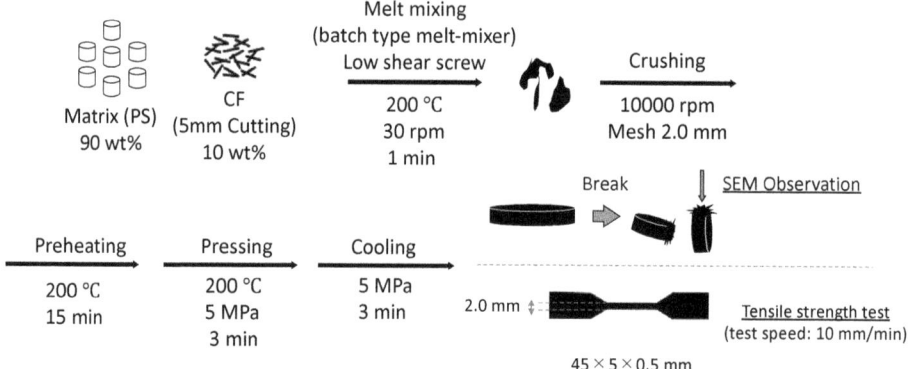

Figure 4. Process flowchart from the preparation of PS composites to SEM observation and evaluation.

3. Results and Discussion

MWCNT was strongly bonded onto CF with an aim to improve the mechanical properties of the composite through a physical anchoring effect, which requires strong bonding between MWCNTs and CFs with the first layer of the reactive Pipozo polymer. It is essential to confirm the reactivity between oxazoline and COOH, together with that between oxazoline and phenol-OH. The chemical reaction between Pipozo and CFs, related with the first layer, is discussed based on the model reaction, together with a quantitative analysis of the acidic functional groups on the CF surface (3-1). The amount of MWCNTs, which is the second layer and is chemically bonded to the first layer, is discussed based on the quantitative analysis (3-2). The mechanical properties of the composite using CF/Pipozo/MWCNT are also discussed with respect to the physical anchoring effect, together with the amount of MWCNTs remaining on the CF surface (3-3). It is important to confirm whether the presence of MWCNTs is maintained or not after melt-mixing of the composite to carefully interpret the effect on the mechanical properties; therefore, the SEM analysis of the surface after melt-mixing is discussed (3-4). The cross section of the composite using CF/Pipozo/MWCNT after freeze fracturing is also discussed with respect to the strong adhesion of the physical anchoring effect (3-5), and the effect of the interfacial properties on the tensile modulus in the case of short fibers is discussed (3-6). The results are thus described and discussed in depth with a focus on the interface formed by the simple layer-by-layer method.

3.1. Model Reaction Using Low-Molecular-Weight Compounds and the Surface Acid Groups of CFs and MWCNTs

COOH and phenol OH are present on the CF surface, both of which can react with Pipozo, which results in the formation of the first layer produced using the layer-by-layer method. As a fundamental experiment, it is important to investigate the reaction conditions, such as the temperature and time, and the resultant reaction rate between COOH and oxazoline, and between phenol OH and oxazoline (summarized in Table 1). We already performed a model reaction between COOH and oxazoline using 2-ethyl-2-oxazoline and nonanoic acid, which indicated that reaction at 100 °C for 3 h provided a complete reaction, i.e., the formation of an amide ester bond [22]. In the present study, the model reaction between phenol OH and oxazoline was conducted using 3-methoxy phenol and phenyl oxazoline as model low-molecular-weight compounds. The reaction rate was evaluated using ^{1}H-NMR, and the results are shown in Figure 5a–e just after mixing (a), and after 1 h (b), 3 h (c), 6 h (d), and 24 h (e). Figure 5 indicates that phenol OH can react with oxazoline at 200 °C for 6 h to form amide ether (>80% of reaction rate). Therefore, the reaction conditions (100 °C for 3 h) for COOH and oxazoline do not allow the reaction between phenol OH and oxazoline.

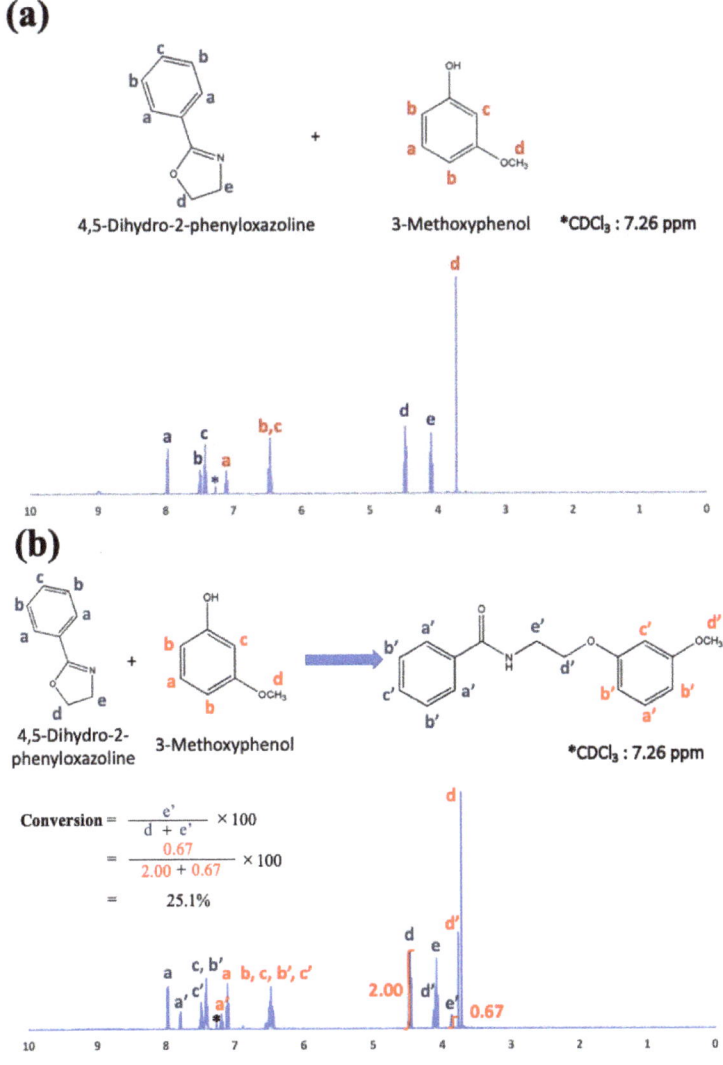

Figure 5. Cont.

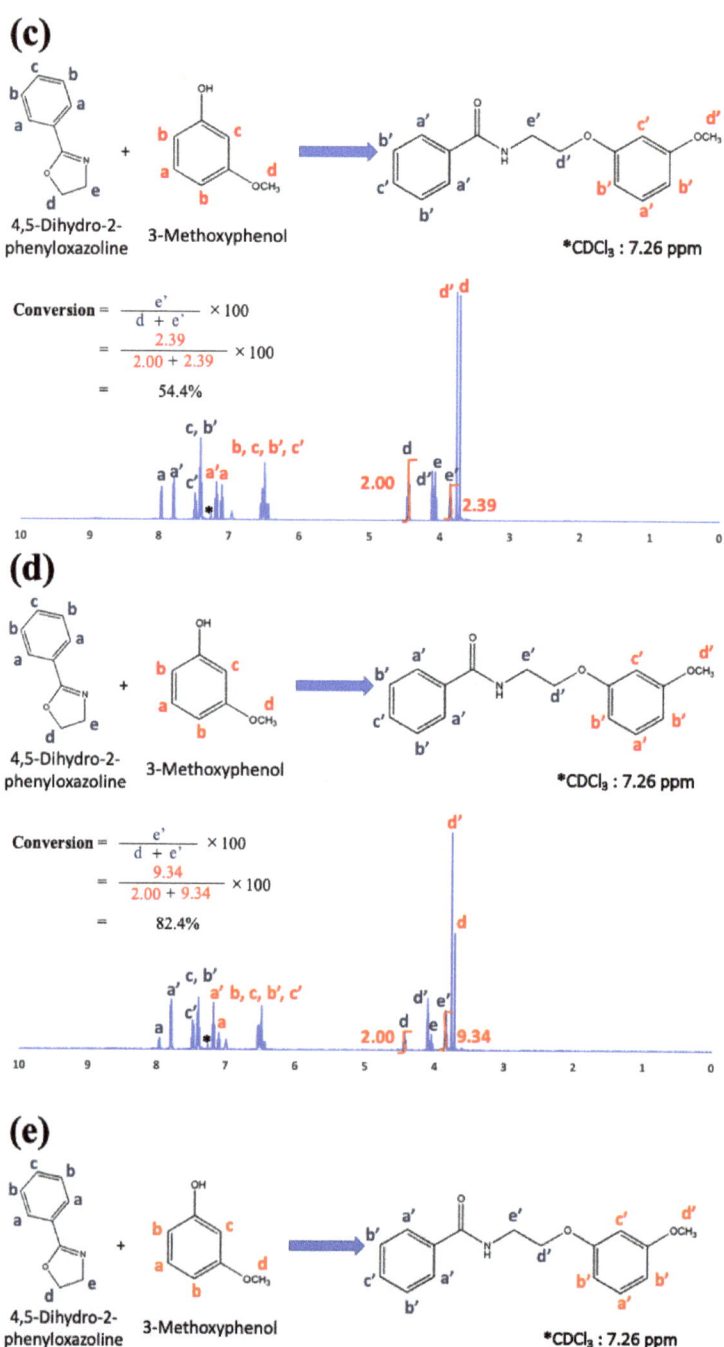

Figure 5. Cont.

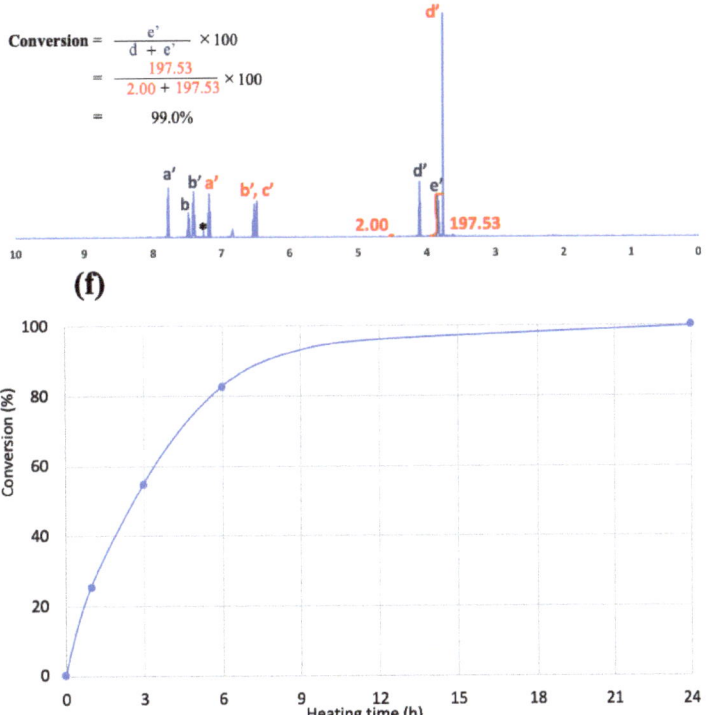

Figure 5. (**a**) ^{1}H-NMR (500 MHz) spectra for the mixture of 4,5-dihydro-2-phenyloxazoline and 3-methoxyphenol immediately after mixing at room temperature, (**b**) after 1 h at 200 °C, (**c**) after 3 h at 200 °C, (**d**) after 6 h at 200 °C, and (**e**) after 24 h at 200 °C. (**f**) Reaction conversion based on ^{1}H-NMR (500 MHz) measurements in (**a–e**) calculated from the integral areas of the d and e' peaks.

It is important to evaluate the acidic functional groups, COOH and phenol OH, on the CF surface by neutralization titration. Let us discuss the accuracy and the reliability of the neutralization titration method by comparison with previous studies using similar commercially available MWCNTs [34–36]. The MWCNTs used in the present study showed a total acid (COOH and phenol OH) content of 7.4×10^{-4} mol/g. Ackermann and Krueger reported a total acid content of 3.0×10^{-4} mol/g [33] and Zhang et al. reported a total acid content of 2.0×10^{-4} mol/g [36]. Based on careful comparison, the total number of acid functional groups was evaluated to be almost identical, which suggests the neutralization titration method is sufficiently accurate. The same method was applied to evaluate the acidic functional groups on the CF surface.

From the neutralization titration method using sodium hydroxide and sodium hydrogen carbonate, the amounts of COOH and phenol OH were determined to be 0.45×10^{-6} mol/g and 1.13×10^{-6} mol/g, respectively. It is important to evaluate the density of acidic functional groups per unit area on the CF surface to confirm the excess amount of oxazoline groups compared with the total acid group content for utilization of the unreacted oxazoline with the MWCNTs. From the 7 μm diameter of the CFs and the density of 1.8 g/cm^3, the unit area was set to be 1×1 nm because the C-C bonding distance is 0.15 nm and the calculation was already performed for the diamond particle surface in our previous work [22]. It was calculated that the number of COOH groups per unit area is ca. 0.8, and that of phenol OH is ca. 2.1, which are similar to that on a diamond particle surface [22]. The number of oxazoline molecules per unit area after monolayer reaction formation on a diamond surface was ca. 10 and that on the CF surface would most probably be similar, which suggests that the number of oxazoline molecules is more than that of the total number of acid groups and that the unreacted available oxazoline should remain after formation of the first layer.

Table 4 summarizes our experimental evidence of the acid groups by the neutralization titration and the results from the references [33,36–39] about the acid groups by the neutralization titration and the oxygen atomic percent by XPS measurements, in which the untreated MWCNTs are used for all results. Regarding the experimental evidence of the presence of the acid functional groups on the untreated MWCNT, we carried out the quantitative analysis of the total acid groups (COOH and phenol OH) according to the established former method using neutralization titration [33,36]. The experimental evidence showed that even the untreated MWCNT has the acid functional groups with 7.4×10^{-4} mol/g. This value is almost in good agreement with the acid functional groups (1.0×10^{-4} mol/g, 2.0×10^{-4} mol/g) from the untreated MWCNT in the former articles [33,36]. In addition, based on the former results from the references [37–39], the evidence of the presence of oxygen on the surface was clearly shown as the atomic percentage (1.4–1.88) even for the commercially available untreated MWCNTs. From our result and the former references, the presence of the acid and the oxygen of the commercially available untreated MWCNTs was demonstrated.

Table 4. Experimental evidence and the former results from the references [33,36–39] about the total acid groups and the oxygen atomic percentage from the neutralization titration and XPS, respectively, using the commercially available untreated MWCNTs.

	Neutralization Titration; Total Acid (-COOH, Phenol-OH) Functional Groups	XPS; Oxygen Atomic % Surface	Product Name of Commercial MWCNT
Experimental value of this study	7.4×10^{-4} mol/g	-	NC7000™ (untreated) from Nanocyl SA
Reference [33]	1.0×10^{-4} mol/g	-	MWCNT(untreated) from FutureCarbon GmbH
Reference [36]	2.0×10^{-4} mol/g	-	MWCNT(untreated) from Cheap Tubes Inc.
Reference [37]	-	1.5 atomic%	NC7000™ (untreated) from Nanocyl SA
Reference [38]	-	1.4 atomic%	NC7000™ (untreated) from Nanocyl SA
Reference [39]	-	1.88 atomic%	NC7000™ (untreated) from Nanocyl SA

3.2. Quantitative Evaluation of MWCNT Amount by the Layer-By-Layer Method

Figure 6 shows SEM micrographs of the CF surface after the layer-by-layer method. Figure 6a shows an SEM micrograph of a CF with the first layer of Pipozo, i.e., CF/Pipozo, which suggests a very smooth surface that originates from the CF surface. This indicates that the first layer of Pipozo is a uniform monolayer structure due to the reactivity of Pipozo, i.e., oxazoline and COOH on the CF surface and the complete washing process of unreacted Pipozo after the first layer reaction. TGA could not detect degradation of the first layer using CF/Pipozo, at least within the detection limit. In our previous study [22] using diamond particles, the uniform monolayer structure was formed by poly-2-vinyl-2-oxazoline, which is similar to Pipozo. TGA analysis could not detect the weight loss of poly-2-vinyl-2-oxazoline due to the limitation of the equipment when using 40 μm-diameter diamond particles, because the surface area is very small. However, when using 1 μm-diameter diamond particles, the weight loss due to the degradation of poly-2-vinyl-2-oxazoline was detectable, which suggests the thickness of the layer was ca. 1–2 nm. The uniformity was demonstrated using MWCNTs as a marker [22]. Therefore, it was considered that a similar uniform monolayer of Pipozo was formed on the CF surface.

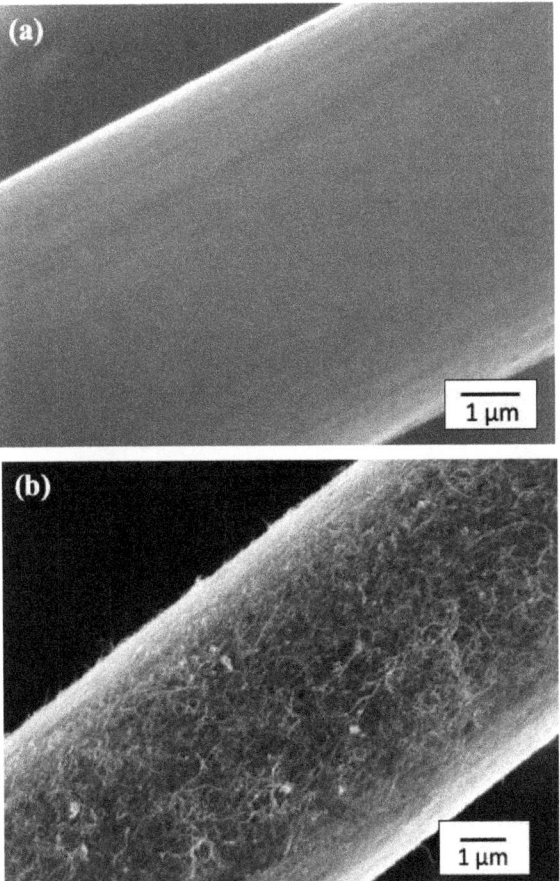

Figure 6. SEM images of the CF surface using the layer-by-layer method. (**a**) CF surface coated with Pipozo (after the first layer, the Pipozo layer was too thin and smooth to be visible by SEM). (**b**) CF surface coated with MWCNTs (after the second layer, and solvent washing several times).

Figure 6b shows an SEM micrograph of MWCNTs, the second layer, chemically bonded with Pipozo, the first layer, chemically bonded with the CF surface. This SEM observation was conducted after a complete washing process to remove unreacted MWCNTs, and suggests that there is a uniform monolayer of Pipozo on the CF surface, which resulted in a very dense and large amount of MWCNTs that formed the uniform coating of the second layer. It should be noted that there are several previous reports [15,16] on MWCNT-coated CFs based on SEM observations; however, all of these reports showed only a small amount of MWCNTs that was much less than observed in the present work. No research to date has shown such a uniform and large amount of MWCNTs chemically bonded on the surface of CFs.

Next, quantitative analysis of the amount of MWCNTs on CFs indicated a substantially large amount with uniformity. For this evaluation, a calibration curve for UV-vis absorption at 800 nm as a function of known MWCNT dispersion concentration was produced. Figure 2 shows the preparation method and how the MWCNTs were removed from the CFs to make a dispersion of MWCNTs. The absorption was measured to be A = 0.205, so that the concentration was determined to be $x = 5.60 \times 10^{-4}$ wt% from the calibration curve.

Here, the solution of PVP dissolved in NMP had the composition (NMP: 9.817 g, PVP: 0.2014 g, Total: 10.0184 g), so that the amount of MWCNTs was calculated to be

5.61×10^{-5} g. The CF/MWCNT amount was 0.0113 g and that of the MWCNTs was 5.61×10^{-5} g; therefore, the MWCNT weight percentage based on the CF/MWCNT (CF plus MWCNT is 100%) was calculated to be 0.50 wt%. This was measured twice in the same way to obtain an average. The amount of MWCNT coating based on the CF was thus evaluated to be 0.44 ± 0.06 wt%.

To achieve a deeper insight into the quantitative reaction by the layer-by-layer method, it is important to determine the mole number of COOH, phenol OH, oxazoline (reacted), and oxazoline (unreacted) in the first layer because the substantial amount of unreacted oxazoline remaining on the first layer is critical to react with the COOH and phenol OH groups of the MWCNT surfaces for formation of the second layer. Our previous study with diamond particles suggested that the total reacted (with COOH and phenol OH) and unreacted oxazoline were 9.1, 2.3, and 6.8, per unit area (1×1 nm), respectively [19,22].

With the various similarities of acidic functional groups (1–2 units) per unit area (1×1 nm), the preparation process and the resultant uniform monolayer between diamond particles and CFs, a noticeable amount of unreacted oxazoline remained in the first uniform layer of Pipozo. This could lead to a substantial amount of MWCNTs on the CFs (0.44 wt%), where the CF surface is not visible due to the coated MWCNTs, which is quite in contrast with the small amount of MWCNTs coated on CFs, where the CF surface was visible in the previous studies [15,16]. Therefore, the first reactive polymer layer has very high reactivity and is quite uniform. The large amount of MWCNTs is considered to be due to physical anchoring of the interfacial effect in the composite, which improves the mechanical properties.

3.3. Mechanical Properties of PS Composites

Figure 7 shows the tensile strength test results for the composites, i.e., tensile strength, tensile modulus, and elongation at break. From Figure 7a, the addition of 10 wt% CF into PS resulted in a 40% increase of the strength and a 70% increase of the modulus. This increase is due to the reinforcement effect of CFs. It is suggested that the increase in the mechanical properties by the addition of CFs is much larger in PS than in polypropylene, which originates from the interface, i.e., the interaction of π–π stacking. There is a substantial amount of MWCNTs (0.4 wt%) at the interface of PS-CF/Pipozo/MWCNT, which leads to large surface unevenness and the physical anchoring effect. The MWCNT layer on the CF surface enhanced the tensile strength with a 20% improvement and the tensile modules with a 20% increase. Figure 7c shows an almost similar elongation at break. The improvement of the tensile strength and modulus is considered to be due to the surface unevenness, i.e., the physical anchoring effect, together with the interaction of π–π stacking.

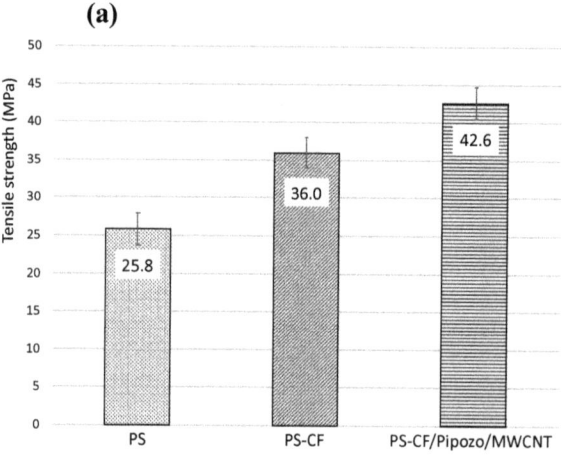

Figure 7. Cont.

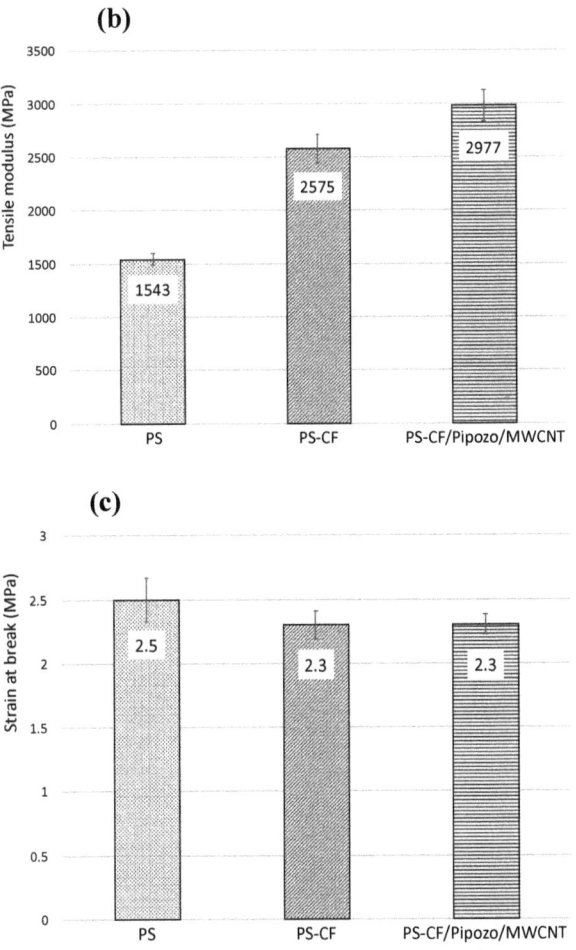

Figure 7. Mechanical properties of various PS composites (PS, PS-CF, PS-CF/Pipozo/MWCNT); (**a**) tensile strength, (**b**) tensile modulus, and (**c**) strain at break.

Careful analysis of the amount of MWCNTs indicated 0.44 wt% MWCNTs on the CFs (therefore, 0.044 wt% MWCNTs based on the total composite). It is notable that only 0.044 wt% MWCNT located at the CF surfaces resulted in such a significant improvement. It is unclear whether the presence of MWCNTs on CFs can be maintained or not after severe melt-mixing, which was the process used for preparation of the present PS composite.

3.4. SEM Observation of MWCNT-Coated CF Surface after Melt-Mixing

There have been several reports on MWCNT-coated CF composites, which are mostly based on thermosetting resin composites, where liquid epoxy penetrates into the CF surface. In the penetration process, the force near the CF surface is limited because the thermosetting resin is a relatively low viscosity liquid. On the other hand, there are a few previous studies on thermoplastic matrix composites [12,13,19]. As an example, Rahmanian et al. produced MWCNTs grown by the CVD method from the CF surface, and composites were prepared using polypropylene [40]. In this case, the CF received a large shear force due to the high viscosity of the molten resin so that MWCNTs could be eliminated from the CFs during melt-mixing. However, there was no investigation of the CF surface by observation after melt-mixing. It is thus necessary to confirm the presence of MWCNTs on CFs after melt-mixing, which can be simply checked by elimination of the PS matrix with the use of an organic solvent, such as dichloromethane (Kanto Chemical Co., Inc., Tokyo, Japan). Elimination of the matrix was performed and the surface of the CF was observed using SEM. Figure 8 shows an SEM image of the CF surface after washing the composite (PS-CF/Pipozo/MWCNT), which revealed that a substantial amount of MWCNTs still remained even after melt-mixing. An additional question arises as to whether the matrix PS resin really penetrates into the MWCNT structure. To answer this question, SEM observations of cross sections after freeze fracturing were conducted.

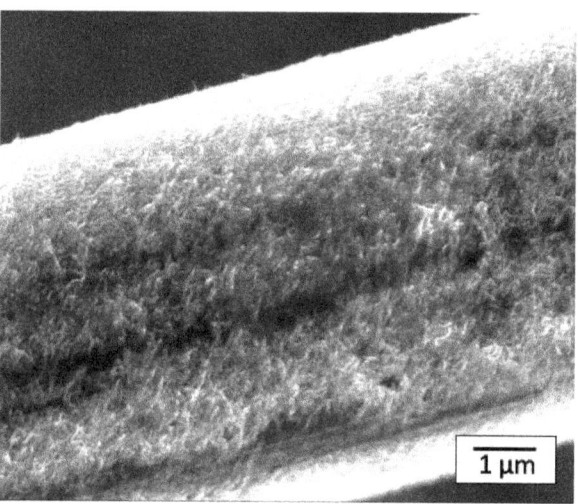

Figure 8. SEM image of the MWCNTs remaining on the CF surface. The sample was prepared using CF/Pipozo/MWCNT (layer-by-layer method, Figure 5), which was melt-mixed with PS and PS was then completely removed with a solvent treatment several times. This demonstrates that the MWCNTs (second layer) were strongly bonded to CF by Pipozo (first layer) and the MWCNTs were not removed by the melt-mixing process.

3.5. SEM Observation of Fracture Cross Sections of the Composite

The improvement of the mechanical properties of the composites can be correlated with the fracture patterns around the surface of CFs. The fracture of the composite with weaker mechanical properties was initiated at the interface, i.e., the weak point, which resulted in a smooth CF surface. On the other hand, that with stronger mechanical properties was initiated in the matrix because of the strongly adhesive interface. There have been no reports on the effect of physical anchoring on the surface of CFs on the fracture phenomenon.

The effect of the uneven structure by the substantial amount of MWCNTs resulted in a strong chemical bonding to the CF surface, as evidenced by the fracture cross section near the CF; Figure 9 shows SEM images of the fracture cross sections of two composites (PS-CF, PS-CF/Pipozo/MWCNT), together with the two types of model structure, interfacial peeling and cohesive failure. The fracture surface of the CF in the PS-CF composite was smooth, which suggests the failure occurred at the interface. On the other hand, for the PS-CF/Pipozo/MWCNT composite, there was PS resin remaining on the surface, which suggested the failure occurred in the matrix, i.e., cohesive failure. This suggests that PS resin penetrated into the MWCNT layer, which resulted in strong adhesion at the interface.

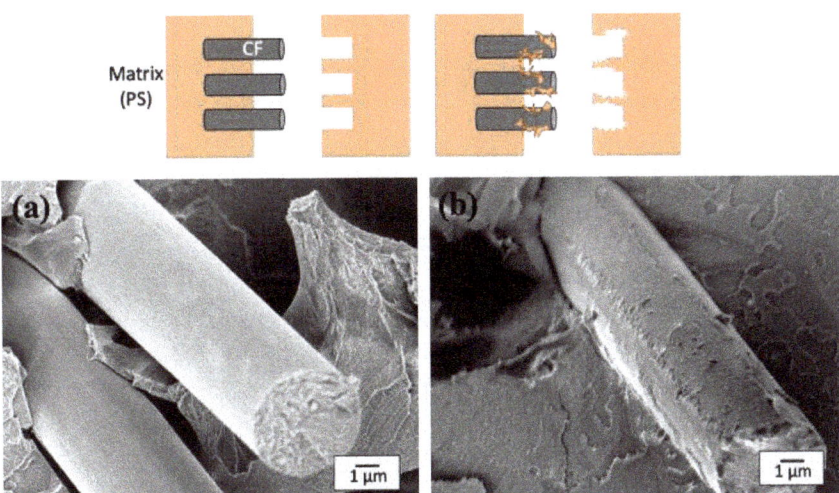

Figure 9. SEM images of freeze-fractured surfaces of the composites after impact tests; (**a**) PS-CF composite (interfacial peeling) and (**b**) PS-CF/Pipozo/MWCNT composite (cohesive failure).

3.6. Effect of Interface on Tensile Properties

The properties of the interface between the matrix and CF are often reported with respect to the microdroplet or interfacial shear strength measurements using long fibers [7,9,11,15,16]. However, these are for long fibers, and the interfacial properties have not been evaluated using short fibers. The strength of the interface between the matrix and CF can be evaluated from the length of the CF and the tensile modulus of the composite [41]. The fiber length in the composite was measured to show that there is no difference in fiber length. For the CF length measurements, approximately 300 fibers were randomly measured and their number and weight averages were calculated. For PS-CF, the number average fiber length was 357.6 ± 142.0 µm and the weight average fiber length was 385.1 µm. For PS-CF/Pipozo/MWCNT, the number average fiber length was 329.3 ± 138.6 µm and the weight average fiber length was 387.4. There was no significant difference in CF length between these two samples. From the theoretical equation, it can be shown that the difference in CF length was due to the effect of the interface of the MWCNTs coated on the CF because there was no difference in CF length, i.e., the increase in the mechanical strength of the composite was due to the action of the MWCNTs at the interface.

The current improvement of the mechanical properties was approximately 20% and further improvement is expected in future work by optimization of the amount of MWCNTs and the structure (e.g., diameter, length, and linearity). These results for the effective modification of CFs by MWCNTs with a simple process to enhance the mechanical properties may be applicable for all types of thermoplastic polymers, especially engineering plastics that will meet the requirements of the automotive industry.

4. Conclusions

MWCNTs were grafted onto the surface of CFs using a reactive polymer including oxazoline. The process, called layer-by-layer, consists of two steps, i.e., coating with the reactive polymer and the MWCNT layer process. A substantial number of MWCNTs were grafted onto CFs by this simple process, which almost completely remained after melt-mixing with PS as a thermoplastic, as confirmed after the elimination of PS with a solvent. This is the first evidential observation of the interfacial structure. The effective surface modification led to an improvement of the mechanical properties, which was supported by SEM observation that showed that fracture was initiated in the matrix instead of the interface. Further improvement will be expected by optimization of the amount of MWCNTs and the structure. This interfacial concept may be applicable to short CF composites based on various thermoplastics, which should meet the requirements for lightweight applications in the automotive industry.

Author Contributions: R.T. performed the measurements, analyzed experimental data, and wrote the manuscript. T.T. supervised the entire project. K.A. and T.G. participated in comprehensive discussion and provided advice and suggestions. All authors have read and agreed to the published version of the manuscript.

Funding: This research received no external funding.

Institutional Review Board Statement: Not applicable.

Informed Consent Statement: Not applicable.

Data Availability Statement: The raw data presented in this study are available on request from the corresponding author.

Acknowledgments: The authors thank Masumi Takamura, Hokuto Chiba, Takuya Nukui and Kiku Ohyama of Yamagata University for helpful discussions and comments on the manuscript.

Conflicts of Interest: The authors declare no conflict of interest.

References

1. Park, H.M.; Park, C.; Bang, J.; Lee, M.; Yang, B. Synergistic effect of MWCNT an carbon fiber hybrid fillers on electrical and mechanical properties of alkali-activated slag composites. *Crystals* **2020**, *10*, 1139. [CrossRef]
2. Petreny, R.; Meszaros, L. Moisture dependent tensile and creep behaviour of multi-wall carbon nanotube and carbon fibre reinforced, injection moulded polyamide 6 matrix multi-scale composites. *J. Mater. Res. Technol.* **2022**, *10*, 689–699. [CrossRef]
3. Yan, X.; Qiao, L.; Tan, H.; Tan, H.; Liu, C.; Zhu, K.; Lin, Z.; Xu, S. Effect of Carbon Nanotubes on the Mechanical, Crystallization, Electrical and Thermal Conductivity Properties of CNT/CCF/PEKK Composites. *Materials* **2022**, *15*, 4950. [CrossRef]
4. Qiao, L.; Yan, X.; Tan, H.; Dong, S.; Ju, G.; Shen, H.; Ren, Z. Mechanical Properties, Melting and Crystallization Behaviors, and Morphology of Carbon Nanotubes/Continuous Carbon Fiber Reinforced Polyethylene Terephthalate Composites. *Polymers* **2022**, *14*, 2892. [CrossRef]
5. Vedrtnam, A.; Sharma, S.P. Study on the performance of different nano-species used for surface modification of carbon fiber for interface strengthening. *Compos. Part A* **2019**, *125*, 105509. [CrossRef]
6. Liu, L.; Jia, C.; He, J.; Zhao, F.; Fan, D.; Xing, L.; Wang, M.; Wang, F.; Jiang, Z.; Huang, Y. Interfacial characterization, control and modification of carbon fiber reinforced polymer composites. *Compos. Sci. Technol.* **2015**, *121*, 56–72. [CrossRef]
7. Yang, L.; Han, P.; Gu, Z. Grafting of a novel hyperbranched polymer onto carbon fiber for interfacial enhancement of carbon fiber reinforced epoxy composites. *Mater. Des.* **2021**, *200*, 109456. [CrossRef]
8. Dharmasiri, B.; Randall, J.D.; Stanfield, M.K.; Ying, Y.; Andersson, G.G.; Nepal, D.; Hayne, D.J.; Henderson, L.C. Using surface grafted poly(acrylamide) to simultaneously enhance the tensile strength, tensile modulus, and interfacial adhesion of carbon fibres in epoxy composites. *Carbon* **2022**, *186*, 367–379. [CrossRef]
9. Peng, O.; Li, Y.; He, X.; Lv, H.; Hu, P.; Shang, Y.; Wang, C.; Wang, R.; Sritharan, T.; Du, S. Interfacial enhancement of carbon fiber composites by poly(amido amine) functionalization. *Compos. Sci. Technol.* **2013**, *74*, 37–42. [CrossRef]
10. Pezegic, T.R.; Anguita, J.V.; Hamerton, I.; Jayawardena, K.D.G.I.; Chen, J.-S.; Stolojan, V.; Ballocchi, P.; Walsh, R.; Silva, S.R.P. Nanotubes as an alternative to polymer sizing. *Sci. Rep.* **2016**, *6*, 37334. [CrossRef]
11. Boroujeni, A.Y.; Tehrani, M.; Nelson, A.J.; Al-Haik, M. Hybrid carbon nanotube-carbon fiber composites with improved in-plane mechanical properties. *Compos. Part B* **2014**, *66*, 475–483. [CrossRef]
12. Sager, R.J.; Klein, P.J.; Lagoudas, D.C.; Zhang, Q.; Liu, J.; Dai, L.; Baur, J.W. Effect of carbon nanotubes on the interfacial shear strength of T650 carbon fiber in an epoxy matrix. *Compos. Sci. Technol.* **2009**, *69*, 898–904. [CrossRef]
13. Wu, D.; Yao, Z.; Sun, X.; Liu, X.; Liu, L.; Zhang, R. Mussel-tailored carbon fiber/carbon nanotubes interface for elevated interfacial properties of carbon fiber/epoxy composites. *Chem. Eng. J.* **2011**, *429*, 132449. [CrossRef]
14. Singh, S.K.; Akhtar, M.J.; Kar, K.K. Hierarchical carbon nanotube-coated carbon fiber: Ultra lightweight, thin, and highly efficient microwave absorber. *Appl. Mater. Interfaces* **2018**, *10*, 24816–24828. [CrossRef]
15. Zhao, M.; Meng, L.; Ma, L.; Ma, L.; Yang, X.; Huang, Y.; Ryu, J.E.; Shankar, A.; Li, T.; Yan, C.; et al. Layer-by-layer grafting CNTs onto carbon fibers surface for enhancing the interfacial properties of epoxy resin composites. *Compos. Sci. Technol.* **2018**, *154*, 28–36. [CrossRef]
16. Wu, G.; Ma, L.; Liu, L.; Wang, Y.; Xie, F.; Zhong, Z.; Zhao, M.; Jiang, B.; Huang, Y. Interfacially reinforced methylphenylsilicone resin composites by chemically grafting multiwall carbon nanotubes onto carbon fibers. *Compos. Part B* **2015**, *82*, 50–58. [CrossRef]
17. Zakaria, M.R.; Akil, H.M.; Kudus, M.H.A.; Ullah, F.; Javed, F.; Nosbi, N. Hybrid carbon fiber-carbon nanotube reinforced polymer composites: A review. *Compos. Part B* **2019**, *176*, 107313. [CrossRef]

18. Gamage, S.J.P.; Yang, K.; Braveenth, R.; Raagulan, K.; Kim, H.S.; Lee, Y.S.; Yang, C.-M.; Moon, J.J.; Chai, K.Y. MWCNT coated free-standing carbon fiber fabric for enhanced performance in EMI shielding with a higher absolute EMI SE. *Materials* **2017**, *10*, 1350. [CrossRef]
19. Zheng, N.; Huang, Y.; Liu, H.-Y.; Gao, J.; Mai, Y.-W. Improvement of interlaminar fracture toughness in carbon fiber/epoxy composites with carbon nanotubes/polysulfone interleaves. *Compos. Sci. Technol.* **2017**, *140*, 8–15. [CrossRef]
20. Salahuddin, B.; Faisal, S.N.; Baigh, T.A.; Alghamdi, M.N.; Islam, M.S.; Song, B.; Zhang, X.; Gao, S.; Aziz, S. Carbonaceous materials coated carbon fibre reinforced polymer matrix composites. *Polymers* **2021**, *13*, 2771. [CrossRef]
21. Miyamoto, M.; Sano, Y.; Kimura, Y.; Saegusa, T. "Spontaneous" Vinyl Polymerization of 2-Vinyl-2-oxazoline. *Macromolecules* **1985**, *18*, 1641–1648. [CrossRef]
22. Goto, T.; Nitta, R.; Nukui, T.; Takemoto, M.; Takahashi, T. Preparation of oxazoline-group-functionalized diamond using poly(2-vinyl-2-oxazoline) based on a model reaction between oxazoline and carboxylic acid. *Diam. Relat. Mater.* **2021**, *120*, 108693. [CrossRef]
23. Kagiya, T.; Matsuda, T. Selective Polymerization of 2-Isopropenyl-2-oxazoline and cross-linking reaction of the polymers. *Polym. J.* **1972**, *3*, 307–314. [CrossRef]
24. Nishikubo, T.; Kameyama, A.; Tokai, H. Synthesis of polymers in aqueous solutions. Selective addition reaction of poly(2-isopropenyl-2-oxazoline) with thiols and carboxylic acids in aqueous solutions. *Polym. J.* **1996**, *28*, 134–138. [CrossRef]
25. Oxazoline-Functional Polymer: EPOCROS™, Nippon Shokubai HP. Available online: https://www.shokubai.co.jp/en/products/detail/pdf/epocros_e.pdf (accessed on 8 July 2022).
26. NC7000 Industrial Multiwall Carbon Nanotubes. Available online: https://www.nanocyl.com/product/nc7000/ (accessed on 8 July 2022).
27. Available online: http://www.toyo-st.co.jp/cgi-bin/toyo-st_.cgi?name=gpps_e&type=pdf (accessed on 8 July 2022).
28. Bauer, M.; Beratz, S.; Ruhland, K.; Horn, S.; Moosburger-Will, J. Anodic oxidation of carbon fibers in alkaline and acidic electrolyte: Quantification of surface functional groups by gas-phase derivatization. *Appl. Surf. Sci.* **2020**, *506*, 144947. [CrossRef]
29. Tiwari, S.; Bijwe, J. Surface Treatment of Carbon Fibers—A Review. *Proc. Technol.* **2014**, *14*, 505–512. [CrossRef]
30. Available online: https://ntrs.nasa.gov/api/citations/19870016001/downloads/19870016001.pdf (accessed on 11 September 2022).
31. Chukov, D.; Nematulloev, S.; Torokhov, V.; Stepashkin, A.; Sherif, G.; Tcherdyntsev, V. Effect of carbon fiber surface modification on their interfacial interaction T with polysulfone. *Results Phys.* **2019**, *15*, 102634. [CrossRef]
32. Ehlert, G.J.; Lin, Y.; Sodano, H.A. Carboxyl functionalization of carbon fibers through a grafting reaction that preserves fiber tensile strength. *Carbon* **2011**, *49*, 4246–4255. [CrossRef]
33. Hanelt, S.; Orts-Gil, G.; Friedrich, J.F.; Meyer-Plath, A. Differentiation and quantification of surface acidities on MWCNTs by indirect potentiometric titration. *Carbon* **2011**, *49*, 2978–2988. [CrossRef]
34. Hu, H.; Bhowmik, P.; Zhao, B.; Hamon, M.A.; Itkis, M.E.; Haddon, R.C. Determination of the acidic sites of purified single-walled carbon nanotubes by acid±base titration. *Chem. Phys. Lett.* **2001**, *345*, 25–28. [CrossRef]
35. Schafer, H.; Hoelderle, M.; Muelhaupt, R. FTi.r. studies of oxazoline functionalized polymer particles. *Polymer* **1998**, *39*, 1259–1268. [CrossRef]
36. Zhang, Z.; Pfefferle, L.; Haller, G.L. Characterization of functional groups on oxidized multi-wall carbon nanobutes by potensiometric titration. *Catal. Today* **2015**, *249*, 23–29. [CrossRef]
37. White, C.M.; Banks, R.; Hamertonc, I.; Watts, J.F. Characterisation of commercially CVD grown multi-walled carbon nanotubes for paint applications. *Prog. Org. Coat.* **2016**, *90*, 44–53. [CrossRef]
38. Taylor-Just, A.J.; Ihrie, M.D.; Duke, K.S.; Lee, H.Y.; You, D.J.; Hussain, S.; Kodali, V.K.; Ziemann, C.; Creutzenberg, O.; Vulpoi, A.; et al. The pulmonary toxicity of carboxylated or aminated multi-walled carbon nanotubes in mice is determined by the prior purification method. *Part. Fibre Toxicol.* **2020**, *17*, 20. [CrossRef]
39. Bhakta, A.K.; Detriche, S.; Kumari, S.; Hussain, S.; Martis, P.; Mascarenhas, R.J.; Delhalle, J.; Mekhalif, Z. Multi-wall Carbon Nanotubes Decorated with Bismuth Oxide Nanocrystals Using Infrared Irradiation and Diazonium Chemistry. *J. Inorg. Organomet. Polym. Mater.* **2018**, *28*, 1402–1413. [CrossRef]
40. Rahmanian, S.; Thean, K.S.; Suraya, A.R.; Shazed, M.A.; Salleh, M.A.M.; Yusoff, H.M. Carbon and glass hierarchical fibers: Influence of carbon nanotubes on tensile, flexural and impact properties of short fiber reinforced composites. *Mater. Des.* **2013**, *43*, 10–16. [CrossRef]
41. Holister, G.S.; Thomas, C. *Fiber Reinforced Materials*; Elsevier Publishing Co., Ltd.: London, UK, 1966; pp. 15–108.

Article

Development of Electrically Conductive Thermosetting Resin Composites through Optimizing the Thermal Doping of Polyaniline and Radical Polymerization Temperature

Kohei Takahashi [1], Kazuki Nagura [1], Masumi Takamura [2], Teruya Goto [1] and Tatsuhiro Takahashi [1,*]

[1] Graduate School of Organic Materials Science, Yamagata University, 4-3-16 Johnan, Yonezawa 992-8510, Japan
[2] Open Innovation Platform, Yamagata University, 4-3-16 Jonan, Yonezawa 992-8510, Japan
* Correspondence: effort@yz.yamagata-u.ac.jp

Abstract: This work developed an electrically conductive thermosetting resin composite that transitioned from a liquid to solid without using solvents in response to an increase in temperature. This material has applications as a matrix for carbon fiber reinforced plastics. The composite comprised polyaniline (PANI) together with dodecyl benzene sulfonic acid (DBSA) as a liquid dopant in addition to a radical polymerization system made of triethylene glycol dimethacrylate with a peroxide initiator. In this system, micron-sized non-conductive PANI particles combined with DBSA were dispersed in the form of conductive nano-sized particles or on the molecular level after doping induced by a temperature increase. The thermal doping temperature was successfully lowered by decreasing the PANI particle size via bead milling. Selection of an appropriate peroxide initiator also allowed the radical polymerization temperature to be adjusted such that doping occurred prior to solidification. Optimization of the thermal doping temperature and the increased radical polymerization temperature provided the material with a high electrical conductivity of 1.45 S/cm.

Keywords: electrically conductive; thermosetting resin; polyaniline

Citation: Takahashi, K.; Nagura, K.; Takamura, M.; Goto, T.; Takahashi, T. Development of Electrically Conductive Thermosetting Resin Composites through Optimizing the Thermal Doping of Polyaniline and Radical Polymerization Temperature. *Polymers* 2022, 14, 3876. https://doi.org/10.3390/polym14183876

Academic Editors: Tomasz Makowski and Sivanjineyulu Veluri

Received: 25 August 2022
Accepted: 14 September 2022
Published: 16 September 2022

Publisher's Note: MDPI stays neutral with regard to jurisdictional claims in published maps and institutional affiliations.

Copyright: © 2022 by the authors. Licensee MDPI, Basel, Switzerland. This article is an open access article distributed under the terms and conditions of the Creative Commons Attribution (CC BY) license (https:// creativecommons.org/licenses/by/ 4.0/).

1. Introduction

Carbon fiber reinforced plastics (CFRPs) prepared by infusing liquid monomers into carbon fiber fabric sheets with subsequent thermosetting are typically lightweight but exhibit high strength. For these reasons, CFRPs have been used as structural materials in aircraft as replacements for various metals. However, aircraft are sometimes subject to lightning strikes [1] and therefore CFRPs that provide lightning strike protection (LSP) are of considerable importance. Specifically, while the metal components of aircraft are electrically conductive and so are not significantly damaged by lightning strikes [2], CFRPs can be severely deteriorated as a result of the decomposition of the CFs and/or the resin [3]. These effects can occur as a consequence of the very low electrical conductivity of typical CFRPs in the thickness direction (which results from the non-conductive matrix resin) and of the Joule heating generated by the large electric current induced by a lightning strike.

Currently, LSP is afforded to the CFRP components of aircraft by applying metal mesh sheets to the surfaces of these items as well as at the interfaces between individual CF layers. However, this increases the mass of the CFRP parts and also the complexity of the fabrication process [4]. For these reasons, the development of electrically conductive thermosetting resins has received considerable attention.

One approach to creating conductive thermosetting resins is to add electrically conductive filler to these materials. There are many types of conductive fillers such as metals and carbon materials, and many studies have been done to add them to thermosetting resins and CFRP [5–9].

When added in the form of particles (either micro- or nano-sized) or as individual molecules, these polymers provide channels for current flow. As an example, the electrically conductive polymer polyaniline (PANI) is often used as a filler for CFRPs because this substance is lightweight, readily synthesized, inexpensive and highly stable compared with other such polymers [10–12]. PANI typically exists in a non-conductive state referred to as emeraldine base (EB) but can be transformed to a conductive emeraldine salt (ES) by the addition of an acid. The acid acts as a dopant and changes the electrical state of the nitrogen atoms contained in the quinoid structure of the macromolecule. PANI is generally not soluble or meltable due to the rigid molecular structure imparted by the benzene rings along each chain as well as the strong molecular interactions in this polymer. However, PANI molecules or nanoparticles can be dispersed in various solvents or liquid resins in conjunction with a liquid acid dopant having a molecular weight of 200 to 500. Typical dopants are do-decyl benzene sulfonic acid (DBSA) and camphor sulfonic acid [10–12].

The thermal properties and LSP characteristics of CFRPs can be improved by incorporating a dispersed PANI/dopant composite in the thermosetting resin [13–15]. However, achieving a suitable degree of electrical conductivity requires a high concentration of this composite, which in turn increases the viscosity of the liquid resin (based on the large surface area of the conductive particles, which leads to significant interparticle friction and interactions). This high viscosity hinders the penetration of the resin into the CF fabric.

To overcome the above difficulty, we have previously proposed and studied the use of PANI/DBSA in conjunction with thermal doping [16–20]. In this process, DBSA (which is a liquid at room temperate) is mixed with micron-sized PANI particles formed from aggregated nanoparticles during the initial fabrication stage. Doping does not occur at this relatively low temperature and this combination produces a low-viscosity dispersion. Rather, the liquid DBSA is only able to penetrate micron-sized aggregated particles and nano-sized particles. With increasing temperature, doping effects appear due to the penetration of the DBSA into the nano-sized particles [16] and the resulting DBSA-doped PANI nano-particles can be dispersed in thermosetting resins [20].

This process is unique in that the DBSA acts as a thermal dopant and also as an initiator for the cationic polymerization of the monomer. The infusion of a liquid thermosetting resin containing PANI/DBSA into stacked layers of CF fabric sheets followed by hot pressing has been found to provide good dispersion of PANI/DBSA nanoparticles or molecules within the individual CF fibers. This process, therefore, generates PANI/DBSA connections (that is, achieves thermal doping) in the thermosetting resin to produce a CFRP having high electrical conductivity in the thickness direction [21]. These materials have thus exhibited excellent LSP properties [22–24]. The CFRP is intended to be applied to parts of aircraft wings and structural materials for the fuselage [3,4]. The CFRP is lightweight because it does not have a metal mesh, so it has the advantages of improving fuel efficiency and reducing carbon dioxide emissions.

Unfortunately, as the room-temperature polymerization of the resin is promoted by the DBSA, which is difficult to inhibit, the viscosity of the mixture increases [25].

This uncontrolled increase in viscosity can be avoided by using different compounds as the dopant and the initiator based on polymerization with a radical initiator. This system also allows the polymerization initiation temperature to be controlled by careful selection of the radical initiator. On this basis, a new design comprising a mixture of PANI, a methyl methacrylate monomer including phosphoric acid as the dopant (P-2M), and a peroxide (Peroxide butyl E) was proposed by Santwana et al. This system provided a constant viscosity of 2500 mPa·s by inhibiting polymerization at room temperature, along with a conductivity of 0.5 ± 0.18 S/m and a flexural modulus of 2.6 ± 0.13 GPa [26] for the pure resin without CFs. When this material was used as the CFRP matrix, the conductivity of the material in the thickness direction was 0.14 ± 0.02 S/cm, which is potentially suitable for LSP [27].

However, this newly developed radical polymerization thermosetting resin system resulted in reduced conductivity, possibly because the material underwent polymerization before the thermal doping process could be completed, since the hour half-life decomposition temperature of Peroxide butyl E has a relatively low value of 119 °C. In addition, the doping effect of phosphoric acid appeared to be less than that of sulfonic acid. On this basis, it would evidently be beneficial to use sulfonic acid as the thermal dopant while controlling the radical polymerization temperature based on using an optimal peroxide.

The present study examined thermal doping by DBSA in conjunction with thermosetting by radical polymerization and assessed the means of increasing the conductivity of the finished product. Experimental trials were carried out using PANI, DBSA, methacrylate (as a bi-functional monomer) and a peroxide. The effect of using DBSA for thermal doping in conjunction with different mixers and the effect of promoting radical polymerization with peroxides having different decomposition temperatures on the conductivity of the resin were investigated. The purpose of the study was to develop a high conductivity thermosetting resin system via the optimization of thermal doping (i.e., by decreasing the thermal doping temperature) and radical polymerization (i.e., by increasing the thermosetting temperature).

2. Materials and Methods

2.1. Raw Materials

Triethylene glycol dimethacrylate (TEGDMA; Tokyo Chemical Industry Co., Ltd., Tokyo, Japan) was used as the liquid monomer in this work (Figure 1a) together with di-t-hexyl peroxide (Peroxide H; Figure 1b) and di-t-butyl peroxide (Peroxide B; Figure 1c; NOF Corp., Tokyo, Japan). These compounds had initiation temperatures (that is, hour half-life temperatures) of 136.2 and 144.1 °C respectively. PANI Emeraldine Base (PANI-EB; Regulus Co., Ltd., Tokyo, Japan), a non-conductive polymer (Figure 1d), was also employed, together with DBSA (Kanto Chemical Co., Inc., Tokyo, Japan) as a liquid compound capable of acting as a dopant in response to a temperature increase (Figure 1e). TEGDMA was employed as the monomer because each molecule contains two reactive acrylate groups that are available for radical polymerization, and so this compound is suitable for the formation of a three-dimensional network in response to initiation by oxygen radicals (-O•) produced by the Peroxide H and Peroxide B as a result of thermal decomposition.

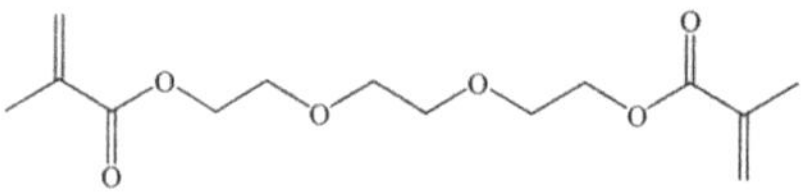

(a) triethylene glycol dimethacrylate (TEGDMA)

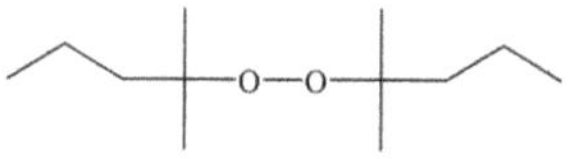

(b) di-t-hexyl peroxide (Peroxide H)
136.2 °C (Hour half-life temperature)

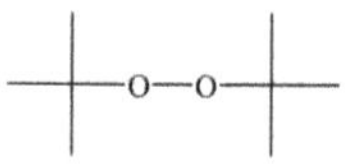

(c) di-t-butyl peroxide (Peroxide B)
144.1 °C (Hour half-life temperature)

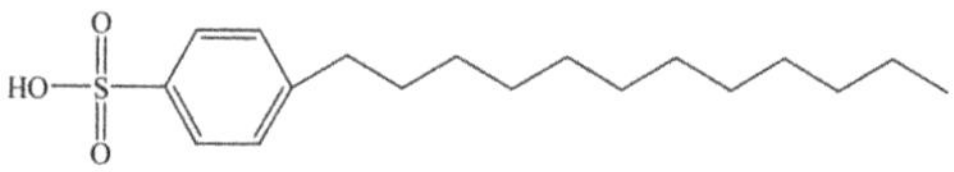

(d) n-dodecylbenzenesulfonic acid (DBSA)

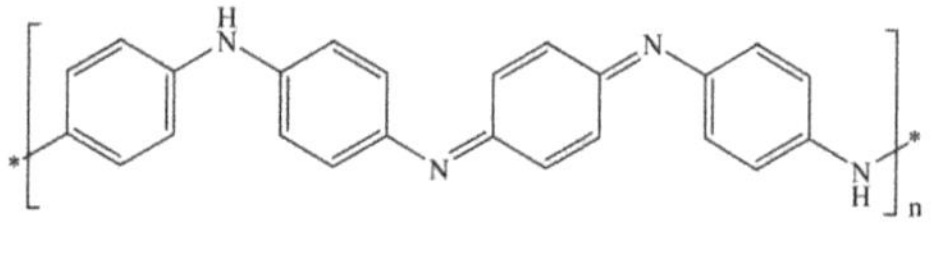

(e) polyaniline emeraldine base (PANI-EB)

Figure 1. Molecular structures of (**a**) TEGDMA, (**b**) di-t-hexyl peroxide, (**c**) di-t-butyl peroxide, (**d**) DBSA and (**e**) PANI-EB.

2.2. Preparation and Evaluation of PANI/DBSA/TEGDMA Composites

The PANI/DBSA/TEGDMA composites were prepared as liquid pastes using the procedures summarized in Figure 2a,b. In this work, either a centrifugal mixer or a bead mill was used to prepare the composite so as to maintain the original PANI particle size or to produce smaller PANI particles, respectively. The intent was to identify the effect of particle size on the thermal doping process.

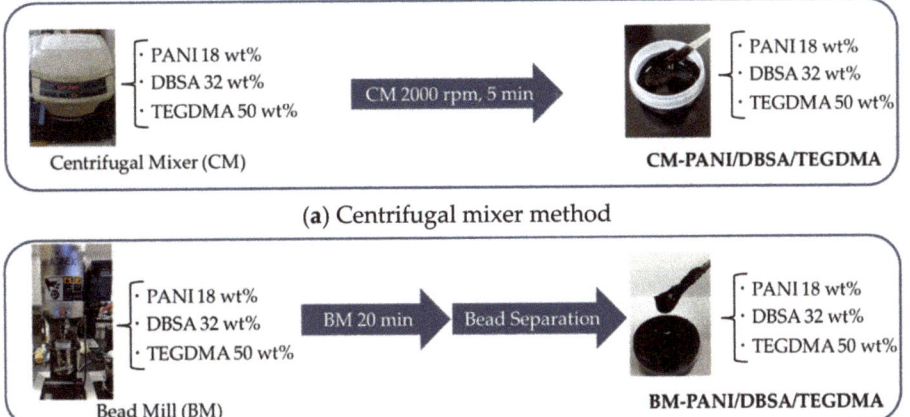

Figure 2. Procedures used to produce PANI/DBSA/TEGDMA composites in conjunction with (**a**) centrifugation and (**b**) bead milling.

(a) Centrifugal mixer method

DBSA and TEGDMA (both liquids) were added to PANI-EB powder to obtain a PANI/DBSA/TEGDMA mass ratio of 18/32/50. The mixture was stirred by hand and then transferred into a bottle and agitated using a centrifugal mixer (ARE-310, THINKY Corp., Tokyo, Japan) at 2000 rpm for 5 min. This procedure generated a liquid paste referred to herein as CM-PANI/DBSA/TEGDMA.

(b) Bead mill method

The DBSA, TEGDMA and PANI-EB were combined in the same manner as described above and stirred first manually and then by a magnetic stirrer for 20 min. The resulting liquid paste was then processed in a bead mill (EasyNano RMB-01 AIMEX Co., Ltd., Tokyo, Japan) at 1000 rpm for 20 min using 160 g of zirconia spheres each having a diameter of 0.5 mm. After milling, the zirconia spheres were removed to give the product, referred to herein as BM-PANI/DBSA/TEGDMA.

(c) Thermal analysis and optical microscopy observations

Doping of the PANI-EB with DBSA in both the CM-PANI/DBSA/TEGDMA and BM-PANI/DBSA/TEGDMA was monitored using differential scanning calorimetry (DSC; DSC Q-200, TA Instruments Japan Inc., Tokyo, Japan) at a heating rate of 5 °C/min from 40 to 190 °C. These trials allowed an evaluation of the effect of the different particle sizes in these specimens. Thermal doping of DBSA into PANI-EB without a solvent is known to reduce the size of PANI particles and aggregates and this effect was evaluated using polarized optical microscopy (BX50, Olympus Corp., Tokyo, Japan) to observe the CM-PANI/DBSA/TEGDMA and BM-PANI/DBSA/TEGDMA while heating the materials from 30 to 150 °C at a rate of 5 °C/min. In situ optical microscopy observations of heated samples were also performed. In these trials, each liquid paste sample was placed between a slide glass and a cover glass to give a constant specimen thickness, after which the material was heated while performing in situ observations.

2.3. Preparation of Composites Including Radical Initiators

Four samples were prepared using combinations of the two different mixing methods and two different peroxides, as summarized in Table 1. In two of these samples, 5 mol% of Peroxide H (based on the moles of TEGDMA in the formulation) was added to the liquid composite. Following this, samples 1 and 2 were processed using the centrifugal mixer at 2000 rpm for 5 min. In the same manner, samples 3 and 4 were made with Peroxide B. Samples 1, 2, 3, and 4 were evaluated by thermal analysis under the same conditions as Section 2.2. (c). Microscopic observations were also evaluated under the same heating conditions, but caution should be exercised as the sample thickness differs from Section 2.2. due to contraction or expansion of the composite during thermosetting.

Table 1. Compositions of test specimens (wt%).

Sample No	PANI	DBSA	TEGDMA	Mixing Method	Peroxide (5 mol% of TEGDMA)
1	18	32	50	Centrifugal mixer (Figure 2a)	Peroxide H
2					Peroxide B
3				Bead mill (Figure 2b)	Peroxide H
4					Peroxide B

2.4. Conductivity Measurements

Samples 1, 2, 3, and 4 prepared in Section 2.3. were heat treated and then measured for conductivity. The liquid composites were poured into metal containers (25 mm in diameter, 1 mm in thickness) and processed using a hot press device (Mini test press, Toyo Seiki Seisaku-sho, Ltd., Tokyo, Japan) at 135 °C and 2 MPa for 1 h to induce radical polymerization. The cured sample produced by hot pressing was allowed to cool to room temperature (approximately 25 °C), and the electrical conductivity was then measured. The conductivity of each specimen was subsequently determined using the four probe method in conjunction with a Loresta-GP MCP-T600 apparatus (Dia Instrument Co., Ltd., Yokohama, Japan). It should be noted that this device actually measured surface resistivity values, from which electrical conductivity data were obtained based on the equation

$$\rho v = \rho s \times t, \quad (1)$$

where ρv is the volume resistivity ($\Omega \cdot cm$), ρs is the surface resistivity (Ω/sq) and t is the specimen thickness (cm) and

$$\sigma = 1/\rho v. \quad (2)$$

To ensure accurate measurements of the conductivity, the surface skin layer on each sample (which comprised an ultrathin layer of non-conductive resin) was carefully removed by hand prior to each measurement.

3. Results

3.1. Effect of the Thermal Doping Temperature

Thermal doping was used to introduce the liquid dopant DBSA into the PANI without using solvents, based on raising the temperature of these materials. The driving force for this doping was the polar interaction between the PANI and DBSA, while the liquid state of the DBSA allowed it to penetrate into the PANI particles. These factors, in turn, were affected by the temperature as well as the PANI particle size and morphology. The latter two parameters could be modified by changing the conditions applied during the polymerization of the PANI. Because of the lack of a solvent system, a temperature increase was required to allow penetration of the liquid dopant, meaning that this was a thermal doping process. This type of system has been frequently reported in the literature [10,11]. It is also known that thermal doping is an endothermic phenomenon that can be followed using DSC.

In the present work, it was important to characterize the thermal doping phenomenon for a simple system without a peroxide initiator. Figure 3 presents the DSC curves obtained from the two liquid PANI/DBSA/TEGDMA composites without peroxide (prepared according to the process detailed in Section 2.2). Figure 3a provides data for the CM-PANI/DBSA/TEGDMA and indicates an exotherm with a peak at 103 °C attributed to the doping of DBSA into the PANI particles [10,11,19]. In contrast, the data in Figure 3b obtained from the BM-PANI/DBSA/TEGDMA sample indicate an exothermic phenomenon at 92 °C. Despite their different locations, the exothermic peaks generated by the two specimens are almost identical. Considering that bead milling produced smaller PANI particles than centrifugal mixing, the decrease in the peak temperature from 103 to 92 °C is attributed to accelerated penetration of the liquid DBSA into the smaller PANI particles.

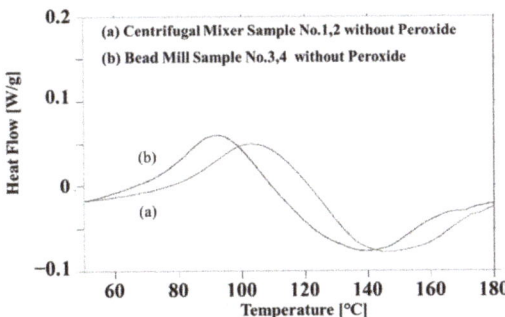

Figure 3. DSC data obtained from "Centrifugal mixer samples 1 and 2 without peroxide" and "Bead mill samples 3 and 4 without peroxide.

Figure 4 presents the resulting images of the PANI/DBSA/TEGDMA samples. Here, the transparent yellow areas are the liquid regions consisting of TEGDMA and DBSA while the black areas are the PANI-EB particles and the greenish areas are the DBSA-doped PANI (that is, PANI in the emeraldine salt state (PANI-ES)). At 40 °C, both the CM-PANI/DBSA/TEGDMA and BM-PANI/DBSA/TEGDMA contained aggregates with sizes in the range of 100–200 μm, although the aggregates in the former were generally larger. Thus, although bead milling reduced the particle size, all materials contained aggregates.

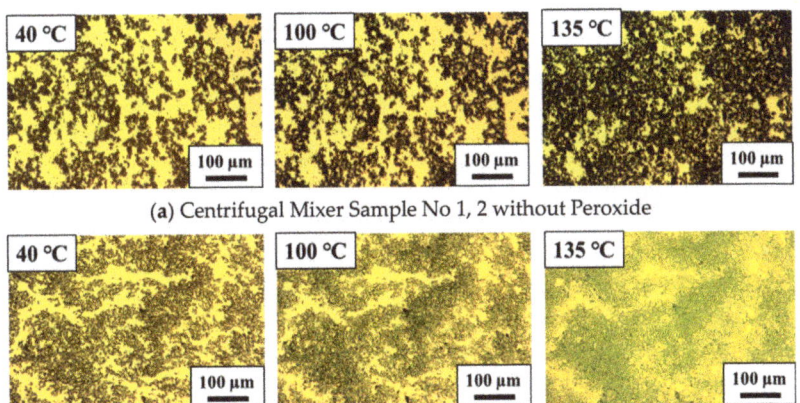

Figure 4. Optical microscopy images of (**a**) centrifugal mixer samples 1 and 2 and (**b**) bead mill samples 3 and 4. All samples were made without a peroxide.

In Figure 4a, the black regions representing initial aggregates in the CM-PANI/DBSA/ TEGDMA are seen to occupy more of the image as the temperature is increased, suggesting that the PANI particles became better dispersed as the DBSA penetrated into the TEGDMA. It should be noted that this behavior did not occur in a mixture composed of only PANI and TEGDMA. In addition, the image acquired at 135 °C indicates that the sample had a slight greenish coloration, suggesting the presence of doped PANI molecules. These images together with the DSC curves confirm that the PANI particles were dispersed during the doping process, such that the aggregates expanded and connected networks were formed. These phenomena would be expected to enhance the conductivity of the material.

Figure 4b presents similar images obtained from the BM-PANI/DBSA/TEGDMA at 40, 100 and 135 °C It is evident that the aggregates became more closely connected with increasing temperature so that the boundaries between particles disappeared. A greenish color appeared at 100 °C corresponding to the peak in the DSC curve for this sample, and became more pronounced at 135 °C Again, this greenish color is attributed to the doping of the PANI molecules.

Figure 5 provides a series of diagrams that summarize the optical microscopy observations. Initially, the CM-PANI/DBSA/TEGDMA and BM-PANI/DBSA/TEGDMA contained aggregates of larger and smaller PANI particles, respectively. At 40 °C, the aggregates were not connected. Increasing the temperature caused these aggregates to increase in size while reducing the sizes of the individual PANI particles, resulting in the connection of the aggregates and inducing electrical conductivity.

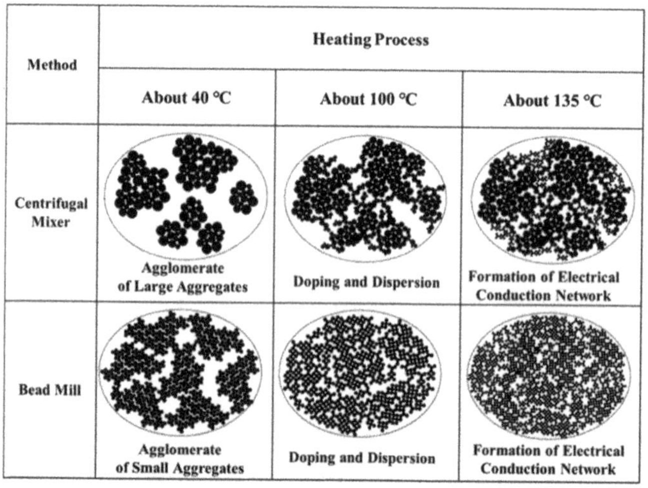

Method	Heating Process		
	About 40 °C	About 100 °C	About 135 °C
Centrifugal Mixer	Agglomerate of Large Aggregates	Doping and Dispersion	Formation of Electrical Conduction Network
Bead Mill	Agglomerate of Small Aggregates	Doping and Dispersion	Formation of Electrical Conduction Network

Figure 5. Diagrams summarizing the effects of the mixing method on thermally-induced dispersion and doping in the PANI/DBSA system.

3.2. Control of the Thermosetting Temperature

Figure 6 summarizes the DSC curves for the four different composites (that is, samples 1 through 4). The exotherms at approximately 100 °C in these plots are ascribed to thermal doping. The exotherms appearing at 150 °C in Figure 6a,b (samples 1 and 3) are related to the radical polymerization of the TEGDMA by Peroxide H. While those at 160 °C in Figure 6a,b (samples 2 and 4) are due to reaction with Peroxide B. Because the hour half-life temperature of Peroxide B is higher than that of Peroxide H, the thermosetting reaction temperatures for samples 2 and 4 were higher than those for samples 1 and 3. These data, therefore, demonstrate the possibility of tuning the thermosetting temperature in the presence of PANI/DBSA by varying the peroxide. Figure 6b shows double exotherm peaks in both curves above 160 °C that corresponds to the radical polymerization, while the exotherms in Figure 6a are single peaks. These results are ascribed to inhibition of the thermosetting reaction based on the trapping of oxygen radicals generated by the peroxides in the smaller particles of PANI produced via bead milling. This effect, in turn, promoted vaporization of the TEGDMA, which has a boiling point of 155 °C. This effect did not occur during radical polymerization of the actual samples at 135 °C.

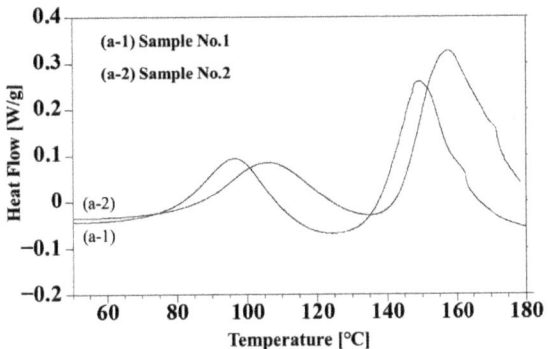

(a) Centrifugal Mixer Sample No 1, 2

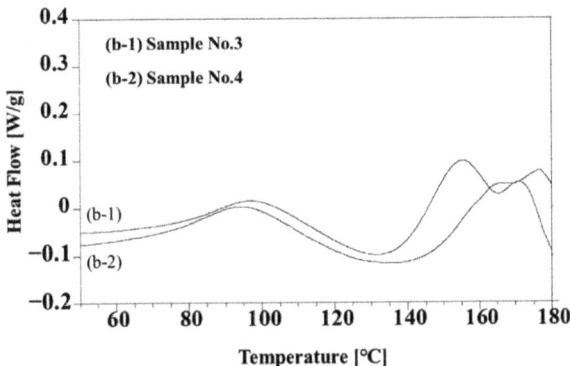

(b) Bead Mill Sample No 3, 4

Figure 6. DSC data obtained from (**a**) Samples 1 and 2 and (**b**) Samples 3 and 4. Samples were processed using peroxides.

3.3. Optimization of Factors Affecting Conductivity

In this work, two key factors (thermal doping as discussed in Section 3.1 and thermosetting as discussed in Section 3.2) were adjusted so as to optimize the conductivity of the proposed new composites. Figure 7 summarizes the conductivity data for the four different composites. Samples 3 and 4, both of which were prepared by bead milling, exhibited higher conductivities, in good agreement with the effective formation of networks of smaller PANI particles. In addition, the conductivity values of Sample No. 2 and No. 4 are approximately an order of magnitude higher than those of Sample No. 1 and No. 3. This difference is attributed to the higher solidification temperature obtained when using Peroxide B. Overall, these data confirm the synergistic effect of decreasing the thermal doping temperature based on bead milling and increasing the thermosetting temperature based on the use of Peroxide B. Optimization of these two factors so as to reduce the thermal doping temperature and increase the radical polymerization temperature provided a high electrical conductivity of 1.45 S/cm.

Electrical Conductivity [S/cm]
1.5
1
0.5
0
0.51
±0.025
0.66
±0.002
1.26
±0.054
1.45
±0.002
Sample No.1
Sample No.3
Sample No.2
Sample No.4

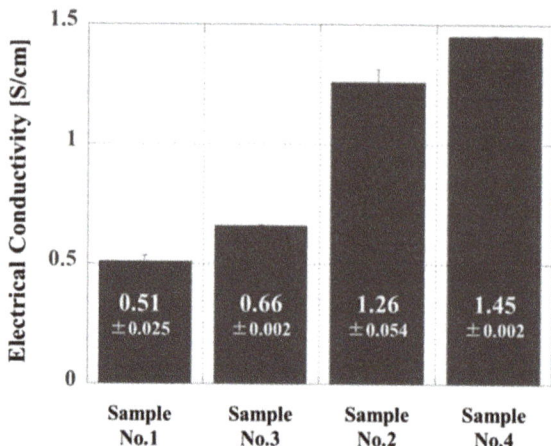

Figure 7. Electrical conductivities of samples 1 and 3 and samples 2 and 4 after heating. The cured sample produced by hot pressing was allowed to cool to room temperature (approximately 25 °C), and the electrical conductivity was then measured.

4. Discussion

Figure 8 summarizes the causes of the large conductivity differences obtained in Section 3.3. It was clear that the thermal doping of PANI/DBSA dominated the conductivity of the composite. Conductivity increases as heat doping progresses, and conductivity decreases as heat doping is hindered. In the case of this radical polymerization system, it was clarified that the thermal decomposition temperature of the peroxide has a great influence on the progress of heat doping.

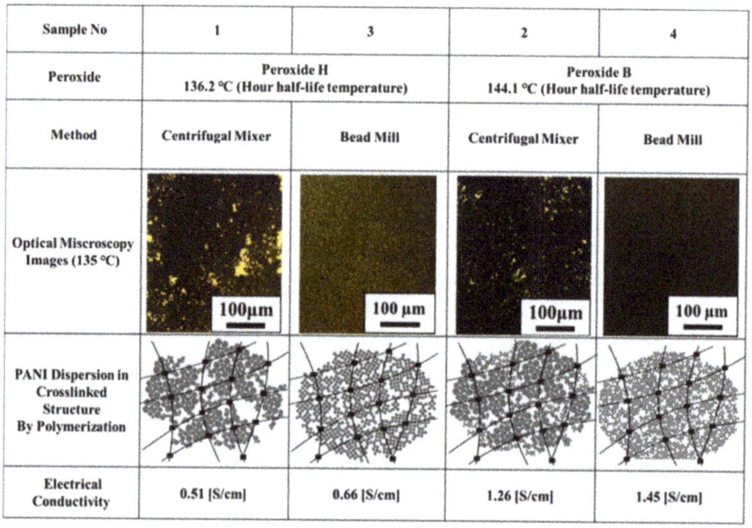

Sample No
1
3
2
4
Peroxide
Peroxide H
136.2 °C (Hour half-life temperature)
Peroxide B
144.1 °C (Hour half-life temperature)
Method
Centrifugal Mixer
Bead Mill
Centrifugal Mixer
Bead Mill
Optical Miscroscopy Images (135 °C)
100μm
100 μm
100μm
100 μm
PANI Dispersion in Crosslinked Structure By Polymerization
Electrical Conductivity
0.51 [S/cm]
0.66 [S/cm]
1.26 [S/cm]
1.45 [S/cm]

Figure 8. Summary of the thermal doping of PANI/DBSA and differences in the thermal decomposition temperatures of the peroxides and effects on conductive network formation.

When peroxide H with a low thermal decomposition temperature was used (Samples 1 and 3), a crosslinked structure was formed by radical polymerization during the heat doping. And it is considered that the change of PANI-ES and the dispersion of PANI were not sufficient, and the electrical conductivity of the composite became low.

When peroxide B, which has a high thermal decomposition temperature, was used (Samples 2 and 4), a crosslinked structure was formed by radical polymerization after the heat doping proceeded. It is considered that sufficient changes in PANI-ES and PANI dispersion occurred, and the electrical conductivity of the composite increased by one order of magnitude. In addition, by lowering the heat doping temperature performed in Section 3.1, even with the same peroxide, the conductivity is higher when the change to PANI-ES and PANI dispersion is more likely to occur.

Since it is found that the thermal decomposition temperature has a greater effect than the heat doping temperature, it is necessary to focus on the investigation of peroxides for further improvement of the conductivity and strength of the composite in the future. [26–29]

5. Conclusions

A new and highly conductive thermosetting resin was developed. This material is a liquid paste at room temperature but undergoes solidification upon heating. This material is made of an electrically conductive polymer produced by thermal doping of PANI with DBSA and a thermosetting resin that undergoes radical polymerization, comprising TEGDMA with a peroxide. Thermal doping of this composite was examined by DSC and in situ optical microscopy in conjunction with processing by centrifugal mixing or bead milling prior to adding a peroxide. Using the bead mill was found to lead to more effective network formation among aggregates that enhanced the conductivity of the material in response to temperature increases. DSC data demonstrated that the temperature associated with thermal doping was reduced after processing by bead milling as a consequence of the more effective penetration of DBSA into smaller PANI particles. The radical polymerization temperature was successfully tuned by varying the peroxide. Optimization of these factors provided a composite with an elevated conductivity of 1.45 S/cm. The high electrical conductivity of this composite is similar to or better than that of PANI-containing thermoset resins used in lightning strike protection CFRP studies [12,13,17,18,21]. Furthermore, in the case of a radical polymerization system, there is an advantage that DBSA and monomers do not react at room temperature and the viscosity does not increase. It can be expected to fabricate larger sizes of CFRP such as aircraft structural materials. However, since the epoxy resin used for CFRP, which is a general structural material, is a cationic polymerizable thermosetting resin, there is a disadvantage that the material of this research cannot be applied immediately. In the future, for the application of CFRP as a structural material, we will investigate the properties including strength and heat resistance as well as the fabrication of electrically conductive CFRP.

Author Contributions: K.T. and K.N. performed the measurements, analyzed experimental data, and wrote the manuscript. T.T. supervised the entire project. M.T. and T.G. participated in comprehensive discussion and provided advice and suggestions. All authors have read and agreed to the published version of the manuscript.

Funding: This research received no external funding.

Institutional Review Board Statement: Not applicable.

Informed Consent Statement: Not applicable.

Data Availability Statement: The raw data presented in this study are available on request from the corresponding author.

Conflicts of Interest: The authors declare no conflict of interest.

References

1. Fisher, F.A.; Plumer, J.A. *Lightning Protection of Aircraft*; 1977 NASA Reference Publication 1008, Document ID 19780003081; US Government Printing Office: Washington, DC, USA, 1977.
2. Louis, M.; Joshi, S.P.; Brockmann, W. An experimental investigation of through-thickness electrical resistivity of CFRP laminates. *Compos. Sci. Technol.* **2001**, *61*, 911–919. [CrossRef]

3. Feraboli, P.; Miller, D. Damage resistance and tolerance of carbon/epoxy composite coupons subjected to simulated lightning strike. *Compos. Part A* **2009**, *40*, 954–967. [CrossRef]
4. Gagne, M.; Therriault, D. Lightning strike protection of composites. *Prog. Aerosp. Sci.* **2014**, *64*, 1–16. [CrossRef]
5. Zhang, D.; Ye, L.; Deng, S.; Zhang, J.; Tang, Y.; Chen, Y. CF/EP composite laminates with carbon black and copper chloride for improved electrical conductivity and interlaminar fracture toughness. *Compos. Sci. Technol.* **2012**, *72*, 412–420. [CrossRef]
6. Kandare, E.; Khatibi, A.A.; Yoo, S.; Wang, R.; Ma, J.; Olivier, P.; Gleizes, N.; Wang, H.C. Improving the through-thickness thermal and electrical conductivity of carbon fibre/epoxy laminates by exploiting synergy between graphene and silver nano-inclusions. *Compos. Part A Appl. Sci. Manuf.* **2015**, *69*, 72–82. [CrossRef]
7. Guo, M.; Yi, X.; Liu, G.; Liu, L. Simultaneously increasing the electrical conductivity and fracture toughness of carbon–fiber composites by using silver nanowires-loaded interleaves. *Compos. Sci. Technol.* **2014**, *97*, 27–33. [CrossRef]
8. Qin, W.; Vautard, F.; Drzal, T.L.; Yu, J. Mechanical and electrical properties of carbon fiber composites with incorporation of graphene nanoplatelets at the fiber–matrix interphase. *Compos. B. Eng.* **2015**, *69*, 335–341. [CrossRef]
9. Yamamoto, N.; Villoria, G.R.; Wardle, B.L. Electrical and thermal property enhancement of fiber-reinforced polymer laminate composites through controlled implementation of multi-walled carbon nanotubes. *Compos. Sci. Technol.* **2012**, *72*, 2009–2015. [CrossRef]
10. MacDiarmid, A.G.; Epstein, A.J. Secondary doping in polyaniline. *Synth. Met.* **1995**, *69*, 85–92. [CrossRef]
11. Han, M.G.; Ki-Cho, S.; Oh, S.G.; Im, S.S. Preparation and characterization of polyaniline nanoparticles synthesized from DBSA micellar solution. *Synth. Met.* **2002**, *126*, 53–60. [CrossRef]
12. Pan, W.; Yang, S.L.; Li, G.; Jiang, J.M. Electrical and structural analysis of conductive polyaniline/polyacrylonitrile composites. *Euro. Polym. J.* **2005**, *41*, 2127–2133. [CrossRef]
13. Jia, W.; Tchoudakov, R.; Segal, E.; Narkis, M.; Siegmann, A. Electrically conductive composites based on epoxy resin with polyaniline-DBSA fllers. *Synth. Met.* **2003**, *132*, 269–278. [CrossRef]
14. Katunin, A.; Krukiewicz, K.; Turczyn, R.; Sul, P.; Lasica, A.; Bilewicz, M. Synthesis and characterization of the electrically conductive polymeric composite for lightning strike protection of aircraft structures. *Compos. Struct.* **2017**, *159*, 773–783. [CrossRef]
15. Katunin, A.; Krukiewicz, K.; Turczyn, R.; Sul, P.; Dragan, K. Lightning strike resistance of an electrically conductive CFRP with a CSA-doped PANI/epoxy matrix. *Compos. Struct.* **2017**, *181*, 203–213. [CrossRef]
16. Kang, Y.; Kim, S.K.; Lee, C. Doping of polyaniline by thermal acid-base exchange reaction. *Mater. Sci. Eng. C* **2004**, *24*, 39–41. [CrossRef]
17. Poussin, D.; Morgan, H.; Foot, J.P. Thermal doping of polyaniline by sulfonic acids. *Polym. Int.* **2003**, *52*, 433–438. [CrossRef]
18. Zilberman, M.; Titelman, I.G.; Siegmann, A.; Haba, Y.; Narkis, M.; Alperstein, D. Conductive Blends of Thermally Dodecylbenzene Sulfonic Acid-Doped Polyaniline with Thermoplastic Polymers. *J. Appl. Polym. Sci.* **1997**, *66*, 243–253. [CrossRef]
19. Levon, K.; Ho, H.K.; Zheng, Y.W.; Laakso, J.; Karana, T.; Taka, T.; Osterholm, E.J. Thermal doping of polyaniline with Dodecylbenzene Sulfonic-Acid without Auxiliary Solvents. *Polymer* **1995**, *36*, 2733–2738. [CrossRef]
20. Goto, T.; Awano, H.; Takahashi, T.; Yonetake, K.; Sukumaran, S.K. Effect of processing temperature on thermal doping of polyaniline without shear. *Polym. Adv. Technol.* **2011**, *22*, 1286–1291. [CrossRef]
21. Kumar, V.; Yokozeki, T.; Goto, T.; Takahashi, T.; Dhakate, S.R.; Singh, B.P. Irreversible tunability of through-thickness electrical conductivity of polyaniline-based CFRP by de-doping. *Compos. Sci. Technol.* **2017**, *152*, 20–26. [CrossRef]
22. Lin, Y.; Gigliotti, M.; Lafarie-Frenot, M.C.; Bai, J.; Marchand, D.; Mellier, D. Experimental study to assess the effect of carbon nanotube addition on the through-thickness electrical conductivity of CFRP laminates for aircraft applications. *Compos. Part B* **2015**, *76*, 31–37. [CrossRef]
23. Kumar, V.; Sharma, S.; Pathak, A.; Singh, B.P.; Dhakate, S.R.; Yokozeki, T.; Okada, T.; Ogasawara, T. Interleaved MCNT buckypaper between CFRP laminates to improve through-thickness electrical conductivity and reducing lightning strike damage. *Compos. Struct.* **2019**, *210*, 581–589. [CrossRef]
24. Kumar, V.; Yokozeki, T.; Karch, C.; Hassen, A.A.; Hershey, C.J.; Kim, S.; Lindahl, J.M.; Barnes, A.; Bandari, Y.K.; Kunc, V. Factors affecting direct lightning strike damage to fiber reinforced composites: A review. *Compos. Part B* **2020**, *183*, 107688. [CrossRef]
25. Kumar, V.; Manomaisantiphap, S.; Takahashi, K.; Goto, T.; Tsushima, N.; Takahashi, T.; Yokozeki, T. Cationic scavenging by polyaniline: Boon or bane from synthesis point of view of its nanocomposites. *Polymer* **2018**, *149*, 169–177. [CrossRef]
26. Pati, S.; Kumar, V.; Goto, T.; Takahashi, T.; Yokozeki, T. Introducing a curable dopant with methacrylate functionality for polyaniline based composites. *Polym. Test.* **2019**, *73*, 171–177. [CrossRef]
27. Pati, S.; Das, S.; Goto, T.; Takahashi, T.; Yokozeki, T. Development of conductive CFRPs using PANI-P-2M thermoset polymer matrix. *Indian J. Eng. Mater. Sci.* **2020**, *27*, 1067–1070.
28. Nanda, A.V.; Raya, S.; Easteala, A.J.; Waterhouse, G.I.N.; Gizdavic-Nikolaidisa, M.; Cooneya, R.P.; Travas-Sejdica, J.; Kilmartina, P.A. Factors affecting the radical scavenging activity of polyaniline. *Synth. Met.* **2001**, *161*, 1232–1237. [CrossRef]
29. Brown, S.C.; Robert, C.; Koutsos, V.; Ray, D. Methods of modifying through-thickness electrical conductivity of CFRP for use in structural health monitoring, and its effect on mechanical properties A review. *Compos. Part A Appl. Sci. Manuf.* **2020**, *133*, 105885. [CrossRef]

Article

Micronized Recycle Rubber Particles Modified Multifunctional Polymer Composites: Application to Ultrasonic Materials Engineering

Vicente Genovés [1], María Dolores Fariñas [1], Roberto Pérez-Aparicio [2], Leticia Saiz-Rodríguez [2], Juan López Valentín [3] and Tomás Gómez Álvarez-Arenas [1,*]

1. Instituto de Tecnologías Físicas y de la Información, Spanish National Research Council (CSIC), 28006 Madrid, Spain
2. Signus Ecovalor, S.L., 28033 Madrid, Spain
3. Instituto de Ciencia y Tecnología de Polímeros, Spanish National Research Council (CSIC), 28006 Madrid, Spain
* Correspondence: t.gomez@csic.es

Citation: Genovés, V.; Fariñas, M.D.; Pérez-Aparicio, R.; Saiz-Rodríguez, L.; Valentín, J.L.; Álvarez-Arenas, T.G. Micronized Recycle Rubber Particles Modified Multifunctional Polymer Composites: Application to Ultrasonic Materials Engineering. *Polymers* **2022**, *14*, 3614. https://doi.org/10.3390/polym14173614

Academic Editors: Sivanjineyulu Veluri and Tomasz Makowski

Received: 15 August 2022
Accepted: 27 August 2022
Published: 1 September 2022

Publisher's Note: MDPI stays neutral with regard to jurisdictional claims in published maps and institutional affiliations.

Copyright: © 2022 by the authors. Licensee MDPI, Basel, Switzerland. This article is an open access article distributed under the terms and conditions of the Creative Commons Attribution (CC BY) license (https://creativecommons.org/licenses/by/4.0/).

Abstract: There is a growing interest in multifunctional composites and in the identification of novel applications for recycled materials. In this work, the design and fabrication of multiple particle-loaded polymer composites, including micronized rubber from end-of-life tires, is studied. The integration of these composites as part of ultrasonic transducers can further expand the functionality of the piezoelectric material in the transducer in terms of sensitivity, bandwidth, ringing and axial resolution and help to facilitate the fabrication and use of phantoms for echography. The adopted approach is a multiphase and multiscale one, based on a polymeric matrix with a load of recycled rubber and tungsten powders. A fabrication procedure, compatible with transducer manufacturing, is proposed and successfully used. We also proposed a modelling approach to calculate the complex elastic modulus, the ultrasonic damping and to evaluate the relative influence of particle scattering. It is concluded that it is possible to obtain materials with acoustic impedance in the range 2.35–15.6 MRayl, ultrasound velocity in the range 790–2570 m/s, attenuation at 3 MHz, from 0.96 up to 27 dB/mm with a variation of the attenuation with the frequency following a power law with exponent in the range 1.2–3.2. These ranges of values permit us to obtain most of the material properties demanded in ultrasonic engineering.

Keywords: multifunctional composites; particle loaded polymers; ultrasonic transducers; ultrasonic materials; damping in composite materials; complex elastic moduli of composites; ultrasonic testing; recycled end-of-life tires rubber

1. Introduction

Two-phase composites made of a polymeric matrix reinforced with particulate matter are widely used in quite diverse fields and applications as the addition of different filler materials and filler volume fractions has proved to be an efficient way to tailor optical, acoustical, mechanical, dielectric, magnetic and thermodynamic properties [1]. A wide variety of particulate fillers has been used [1]: metals (tungsten, iron, etc.), semimetals (silicon), oxides (alumina, cerium oxide, silicone oxide, titanium oxide, zirconia, etc.), ceramics (barium strontium titanate, PZT, silicon carbide, etc.), polymers (elastomers, rubbers [2,3], resins, etc.) and other more complex materials like different recycled materials (carbon black, fly ash, toner waste, bio-agricultural waste, and recycled rubber granulates and powders from end-of-life tires [4–7]. There is also growing interest and applications for the case of nanoparticle-loaded polymers (see [8] for a review).

Applications of particle-loaded polymers are as numerous as the different types of materials mentioned above [9]. Examples can be found in the design of high damping materials [10], in the construction industry and in the fabrication of precision machine tools [5], in the fabrication of electronic components for encapsulation and EMI shielding [11] and in aerospace applications or for automobiles, ships, and different electronic devices (where these composites are used to improve brittleness and impact strength of epoxy resins [3]). Rubber-reinforced composites have been used for improving mechanical properties in applications such as sound absorbing panels and impact protection slabs [3]. Hollow glass spheres are used to increase thermal and acoustic insulation of paintings and coatings, and to produce buoyancy materials [12]. Recycled rubber granulates and powders from end-of-life tires (ELT) also have well-established applications in the industry such as sport surfaces, road safety elements, modification of bituminous mixtures, insulation, brake pads or even for the manufacture of new tires. However, new applications for ELT-derived products are still demanded [13,14]. Recent developments in recycling processes open up new perspectives for ELT rubber. In this way, cryogenic milling allows reducing the particle size to about 100 μm, then, the resultant micronized rubber can be used in high added value applications.

In the ultrasonic field, these composites are normally called 0-3 connectivity composites [15], or 0-3/3-3 in the case of composites with very high particle volume fraction [16,17]. They are used in two different applications: (i) as components of piezoelectric transducers, either as active (piezoelectric) materials [18], or as passive materials (mainly backing blocks and matching layers [19]) to further extend the functionalities of the piezoelectric component (in terms of bandwidth, resolution and sensitivity), and (ii) to produce phantoms of human tissues for testing of ultrasonic image systems and procedures for medical applications. The design and manufacture of active 0-3 connectivity piezocomposite materials for ultrasonic transducers have been widely investigated [20–23]. This allowed achieving lower impedance piezoelectric composites, and ultrasonic transducers with larger bandwidth, or higher center frequency [24], as well as piezoelectric paints [25] and cements [26]. Passive 0-3 connectivity composites have been largely used to produce damping backing blocks [27–34] or matching layers, where the use of nanoparticles allow for the possibility to make matching layers for high frequency applications [35,36]. Sometimes, the use of two different types of particulate matter has also been reported [29,37]. Another use of passive 0-3 connectivity composites is to produce human tissue phantoms (with reduced cost, longer lifetime, and able to replicate more complex anatomical structures) that contribute to facilitated advances in ultrasonic diagnostic and therapeutic strategies [38,39].

In general, in these types of applications, it is necessary to precisely control the acoustic impedance of the composite, the ultrasonic damping and the contribution of the scattering. In the case of backing blocks, a large impedance is normally required (so that the vibration in the piezoelectric element can be damped and hence the bandwidth of the transducer enlarged, and the axial resolution improved), while a large attenuation and a very reduced backscattering is required to reduce noise in the transducer and eliminate any echo from the backing block back surface. In the case of impedance matching layers, impedance must be precisely tuned to an intermediate value between that of the piezoelectric component and the insonicated medium, while attenuation has to be minimized. Finally, for tissue mimicking phantoms, it is necessary to tune the impedance, the attenuation coefficient and its variation with the frequency and the backscattering of the material to that of the tissues. A similar problem is found in the design of damping composite materials, where two functionalities are involved: the load capability and the damping efficiency, and where the proposed figure of merit is the product $|E^*|\tan\delta$, where $|E^*|$ is the modulus of the complex elastic modulus (E^*) and $\tan\delta = Im(E^*)/Real(E^*)$ [10].

Several strategies have been attempted to increase the damping, to control the scattering in composite materials at ultrasonic frequencies and to decoupled damping and impedance modifications. They can be classified in three main groups depending on the composite component that is modified to obtain the desired properties: (i) the particles, (ii) the matrix, and iii) both. Lutsch (1962) [27] introduced large rubber particles achieving a moderate attenuation (up to 0.8 dB/mm at 1 MHz), with intermediate acoustic impedance

(~8 MRayl). Similarly, Ju-Zhen [29], added cerium oxide particles to tungsten-loaded epoxy to increase attenuation. Scattering by the filler particles in the backing block can be enhanced by increasing the particle concentration or the mean particle size or the impedance mismatch between the particles and the matrix, but this must be done carefully as the backscattering can be an unacceptable source of noise. Cho et al. [40] proposed a fabrication process that permits the increase of the load of particles, achieving attenuation values between 3 to 5 dB/mm at 3 MHz, with acoustic impedance between 4 and 4.6 MRayl.

Increase of the composite damping by using more attenuating polymer matrices can be achieved by either using a different polymer or by modifying the properties of the selected one (by adding a plasticizer, by blending different formulations, or by lowering the cross-link density [3]). In this sense, State et al. [37] achieved attenuation values between 35 and 40 dB/mm at 8 MHz, and an almost linear variation with frequency, for tungsten and alumina-loaded polyurethanes (compared with reference values of about 19 dB/mm at 8 MHz for similar alumina-loaded epoxy composites), with impedance of about 2.6–3.3 MRayl.

El-Tantawy and Sung [41] tried a combined approach: large Ti particles, and a modification of the polymer by adding a plasticizer (glycerol) and a coupling agent (silane). Achieved attenuation values were between 2.2–3.9 dB/mm at 3 MHz, with acoustic impedances between 2 and 7.8 MRayl. Another approach consists of adding liquid rubber [42,43] that has also been used to reinforce the brittle character of thermosetting epoxies [3]. There are two different ways for the modification of the composite properties by adding liquid rubber. In the first case, rubber-epoxy separation occurs during the epoxy solidification giving rise to rubber domains, mainly as a result of the decrease in configurational entropy due to the increase in molecular weight as the epoxy cures [3]. In the second case, this separation does not take place [42]. In the case where rubber domains appear in the composite, impedance decreases (from 11.7 MRayl to 8.7 MRayl), however, attenuation dramatically increases, from 1 dB/mm to 6 dB/mm at 2 MHz. The advantages are that the viscosity increase of the epoxy + rubber mixture is moderate; this allows for a high load of particles and the obtained attenuation values are very high.

In this paper we analyze the possibility of tailoring, in a decoupled way, the load capability (determined by the composite impedance), the ultrasonic attenuation coefficient, its variation with the frequency, and the contribution of scattering of particle polymers by using two different types of particles: small particles of heavy metal (intended to composite impedance, with reduced scattering and increase EMI shielding) and micronized rubber powder from ELTs (intended to increase attenuation with reduced scattering and reduced impedance modification). A fabrication route compatible with transducer manufacturing is also proposed as this can be one of the applications of these composites: to enhance the functionality of the piezoelectric layer in piezoelectric transducers (in terms of bandwidth, resolution and sensitivity). This involves avoiding high temperatures and pressures in the fabrication of the composites that may compromise the piezoelectric response of the active component of the transducers. Finally, a modelling approach to calculate the complex elastic constants of the composites, the variation of the attenuation with the frequency and the relative contribution of the scattering is also proposed.

2. Materials and Methods

2.1. Materials

2.1.1. Raw Materials

The employed raw materials are summarized in Table 1.

Disk samples with a 30 mm diameter and thickness between 1.7 and 4 mm (Table 2) were manufactured following the procedure described in Section 2.2.2. In addition, a rubber plate sample was also produced by press molding of the ELT rubber powder at 160 °C and 200 bar for 30 min. Density is worked out from weight and size (disk diameter and thickness) measurements.

Table 1. Raw materials.

Material Description	Commercial Name	Supplier	Other Info
Epoxy resin	EpoxAmite 101	SmoothOn (Macungie, PA, USA)	Viscosity: 1000 cps Pot life: 11 min
Polyurethane resin	Urebond	SmoothOn (PA, USA)	Viscosity: 5400 cps Pot life: 5 min
Polyurethane resin	Liquid Plastic	SmoothOn (PA USA)	Viscosity: 80 cps Pot life: 3 min
Tungsten powder	–	Alfa Aesar (Haverhill, MA, USA)	Tungsten 99.9% Particle size 12 μm
Cryogenic rubber powder	MicroDyne 75-TR	Lehigh Technologies (Atlanta, GA, USA)	Particle size < 70 μm (90% wt has a smaller size than 70 μm)

Table 2. Disk samples (30 mm diameter) of raw materials for ultrasonic characterization.

Sample Denomination	Material	Thickness (mm)
Ep-00	EpoxAmite	4.25 ± 0.13
PU1-00	Urebond	4.82 ± 0.15
PU2-00	Liquid Plastic	4.45 ± 0.13
Rb-00	Rubber	2.73 ± 0.14

The properties of these materials are shown in Table 3; in this case, the method to measure these properties is the one used to measure the composite samples. This method is described in Section 2.2.2. To account for the variation of the ultrasonic attenuation (α) with the frequency (f), we used a power law: $\alpha = \alpha_0 (f/f_0)^n$. The exponent "$n$" is also listed in Table 3.

Table 3. Ultrasonic properties of the raw materials.

Sample	Density (kg/m³)	Ultrasound Velocity (m/s)	α @ 3 MHz (dB/mm)	n
Ep-00	1150 ± 33	2620 ± 75	0.69 ± 0.02	0.90
PU1-00	1125 ± 32	1700 ± 50	1.34 ± 0.04	1.13
PU2-00	1130 ± 30	2195 ± 66	2.66 ± 0.08	1.11
Rb-00	1100 ± 50	1200 ± 60	16.43 ± 0.80	0.79

Given the novelty in the use of recycled rubber from ELT for ultrasonic applications and the lack and variability of ultrasonic data (in particular for "n") for polymers, Figure 1 shows the measured attenuation versus frequency and the fitting with the proposed power law that permits the estimation of the exponent "n".

Finally, Table 4 summarizes the main properties of the tungsten powder.

Table 4. Properties of the Tungsten powder.

Property	Value
Longitudinal velocity (m/s)	5200
Bulk modulus (GPa)	305
E (GPa)	414
Longitudinal attenuation @ 5 MHz (dB/m)	66.5
"n"	1.0
Shear Modulus (GPa)	162.3
Density (kg/m³)	19,300

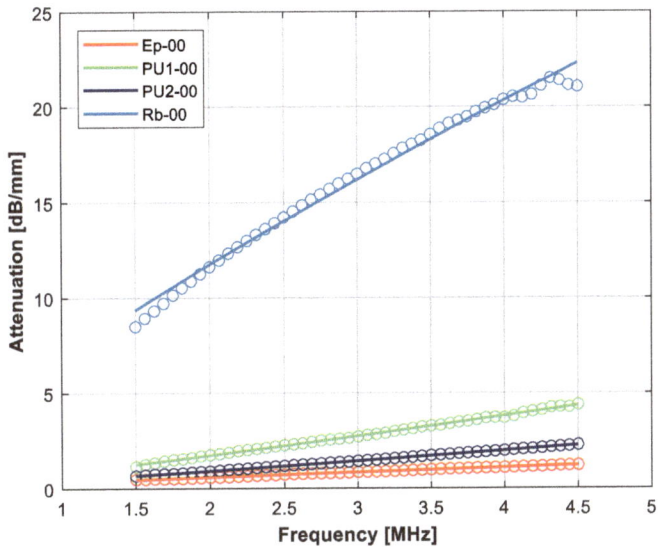

Figure 1. Variation in the ultrasonic attenuation coefficient with the frequency for the raw materials. Open circles: measurements. Solid line: fitting using a power law.

2.1.2. Composite Material Samples

Following the methods described in Section 2.2.1, several samples were produced (see Tables 5–8). The mass fraction of each component used in the mixture (μ_i) is defined a (Equation (1)):

$$\mu_i = m_i / \sum_{i=1}^{N} m_i, \tag{1}$$

where m_i is the mass of each component in the mixture. This mass fraction data together with the density of each component (ρ_i) is used to calculate the volume fraction of each component (Equation (2)):

$$\phi_i = (\mu_i / \rho_i) / \sum_{i=1}^{N} \mu_i / \rho_i, \tag{2}$$

Table 5. Rubber-powder-loaded epoxy resin composites.

Sample Denomination	Rubber Volume Fraction (%)	Sample Thickness (mm)	Density Deviation (%)
Ep-R-02	2	4.5 ± 0.13	−0.32
Ep-R-04	4	4.1 ± 0.12	−0.39
Ep-R-10	10	3.9 ± 0.12	−2.32
Ep-R-15	15	4.7 ± 0.14	−2.49
Ep-R-20	20	4.2 ± 0.13	−4.39
Ep-R-30	30	3.9 ± 0.12	−1.92
Ep-R-35	35	4.1 ± 0.12	−0.32

Hence, we calculate the nominal density of the final composite (ρ^*_{comp}) (Equation (3)):

$$\rho^*_{comp} = \sum_{i=1}^{N} \phi_i \rho_i, \tag{3}$$

Table 6. Tungsten powder + rubber-powder-loaded epoxy resin composites.

	Sample Denomination	Tungsten/Rubber Volume Fraction (%)	Sample Thickness (mm)	Density Deviation (%)
Low W load	Ep-WR-14-7	14.1/7.3	2.96 ± 0.14	−2.44
	Ep-WR-13-13	13.9/13.8	2.82 ± 0.16	−3.84
	Ep-WR-16-16	15.9/16.1	2.67 ± 0.15	−1.73
	Ep-WR-15-27	15.2/26.6	3.27 ± 0.18	−2.93
	Ep-WR-15-37	15.2/37.8	4.35 ± 0.27	−4.35
High W load	Ep-WR-48-03	48/3.2	3.8 ± 0.24	−1.54
	Ep-WR-47-04	47.1/4.4	3.2 ± 0.21	−2.38
	Ep-WR-41-08	41.1/8.3	1.8 ± 0.15	−1.62
	Ep-WR-38-12	38.1/12.5	4.0 ± 0.26	−1.97

Table 7. Samples of tungsten-loaded epoxy resin composites.

Sample Denomination	Tungsten Volume Fraction (%)	Sample Thickness (mm)	Density Deviation (%)
Ep-W-16	16.0	2.4 ± 0.12	−5.32
Ep-W-52	52.2	2.6 ± 0.13	2.81
Ep-W-53	53.2	2.7 ± 0.13	−3.55

Table 8. Samples of tungsten-powder-loaded polyurethane composites.

Sample Denomination	Tungsten Volume Fraction (%)	Sample Thickness (mm)	Density Deviation (%)
PU1-W-45	45.9	1.88 ± 0.11	−7.5
PU1-W-37	37.0	3.9 ± 0.23	1.7
PU2-W-52	51.7	2.63 ± 0.16	−11.7
PU2-W-37	37.0	4.8 ± 0.29	−3.0

This nominal density is compared with the actual density (ρ_{comp}) of the fabricated samples, which is obtained from the sample dimensions (diameter and thickness) and mass. From this comparison we work out the density deviation, (Equation (4)):

$$\text{Density deviation (\%)} = \frac{\rho_{comp} - \rho^*_{comp}}{\rho^*_{comp}} \times 100, \quad (4)$$

Table 5 describes the fabricated rubber-powder-loaded epoxy resin samples. The purpose of this series is to confirm the capability to efficiently fabricate these samples with the proposed method, to verify that the attenuation coefficient increases with the rubber load while scattering is not significantly increased, and to check the efficiency of the proposed model to predict the attenuation in these composites. The volume fraction ranges from 2% up to 35%. Just for comparison purposes, it can be mentioned that the rubber concentration values employed for epoxy toughening are, normally, 5–25 wt% [3]. The thickness of the sample has no relation with the composition and depends on the mold used (we used different molds with the same diameter but different height) and the degree of polishing.

Table 6 summarizes the fabricated tungsten + rubber-loaded epoxy samples. Two series of samples have been produced, with low and high tungsten load, respectively. The first series is intended to obtain composites with acoustic impedance around 5 MRayl. In this case, the load of tungsten particles is reduced (about 15%), hence it is possible to add a relatively large amount of rubber load (up to 40%). The second series is intended to obtain composites with acoustic impedance close to 17 MRayl. In this case, the load of tungsten particles is larger (38–50%) so that it is only possible to add a more reduced amount of rubber load (up to 12%). The limit of the maximum particle load is determined by the epoxy viscosity, the pot life, the particle size and shape and the mixing technique.

For comparison purposes, some samples of tungsten-loaded epoxy resin and tungsten-loaded polyurethane were also fabricated and tested. These samples are described in Tables 7 and 8.

2.1.3. Equipment Used

Samples were weighed using a precision analytical laboratory balance, Nahita Blue, diameter of the samples was measured using a caliber and the thickness was measured using a micrometer (Mitutoyo, Spain). Samples were polished using a Saphir 250 A1-ECO polisher (Neurtek, Madrid, Spain) and post cured in a JP-Selecta oven (Barcelona, Spain). Mixing was performed using a Hauschild high speed orbital mixer (Haushild, Hamm, Germany). For the ultrasonic measurements, an ultrasonics, pulser-receiver: DPR300 (JSR Ultrasonics, Pittsford, NY, USA), an oscilloscope (Tektronix, DPO5054, Tektronix, Beaverton, OR, USA), and one pair of flat water immersion wide band transducers centered at 3.5 MHz (Olympus V383-SU, 3.5 MHz, Olympus, Allentown, PA, USA) were used.

2.2. Experimental Methods

2.2.1. Fabrication of Composite Samples

The key elements in order to establish the fabrication route for these composites are: (i) the capability to produce from moderate to highly loaded composites with a good mixture of components, complete curing of the sample, good adhesion between matrix and particles, and without internal cavities or air trapped; (ii) the compatibility of this process with the transducer manufacturing process (this means to avoid high temperatures, pressures and use of solvents that could affect the piezoelectric material and the electrical connections at the piezoelement and to ensure an interface between the piezo and the composite free of defects and with good adhesion).

Conventional approaches using vacuum degassing or ultrasonic cavitation baths are, in most cases, not efficient with very high particle loads as the viscosity of the mixture becomes very large. In this paper, we propose to fabricate the composite materials by simultaneously mixing and degassing the mixture using a high-speed orbital mixer. The mixing/degassing protocol took place during 3 mins with two stages; first stage with a velocity of 1800 rpm for 2 mins and then, a second stage where velocity was increased (10 s ramp) up to 2500 rpm for the rest of the process. This process was revealed to be fully compatible with transducer manufacturing (i.e., direct fabrication of the composite on the piezoelectric element), is efficient in achieving high particle loads without trapped air, and has good mixing, good homogeneity, and no curing problems. In addition, as the processing times are reduced (3 mins), this also allows for the use of polymers with reduced pot life that cannot be used when large degassing times (as in vacuum degassing) are required.

First, component A of the polymer is mixed with the desired particle load. The rubber is first added and mixed/degassed, then the tungsten is added and then mixed/degassed again. After this, component B of the polymer is added, and mixed and degassed. After mixing, samples were cured for 24 hours at 25 °C in a cylindrical mold, then demolded and post cured for 1 h at 80 °C in a JP-Selecta oven (JPS, Barcelona, Spain). To avoid any manipulation of the sample after mixing, the cylindrical mold is the same recipient that was used for mixing. Once the samples cooled down, they were demolded and polished, using an automatic Saphir 250 A1-ECO polisher (Neurtek, Madrid, Spain), to achieve uniform, flat and parallel surfaces. Thickness variation for each sample was kept under 200 μm. Polished surfaces of the composites were observed with an optical microscope to verify the dispersion of the fillers. The fabrication route is also shown in Figure 2.

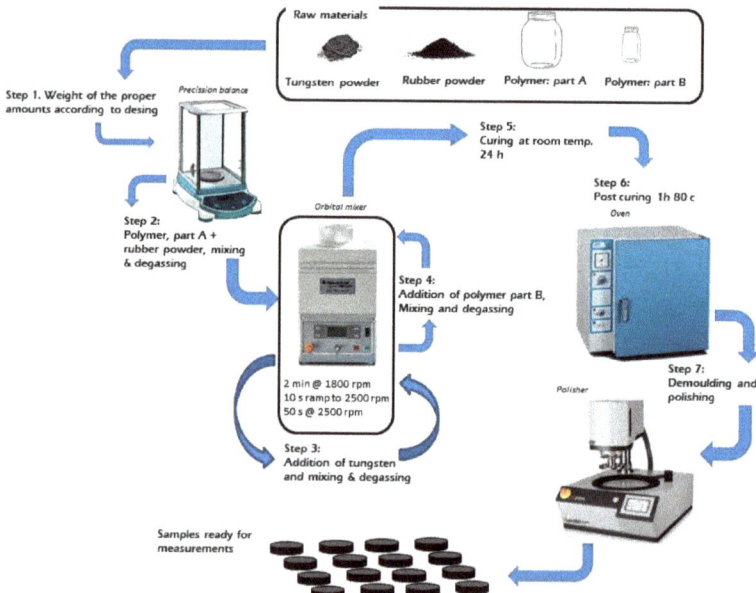

Figure 2. Schematic diagram of the fabrication procedure.

As a first and simple verification of the fabrication process, the density of the samples was measured and compared with the nominal density estimated from the amount of the different materials added to the mixer. In addition, the measured ultrasonic velocity and attenuation and comparison with theoretically predicted values are also used to detect problems like poor compatibility between filler and matrix, the presence of trapped air or lack of homogeneity. To test the homogeneity of the samples, in a few cases, thicker disks were fabricated and cut into two thinner disks that were polished and measured to determine if there is any gradient of properties. No significant differences were observed. In addition, no curing problems were observed in any of the cases presented here. All fabricated samples could be machined and polished.

2.2.2. Ultrasonic Measurements

Ultrasonic measurements were performed using a pulser-receiver (JSR Ultrasonics, DPR300, Pittsford, NY, USA), an oscilloscope (Tektronix, DPO5054, Tektronix, Beaverton, OR, USA), one pair of flat-water immersion wide band transducers centered at 3.5 MHz (Olympus V383-SU, 3.5 MHz Olympus, Allentown, PA, USA) and a custom made tank (45 × 45 × 120 mm^3 PMMA) (see Figure 3) that easily allows for transducers and sample positioning.

All measurements were performed in distilled and degassed water at 22 °C. Samples were first immersed in water and vacuum degassed to ensure that no air bubble is trapped on the sample surface, as this will strongly affect the measured attenuation. Then, temperature was stabilized at 22 °C and measurements performed. All measurements were performed at normal incidence. The through transmitted signal without a sample in between is used as reference or calibration. Then, the sample is put in between the transducers. Acquired signals were transferred to MATLAB where Fast Fourier Transform (FFT) was extracted to obtain the magnitude and phase spectra of the transmission coefficient. Samples were measured at several points (up to 5) to verify the homogeneity of the samples. Phase spectra allow the determination of the velocity in the sample if sample thickness and velocity in the water are known [44,45], while magnitude spectra permit the determination of attenuation and variation of the attenuation with the frequency if impedance of the water and attenuation in the water are known [45]. In addition, the variation in the attenuation coefficient with the frequency is quantified using a power law (Equation (5))

$$\alpha = \alpha_0 \left(\frac{f}{f_0}\right)^n, \quad (5)$$

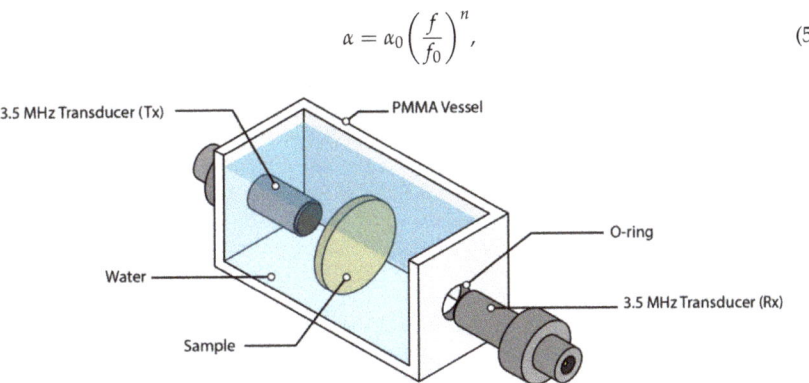

Figure 3. Ultrasonic water-immersion through-transmission layout.

2.2.3. Other Measurements

Weight (precision analytical laboratory balance, Nahita Blue), diameter (caliber) and thickness (micrometer, Mitutoyo) of all samples were measured and material density worked out. As the density of a composite material (ρ_{comp}) can also be obtained from the density of its i-constituents (ρ_i) and their volumetric fraction (ϕ_i) (Equation (2)), the agreement between measured and calculated density can be used to verify that the proper proportions of components were effectively added, their correct mixture during fabrication and the lack of any trapped air. In this comparison, it is considered that the main source of error is the variability of the sample thickness.

2.3. Theoretical Methods: Moelling Composite Properties

The purpose of the composite modelling is to make possible the prediction of the properties of the composites when the properties of the constituent materials and the volume fraction of the components are known. In this particular case, we are interested in the capability to predict the composite impedance, the ultrasonic attenuation coefficient and its variation with the frequency. So far, efforts have been focused on models that permit the calculation of the elastic moduli of two-phase composites from the elastic moduli of constituent materials and the volume fraction. With these moduli and the composite effective density, it is possible to work out the ultrasound velocity and the acoustic impedance. Reviews of the different modelling approaches can be seen in [16,22,42,46]. The simplest approach is the use of averaging models or mixture rules like the well-known Voigt and Reuss models (or series and parallel model). The main advantage is that these models are very simple and provide predictions for the entire volume fraction range. They provide an upper and lower limit for the composite properties depending on how the two phases are distributed in the space. The main drawback of these mixture rules is that when properties of the two constituent phases are very different (for example epoxy resin and tungsten), the upper and lower bounds are too separated such that they have no predictive value. The approach of Hashin and Shtrikman provides an improvement as upper and lower bounds are closer [46,47]. However, when composite components are too different, the separation between these two limits is still too big and they may have little predictive value [17,46].

According to the Hashin-Shtrikman (HS) approach [47] for a two-phase composite, the modulus of compressibility (K_{comp}) and stiffness (G_{comp}) of the composite are given by (Equations (6) and (7)):

$$K_{comp}^{L,U} = K_{1,2} + \frac{v_{2,1}}{1/(K_{2,1} - K_{1,2}) + 3v_{1,2}/(3K_{1,2} + 4G_{1,2})'} \quad (6)$$

and

$$G_{comp}^{L,U} = G_{1,2} + \frac{v_{2,1}}{1/(G_{2,1} - G_{1,2}) + 6(K_{1,2} + 2G_{1,2})v_{1,2}/[5G_{1,2}(3K_{1,2} + 4G_{1,2})]}, \quad (7)$$

where the subscripts 1 and 2 refer to the two components and v is the volume fraction, $K_2 > K_1$ and $G_2 > G_1$ and superscripts L and U stand for the Upper and the Lower limits of the Hashin-Strikman model. In particular, L is the exact solution for the composite made of matrix of phase "one" material in which spherical inclusions of phase "two" material are distributed in a particular way. In addition, U is the exact solution for matrix of phase "two" material in which spherical inclusions of phase "one" material are distributed in a particular way.

The coherent potential approximation (CPA) model is based on the scattering theory. This model predicts the modulus of compressibility (K_{comp}) and stiffness (G_{comp}) of two phase composites, assuming that the inclusions are spherical, the wavelengths are much longer than the size of the inclusions, and multiple scattering effects are negligible (Equations (8)–(10)):

$$\frac{1}{K_{comp} + \frac{4}{3}G_{comp}} = \frac{v_1}{K_1 + \frac{4}{3}G_{comp}} + \frac{v_2}{K_2 + \frac{4}{3}G_{comp}}, \quad (8)$$

$$\frac{1}{G_{comp} + F} = \frac{v_1}{G_1 + F} + \frac{v_2}{G_2 + F}, \quad (9)$$

$$F = \frac{G_{comp}}{6} \frac{9K_{comp} + 8G_{comp}}{K_{comp} + 2G_{comp}}, \quad (10)$$

where subscripts 1 and 2 refer to the matrix and the inclusions, respectively, and v is the volume fraction.

In order to predict both the attenuation coefficient and the acoustic impedance of two-phase composites we make use of the correspondence principle of viscoelasticity, where complex elastic modulus (K^*, G^*) are used to take into account the dynamic damping in the material (Equation (11)):

$$K \rightarrow K^*, \quad G \rightarrow G^*, \quad (11)$$

Complex wave number for longitudinal and shear waves is defined from the angular frequency (ω), and the attenuation coefficient ($\alpha_{L,S}$), (Equation (12)):

$$k_{L,S}^* = \frac{\omega}{v_{L,S}} - i\alpha_{L,S}, \quad (12)$$

where the subscript L and S denote longitudinal and shear wave, respectively. The complex wave velocity ($v_{L,S}^*$) is given then by (Equation (13)):

$$v_{L,S}^* = \frac{\omega}{k_{L,S}^*}, \quad (13)$$

and complex elastic moduli (K^*, G^*) are obtained from (Equation (14))

$$K^* + \frac{4}{3}G^* = (v_L^*)^2 \rho_{comp}, G^* = (v_S^*)^2 \rho_{comp}, \quad (14)$$

These complex elastic moduli of the composite components can be used with Equations (6) and (7) or (8)–(10) to calculate the complex elastic moduli of the composite and hence the attenuation coefficient. This can be repeated for several frequencies, so that variation in the attenuation coefficient with the frequency can be calculated. This method to calculate the attenuation coefficient in the composite takes into account the value of attenuation in each component and the volume fraction, but not the contribution of scattering losses. In this sense, this prediction can be expected to be more accurate in the case of epoxy resins loaded with rubber particles, as the scattering of the rubber particles in the epoxy matrix is expected to be reduced, but the increase in the attenuation due to the attenuation in the rubber is expected to be large. On the contrary, actual attenuation can be expected to be larger than the calculated one in tungsten-loaded epoxy resins as the attenuation in the tungsten is expected to be reduced but the contribution of the scattering can be expected to be significant.

In order to predict both the attenuation coefficient and the acoustic impedance of three phase composites, we propose the following approach. First, we model the composite made of epoxy resin and micronized rubber. As elastic moduli and density of these two materials are not very different, the Hashin-Strikman approach can be used to get a sensible prediction of the composite properties. In addition, we also calculate composite properties using the CPA model. Then, we consider this composite as the matrix material and we add the tungsten particles. We apply the H-S and the CPA approach to calculate the properties of this three-phase composite.

3. Results

The characterization of the raw materials verifies the work hypothesis (see Table 3): the rubber sample presents a very high attenuation coefficient (16.4 dB/mm at 3 MHz), compared with epoxy resin (0.7 dB/mm at 3 MHz) and polyurethane (1.3–2.7 dB/mm at 3 MHz) and also a lower variation with frequency (n = 0.79), which is good to retain high damping at low frequencies.

In order to first test the efficiency of the fabrication technique, the expected and the actual density of the samples are always calculated as well as the deviation between them. The main source of error in the estimated density is due to thickness variability of the samples (within 200 µm). Density error due to this thickness variability is, for example, about 3% for rubber-loaded resin samples and about 6% for tungsten-loaded epoxy resin samples. Density deviations in Tables 5 and 6 can be explained by these errors. On the contrary, some of the tungsten-loaded polyurethane samples in Table 7 (-7.5% and -11.7%), present a larger deviation that, most likely, suggests the presence of some trapped air in the composite or some mixing deficiencies.

Table 9 summarizes the measured ultrasonic properties (velocity, attenuation at 3 MHz and variation with the frequency assuming a power law) for the rubber-loaded epoxy composites. Figure 4 shows the measured variation in the attenuation coefficient with the frequency, and Figure 5 shows the comparison between the measured and the calculated acoustic impedance and attenuation coefficient at 3 MHz in the micronized-rubber-loaded epoxy composites (using the upper and lower HS bounds and CPA models).

Table 9. Measured ultrasonic properties of the rubber-loaded epoxy composites.

Sample	Longitudinal Ultrasound Velocity (m/s)	Impedance (MRayl)	α @ 3 MHz (dB/mm)	n
Ep-R-02	2567 ± 77	2.94 ± 0.09	0.96 ± 0.03	1.7
Ep-R-04	2536 ± 75	2.90 ± 0.09	1.12 ± 0.03	1.8
Ep-R-10	2423 ± 70	2.71 ± 0.08	1.39 ± 0.04	1.9
Ep-R-15	2298 ± 67	2.56 ± 0.08	1.66 ± 0.05	2.0
Ep-R-20	2202 ± 66	2.40 ± 0.07	1.97 ± 0.06	1.9
Ep-R-35	2111 ± 62	2.35 ± 0.07	2.35 ± 0.07	1.7

Table 10 presents the measured impedance, and ultrasound velocity and attenuation for the rubber + tungsten-loaded epoxy composites, Figures 6 and 7 show the variation in the attenuation coefficient with the frequency, and Figure 8 presents measured and calculated (upper and lower HS bounds and CPA models) attenuation and impedance vs. rubber volume fraction. For theoretical calculations, it is assumed that tungsten volume fraction is either 45% or 14%.

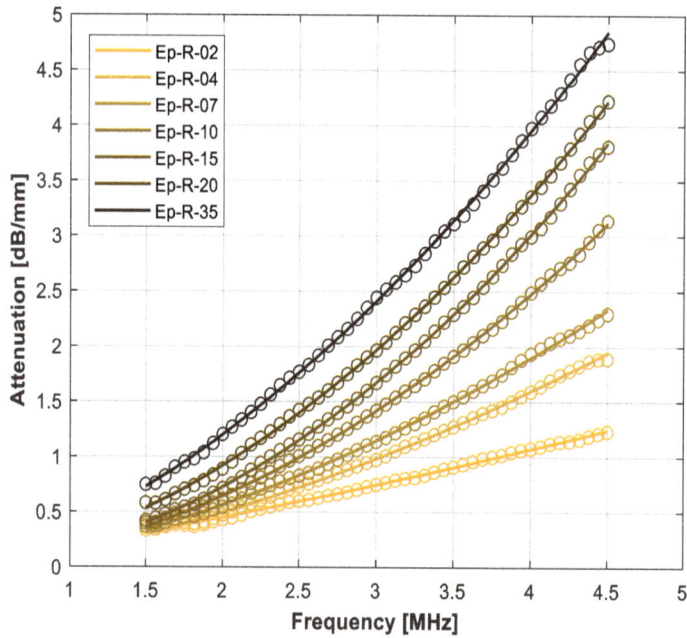

Figure 4. Variation in the ultrasonic attenuation coefficient with the frequency for the rubber-loaded epoxy resin composites. Open circles: measurements. Solid line: fitting using a power law (Equation (1)).

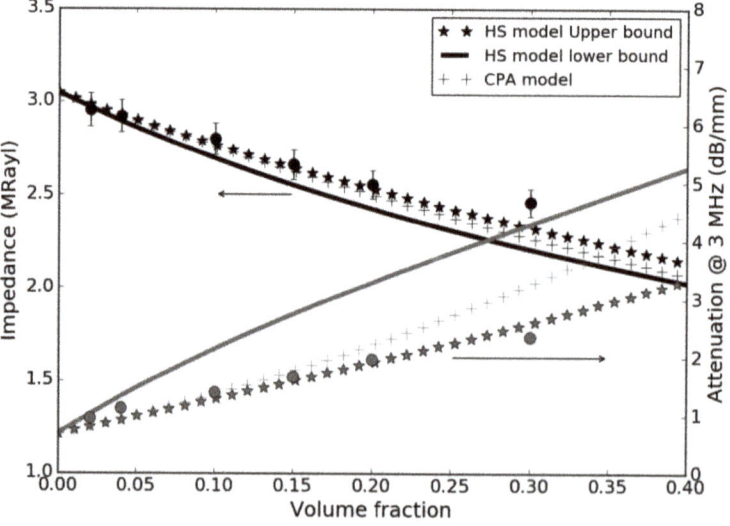

Figure 5. Variation in the impedance (black) and the ultrasonic attenuation coefficient at 3 MHz (gray) with the rubber volume fraction in the rubber-loaded epoxy composites. Circles: experimental measurements. Solid line: HS model, lower bound; Starts: HS, upper bound, +: CPA model.

Table 10. Measured ultrasonic properties of the tungsten + rubber-loaded epoxy composites.

	Sample	Longitudinal Ultrasound Velocity (m/s)	Impedance (MRayl)	α @ 3 MHz (dB/mm)	n
Low W-load	Ep-WR-14-7	1562 ± 94	5.57 ± 0.3	3.80 ± 0.2	2.3
	Ep-WR-13-13	1311 ± 81	4.66 ± 0.3	5.46 ± 0.3	2.3
	Ep-WR-16-16	1463 ± 87	5.77 ± 0.3	6.69 ± 0.4	2.3
	Ep-WR-15-27	1166 ± 69	4.40 ± 0.3	10.01 ± 0.6	2.0
	Ep-WR-15-37	1127 ± 69	4.17 ± 0.3	13.95 ± 0.8	3.2
High W-load	Ep-WR-48-03	1618 ± 95	15.64 ± 0.9	1.18 ± 0.07	2.5
	Ep-WR-47-04	1594 ± 94	15.01 ± 0.9	2.66 ± 0.1	2.0
	Ep-WR-41-08	1507 ± 90	12.70 ± 0.9	3.19 ± 0.2	2.1
	Ep-WR-38-12	1459 ± 89	11.48 ± 0.9	3.36 ± 0.2	2.2

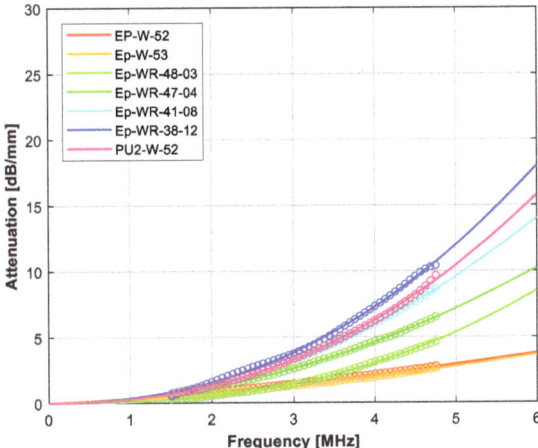

Figure 6. Variation in the ultrasonic attenuation coefficient with the frequency for the rubber + tungsten-loaded epoxy resin composites. High impedance series. Open circles: measurements. Solid line: fitting using a power law.

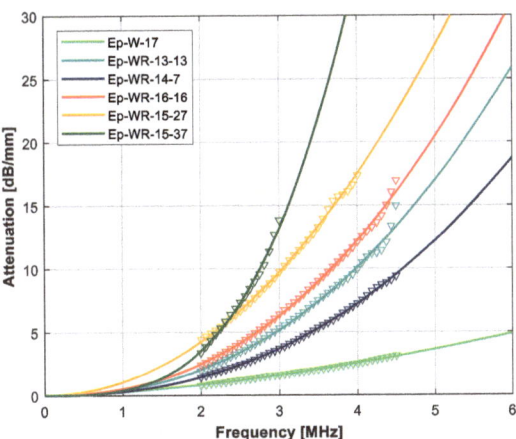

Figure 7. Variation in the ultrasonic attenuation coefficient with the frequency for the rubber + tungsten-loaded epoxy resin composites. Low impedance series. Open circles: measurements. Solid line: fitting using a power law.

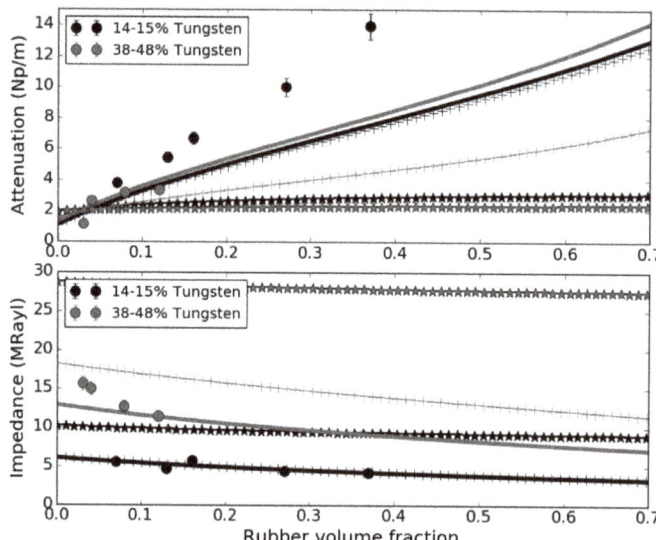

Figure 8. Attenuation (**up**) and Impedance (**bottom**) vs. rubber volume fraction. Gray: tungsten volume fraction: 38–48%. Black: tungsten volume fraction 14–15%. Circles: Experimental measurements; Solid line: HS model, lower bound; Starts: HS model, upper bound; +: CPA model.

Finally, Figure 9 shows the variation in the ultrasonic attenuation at 3 MHz with rubber volume fraction for tungsten + rubber + epoxy composite (for both the low and the high impedance series).

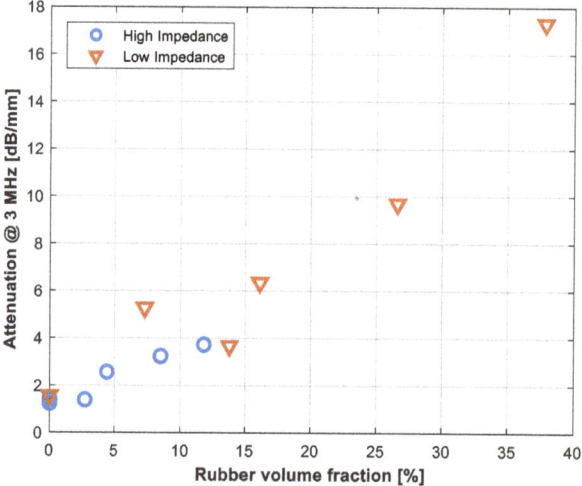

Figure 9. Ultrasonic attenuation coefficient at 3 MHz vs. rubber volume fraction. Results for both the high and the low impedance series.

For completeness, Tables 11 and 12 show the measured properties of the tungsten-loaded epoxy and polyurethane, respectively.

Table 11. Measured ultrasonic properties of the tungsten-loaded epoxy composites.

Sample	Longitudinal Ultrasound Velocity (m/s)	Impedance (MRayl)	α @ 3 MHz (dB/mm)	n
Ep-W-17	1695	6.77	1.44	1.7
Ep-W-52	1706	18.48	1.53	1.4
Ep-W-53	1666	17.22	1.35	1.6

Table 12. Measured ultrasonic properties of the tungsten-loaded polyurethane composites.

Sample	Longitudinal Ultrasound Velocity (m/s)	Impedance (MRayl)	α @ 3 MHz (dB/mm)	n
PU1-W-45	788	6.90	24.44	1.2
PU1-W-37	863	6.90	27.02	1.05
PU2-W-52	1139	10.58	3.97	2.3
PU2-W-37	1259	9.59	6.69	1.05
PU1-W-45	788	6.90	24.44	1.2

4. Discussion

As a first test of the capability of the rubber powder to increase the ultrasonic attenuation coefficient, samples with rubber volume fraction from 2 to 35% were fabricated and measured. As expected, the rubber load produces a moderate decrease of the acoustic impedance (up to 22%), and a remarkable increase in the attenuation coefficient (up to 236%). Compared with the unloaded epoxy, the variation with frequency changes notably (from n = 0.9 to n~1.8), probably due to the contribution of either the scattering produced by the rubber particles or their viscoelastic response. Compared with attenuation in polyurethane (Table 3), this approach permits the achievement of similar attenuation values but a wider range of variation and allows for a precise tailoring of the attenuation in the composite. As expected, the difference in the predictions made with HS upper and lower bounds are reduced (see Figure 5). The HS-UB model provides the closest predictions to the experimental data for both impedance and attenuation.

As it is already well known, loading the polymer with tungsten powder permits the increase of the acoustic impedance of the composite. This result is also reproduced here, see data in Tables 11 and 12. Compared with epoxy resin, the use of polyurethanes provides composites with relatively lower acoustic impedance, mainly due to the lower acoustic impedance of the polyurethanes, which is a negative feature for backing blocks. In addition, in some cases, the larger viscosity or shorter pot life time of the polyurethane complicate the composite mixing, and this limits the maximum possible load of particles, so this also contributes to further reduce the impedance of these composites. For tungsten-loaded epoxy and polyurethane composites (Tables 11 and 12), the measured impedance is close to the impedance predicted by the HS-LB model, with the exception of PU1-W-45 that presents a much smaller value, that is outside the HS bounds, so these suggest the presence of some air trapped.

Measurements also confirm that by loading the polymer with tungsten particles, the attenuation increases. The HS-LB model also provides good predictions for the attenuation in the composites with the exception of PU1 composites where measured attenuation is much larger than expected. This can be due to the presence of small air bubbles.

For the proposed solution for an intermediate impedance backing block with enhanced attenuation (epoxy + rubber + tungsten, impedance 5–7 MRayl), the measurements confirm the possibility to significantly increase the attenuation in the composite material by adding a third phase of rubber powder. As the required load of tungsten is reduced (about 13–15%), a higher load of rubber powder is possible (up to 37%). Measurements reveal that the attenuation coefficient can be increased (from 1.44 dB/mm, up to 13.95 dB/mm) with a moderate impedance reduction (from 6.7 to 4.2 MRayl). Comparison with calculated values (Figure 8) reveals that measured attenuation is larger than expected while the impedance is very close to the HS-LB approach. This behavior can be the result of the contribution of the scattering of the particles or can be due to a poor bounding between the particles and the matrix. On the other hand, for the proposed solution for a high impedance backing with enhanced attenuation (epoxy + rubber + tungsten, impedance 10-15 MRayl), the measurements confirm the possibility to increase, in a moderate way, the attenuation in the composite material by adding a third phase of rubber powder. As the required load of tungsten is larger (about 30–50%), only a smaller load of rubber powder is possible in this case (up to 12%). Measurements reveal that the attenuation coefficient can be increased (from 1.4 dB/mm up to 3.4 dB/mm) while the impedance decreases (from 18 to 12 MRayl). In this case, the predicted attenuation using the HS-LB provides a good matching into the measured values, while the measured impedance values are within the CPA and the HS-LB model predictions.

Table 13 shows a comparison of attenuation and impedance values obtained for rubber + tungsten-loaded epoxy and main reference values obtained from previous published works. The largest attenuation value is obtained for polyurethane composites (13–15 dB/mm) [37], though this approach is only valid for very low impedance materials (<3.5 MRayl). Very similar values are obtained in this work (13.95 dB/mm) but in this case with a larger impedance (4.17 MRayl), which make this approach more interesting for backing blocks. For high impedance materials (>10 MRayl), the best solution is the one provided in this work while for intermediate impedance materials (5–10 MRayl), the solution of this paper provides similar results to that in [44].

Table 13. Comparative of attenuation coefficient and acoustic impedance values in particle-loaded polymer composites.

Reference	Attenuation Coeff @ 3 MHz (dB/mm)	Acoustic Impedance (MRayl)
Lutsch [27]	2.4	~8
Tiefensee et al. [36]	5	4-7
State et al. [37]	13–15	2.6–3.3
	7.1	
Cho et al. [40]	3–5	4–5
Nguyen et al. [42]	10	8.7
	1.8	11.7
Abas et al. [32]	1	7
Nguyen et al. [46]	9	8.7
El-Tantawy &. Sung [41]	2.2–3.9	2–7.8
Wang et al. [48]	~5.0	3–5.5
This work		
Rubber + epoxy	0.96–2.35	2.3–2.9
Tungsten + epoxy	1.4–1.5	6.8–18.5
Tungsten + polyurethane	4–27	6.9–10.5
Rubber + Tungsten (low) + epoxy	4–14	4–5
Rubber + Tungsten (high) + epoxy	1.2–3.4	12–16

5. Conclusions

This work presented a study of the possibilities to control the damping in tungsten-loaded polymers for backing blocks in ultrasonic transducers by adding micronized rubber powder obtained from end-of-life tires (ELT). A fabrication technique based on a high-speed orbital mixer is proposed as it is compatible with transducers' fabrication restrictions and allows in situ fabrication and curing of the backing block directly on the piezoelectric element, so that interface problems can be minimized. It can be concluded that is it possible to fabricate composites in the impedance range of 4–5 MRayl with attenuation coefficient in the range 4–14 dB/mm (at 3 MHz), and higher impedance composites (12–16 MRayl), with attenuation in the range 1.2–3.4 dB/mm. In fact, it has been shown that the attenuation coefficient in the composite increases almost linearly with the load of rubber regardless of the load of tungsten (Figure 9).

The inclusion of rubber powder in tungsten-loaded polymers permits the tailoring of several functionalities of the composite. In the case of materials for ultrasonic transducers these functionalities comprise the capability to simultaneously and independently control the impedance, ultrasound velocity, ultrasound attenuation, variation of ultrasound attenuation with the frequency, the scattering strength and even the EMI shielding. In particular, it is possible to obtain materials with acoustic impedance in the range 2.35–15.6 MRayl, ultrasound velocity in the range 790–2570 m/s, attenuation at 3 MHz, from 0.96 up to 27 dB/mm with a variation of the attenuation with the frequency following a power law with exponent in the range 1.2–3.2. In addition, the inclusion of these materials in piezoelectric transducers, as backing blocks or matching layers, permits the further control of different properties and functions of the transducer, such as the bandwidth, the ringing, the amplitude and time for the appearance of the backing block back echo and the sensitivity.

For the fabrication of backing materials for ultrasonic piezoelectric transducers based on ceramics and single crystals, the best results correspond to Ep-WR-48-03, Ep-WR-47-04, Ep-WR-41-08 and Ep-WR-38-12 that combine large impedance (good damping that contributes to enlarge the bandwidth, reducing the ringing and improving axial resolution) and high attenuation and low velocity (that contribute to eliminate the backing block back echo). For the fabrication of backing materials for ultrasonic piezoelectric transducers based on 1-3 composites (ceramic volume fraction between 30 and 70%), the best option can be Ep-WR-15-27 and Ep-WR-15-37 that combine moderate impedance, large attenuation and low ultrasonic velocity.

Therefore, the materials and the manufacturing route proposed provide an interesting alternative to tailor the properties of ultrasonic transducers' backing blocks. In addition, this solution also offers an application for the use of micronized rubber recycled from ELT.

6. Patents

Title: "A manufacturing method of a passive composite material for an ultrasonic transducer". Number: P202230611. Date: 5 July 2022.

Title: "Use of composite material as an artificial tissue or organ for testing the performance of an ultrasound diagnosis apparatus". Number: P202230610. Date: 5 July 2022.

Author Contributions: Conceptualization, V.G., J.L.V., L.S.-R. and T.G.Á.-A.; methodology, V.G., J.L.V., R.P.-A. and T.G.Á.-A.; software, V.G.; validation, V.G. and T.G.Á.-A.; formal analysis, V.G. and T.G.Á.-A.; investigation, V.G., M.D.F., J.L.V. and T.G.Á.-A.; resources, R.P.-A., L.S.-R., J.L.V. and T.G.Á.-A.; data curation, V.G.; writing—original draft preparation, V.G. and T.G.Á.-A.; writing—review and editing, T.G.Á.-A., M.D.F. and R.P.-A.; visualization, V.G.; supervision, T.G.Á.-A. All authors have read and agreed to the published version of the manuscript.

Funding: This research was funded by Ministerio de Economía y Competitividad (DPI2016-78876-R and MAT2017-87204-R), CSIC (201860E045) and Signus Ecovalor, S.L. Author J.L.V. is member of the SusPlast platform of CSIC.

Data Availability Statement: Data are available upon request to the corresponding author.

Conflicts of Interest: The authors declare no conflict of interest. The funders had no role in the design of the study; in the collection, analyses, or interpretation of data; in the writing of the manuscript; or in the decision to publish the results.

References

1. Rothon, B. (Ed.) *Fillers for Polymer Applications*; Springer: Berlin/Heidelberg, Germany, 2017; ISBN 978-3-319-28116-2.
2. Bagheri, R.; Marouf, B.T.; Pearson, R.A. Rubber-toughened epoxies: A critical review. *Polym. Rev.* **2009**, *49*, 201–225. [CrossRef]
3. Kargarzadeh, H.; Ahmad, I.; Abdullah, I. Chapter 10: Mechanical Properties of Epoxy-Rubber Blends. In *Handbook of Epoxy Blends*; Springer International Publishing: Cham, Switzerland, 2015.

4. Ismail, H.; Omar, N.F.; Othman, N. Effect of carbon black loading on curing characteristics and mechanical properties of waste tyre dust/carbon black hybrid filler filled natural rubber compounds. *J. Appl. Polym. Sci.* **2011**, *121*, 1143–1150. [CrossRef]
5. Colom, X.; Carrillo, F.; Cañavate, J. Composites reinforced with reused tyres: Surface oxidant treatment to improve the interfacial compatibility. *Compos. Part A Appl. Sci. Manufac.* **2007**, *38*, 44–50. [CrossRef]
6. Valášek, P. Mechanical Properties of Epoxy Resins Filled with Waste Rubber. *Manuf. Technol.* **2014**, *14*, 632–637. [CrossRef]
7. Ramarad, S.; Khalid, M.; Ratnam, C.T.; Chua, A.I.; Rashmi, W. Waste tire rubber in polymer blends: A review on the evolution, properties and future. *Prog. Mater. Sci.* **2015**, *72*, 100–140. [CrossRef]
8. Woldemariam, M.H.; Belingardi, G.; Koricho, E.G.; Reda, D.T. Effects of nanomaterials and particles on mechanical properties and fracture toughness of composite materials: A short review. *AIMS Mater. Sci.* **2019**, *6*, 1191–1212. [CrossRef]
9. Móczó, J.; Pukánszky, B. Particulate Fillers in Thermoplastics. In *Encyclopedia of Polymers and Composites*; Palsule, S., Ed.; Springer: Berlin/Heidelberg, Germany, 2015.
10. Lakes, R.S. High damping composite materials: Effect of structural hierarchy. *J. Compos. Mater.* **2002**, *36*, 287–297. [CrossRef]
11. Sankaran, S.; Deshmukh, K.; Ahamed, M.B.; Khadheer Pasha, S.K. Recent advances in electromagnetic interference shielding properties of metal and carbon filler reinforced flexible polymer composites: A review. *Compos. Part A Appl. Sci. Manufact.* **2018**, *114*, 49–71. [CrossRef]
12. Quesenberry, M.J.; Madison, P.H.; Jensen, R.E. Characterization of Low Density Glass Filled Epoxies, Army Research Laboratory, March 2003, ATRL-TR-2938. Available online: https://apps.dtic.mil/sti/pdfs/ADA412137.pdf (accessed on 26 August 2022).
13. Valentín, J.L.; Pérez-Aparicio, R.; Fernandez-Torres, A.; Posadas, P.; Herrero, R.; Salamanca, F.M.; Navarro, R.; Saiz-Rodríguez, L. Advanced Characterization of Recycled Rubber from End-of-life Tires. *Rubber Chem. Technol.* **2020**, *93*, 683–703. [CrossRef]
14. Valentín, J.L.; Saiz-Rodríguez, L.; Pérez-Aparicio, R. Guía para el empleo de caucho reciclado procedente del neumático en la industria del caucho. *Rev. De Plaásticos Mod.* **2021**, *122*, 4.
15. Newnham, R.E.; Skinner, D.P.; Cross, L.E. Connectivity and piezoelectric-pyroelectric composites. *Mat. Res. Bull.* **1978**, *13*, 525–536. [CrossRef]
16. Levassort, F.; Lethiecq, M.; Desmare, R. Effective electroelastic moduli of 3-3(0-3) piezocomposites. *IEEE Trans. Ultrason. Ferroelectr. Freq. Control* **1999**, *46*, 1028–1034. [CrossRef] [PubMed]
17. Gómez Alvarez-Arenas, T.E.; Mulholland, A.J.; Hayward, G.; Gomatam, J. Wave propagation in 0-3/3-3 connectivity composites with complex microstructure. *Ultrasonics* **2000**, *38*, 897–907. [CrossRef]
18. Newnham, R.E.; Bowen, L.J.; Klicker, K.A.; Cross, L.E. Composite piezoelectric transducers. *Mater. Des.* **1980**, *2*, 93–106. [CrossRef]
19. Haifeng Wang, K.S.; Ritter, T.; Cao, W. High frequency properties of passive materials for ultrasonic transducers. *IEEE Trans. Ultrason. Freq. Control* **2001**, *48*, 78–84. [CrossRef]
20. Banno, H.; Saito, S. Piezoelectric and dielectric properties of composites of synthetic rubber and PbTiO or PZT. *Jpn. J. Appl. Phys.* **1983**, *22*, 67–69. [CrossRef]
21. Gururaja, T.R.; Xu, Q.C.; Ramachandran, A.R.; Halliyal, A.; Newnham, R.E. Preparation and piezoelectric properties of fired 0–3 composites. In Proceedings of the IEEE 1986 Ultrasonics Symposium, Williamsburg, VA, USA, 17–19 November 1986; pp. 703–706. [CrossRef]
22. Levassort, F.; Lethiecq, M.; Certon, D.; Patat, F. A matrix method for modelling electroelastic moduli of 0-3 piezo-composites. *IEEE Trans. Ultrason. Ferroelect. Freq. Control* **1997**, *44*, 445–452. [CrossRef]
23. Gómez Álvarez-Arenas, T.E.; Montero, F.; Levassort, F.; Lethieq, M.; James, A.; Ringgard, E.; Millar, C.E.; Hawkins, P. Ceramic powder–polymer piezocomposites for electroacoustic transduction: Modeling and design. *Ultrasonics* **1998**, *36*, 907–923. [CrossRef]
24. Lau, S.; Li, X.; Zhang, X.; Zhou, Q.; Shung, K.K.; Ji, H.; Ren, W. High frequency ultrasonic transducer with KNN/BNT 0–3 composite active element. In Proceedings of the 2010 IEEE International Ultrasonics Symposium, San Diego, CA, USA, 11–14 October 2010; pp. 76–79. [CrossRef]
25. Hanner, K.A.; Safari, A.; Newnham, R.E.; Runt, J. Thin-film 0–3 polymer piezo- electric ceramic composites-piezoelectric paints. *Ferroelectrics* **1989**, *100*, 255–260. [CrossRef]
26. Wang, F.; Wang, H.; Song, Y.; Sun, H. High piezoelectricity 0–3 cement-based piezoelectric composites. *Mater. Lett.* **2012**, *76*, 208–210. [CrossRef]
27. Lutsch, A. Solid Mixtures with Specified Impedances and High Attenuation for Ultrasonic Waves. *J. Acoust. Soc. Am.* **1962**, *34*, 131–132. [CrossRef]
28. Lees, S.; Gilmore, R.S.; Kranz, P.R. Acoustic Properties of Tungsten-Vinyl Composites. *IEEE Trans. Sonics Ultrason.* **1973**, *20*, 1–10. [CrossRef]
29. Ju-Zhen, W. Backing Material for the Ultrasonic Transducer. USA Patent 4800316, 24 June 1989.
30. Sayers, C.M.; Tait, C.E. Ultrasonic properties of transducer backings. *Ultrasonics* **1984**, *22*, 57–60. [CrossRef]
31. Grewe, M.G.; Gururaja, T.R.; Shrout, T.R.; Newnham, R.E. Acoustic properties of particle/polymer composites for ultrasonic transducer backing applications. *IEEE Trans. Ultrason. Ferroelect. Freq. Control* **1990**, *37*, 506–514. [CrossRef] [PubMed]
32. Abas, A.A.; Ismail, D.M.P.; Sani, S.; Noorul, M.; Ahmed, I. Effect of Backing layer Composition on Ultrasonic Probe Bandwith. In Proceedings of the RnD Seminar 2010: Research and Development Seminar, Bangi, Malaysia, 12–15 October 2010.
33. Hidayat, D.; Syafei, N.S.; Wibawa, B.M.; Taufik, M.; Bahtiar, A.; Risdiana, R. Metal-Polymer Composite as an Acoustic Attenuating Material for Ultrasonic Transducers. *KEM* **2020**, *860*, 303–309. [CrossRef]

34. Zhang, W.; Jia, H.; Gao, G.; Cheng, X.; Du, P.; Xu, D. Backing layers on electroacoustic properties of the acoustic emission sensors. *Appl. Acoust.* **2019**, *156*, 387–393. [CrossRef]
35. Zhou, Q.; Cha, J.H.; Huang, Y.; Zhang, R.; Cao, W.; Shung, K.K. Alumina/epoxy nanocomposite matching layers for high-frequency ultrasound transducer application. *IEEE Trans. Ultrason. Ferroelect. Freq. Control* **2009**, *56*, 213–219. [CrossRef]
36. Tiefensee, F.; Becker-Willinger, C.; Heppe, G.; Herbeck-Engel, P.; Jakob, A. Nanocomposite cerium oxide polymer matching layers with adjustable acoustic impedance between 4 MRayl and 7 MRayl. *Ultrasonics* **2010**, *50*, 363–366. [CrossRef]
37. State, M.; Brands, P.J.; van de Vosse, F.N. Improving the thermal dimensional stability of flexible polymer composite backing materials for ultrasound transducers. *Ultrasonics* **2010**, *50*, 458–466. [CrossRef]
38. Cafarelli, A.; Verbeni, A.; Poliziani, A.; Dario, P.; Menciassi, A.; Ricotti, L. Tuning acoustic and mechanical properties of materials for ultrasound phantoms and smart substrates for cell cultures. *Acta Biomater.* **2017**, *49*, 368–378. [CrossRef]
39. Culjat, M.O.; Goldenberg, D.; Tewari, P.; Singh, R.S. A review of tissue substitutes for ultrasound imaging. *Ultrasound Med. Biol.* **2010**, *36*, 861–873. [CrossRef] [PubMed]
40. Cho, E.; Park, G.; Lee, J.-W.; Cho, S.-M.; Kim, T.; Kim, J.; Choi, W.; Ohm, W.-S.; Kang, S. Effect of alumina composition and surface integrity in alumina/epoxy composites on the ultrasonic attenuation properties. *Ultrasonics* **2016**, *66*, 133–139. [CrossRef]
41. El-Tantawy, F.M.; Sung, Y.K. A novel ultrasonic transducer backing from porous epoxy resin-titanium-silane coupling agent and plasticizer composites. *Mater. Lett.* **2004**, *58*, 154–158. [CrossRef]
42. Nguyen, N.T.; Lethiecq, M.; Karlsson, B.; Patat, F. Highly attenuative rubber modified epoxy for ultrasonic transducer backing applications. *Ultrasonics* **1996**, *34*, 669–675. [CrossRef]
43. Thomas, R.; Ahmad, I.; Ahmad, S.; Koshy, S. Blends and IPNs of Natural Rubber with Thermo-Setting Polymers. In *Natural Rubber Materials*; Thomas, S., Chan, C.H., Pothan, L., Rajisha, K.R., Maria Hanna, J., Eds.; Royal Society of Chemistry: Cambridge, UK, 2014; pp. 336–348.
44. Sachse, W.; Pao, Y.H. On the determination of phase and group velocities of dispersive waves in solids. *J. Appl. Phys.* **1978**, *49*, 4320–4327. [CrossRef]
45. Kline, R.A. Measurement of attenuation and dispersion using an ultrasonic spectroscopy technique. *J. Acoust. Soc. Am.* **1984**, *76*, 498–504. [CrossRef]
46. Nguyen, T.N.; Lethiecq, M.; Levassort, F.; Pourcelot, L. Experimental verification of the theory of elastic properties using scattering approximations in (0-3) connectivity composite materials. *IEEE Trans. Ultrason. Ferroelectr. Freq. Control* **1996**, *43*, 640–645. [CrossRef]
47. Hashin, Z.; Shtrikman, S. A variational approach to the theory of the elastic behavior of multiphase materials. *J. Mech. Phys. Solids* **1963**, *11*, 127–140. [CrossRef]
48. Wang, H.; Ritter, T.A.; Cao, W.; Shung, K.K. Passive materials for high frequency ultrasound transducers. In Proceedings of the SPIE Conference on Ultrasonic Transducer Engineering, San Diego, CA, USA, 24–25 February 1999; Volume 3664, pp. 35–42. [CrossRef]

Article

A Thermo-Mechanically Robust Compliant Electrode Based on Surface Modification of Twisted and Coiled Nylon-6 Fiber for Artificial Muscle with Highly Durable Contractile Stroke

Sungryul Yun [1,*], Seongcheol Mun [1], Seung Koo Park [2], Inwook Hwang [1] and Meejeong Choi [1]

1. Tangible Interface Creative Research Section, Electronics and Telecommunications Research Institute (ETRI), Daejeon 34129, Korea
2. Human Enhancement & Assistive Technology Research Section, Electronics and Telecommunications Research Institute (ETRI), Daejeon 34129, Korea
* Correspondence: sungryul@etri.re.kr

Citation: Yun, S.; Mun, S.; Park, S.K.; Hwang, I.; Choi, M. A Thermo-Mechanically Robust Compliant Electrode Based on Surface Modification of Twisted and Coiled Nylon-6 Fiber for Artificial Muscle with Highly Durable Contractile Stroke. *Polymers* 2022, *14*, 3601. https://doi.org/10.3390/polym14173601

Academic Editors: Tomasz Makowski and Sivanjineyulu Veluri

Received: 5 August 2022
Accepted: 29 August 2022
Published: 31 August 2022

Publisher's Note: MDPI stays neutral with regard to jurisdictional claims in published maps and institutional affiliations.

Copyright: © 2022 by the authors. Licensee MDPI, Basel, Switzerland. This article is an open access article distributed under the terms and conditions of the Creative Commons Attribution (CC BY) license (https://creativecommons.org/licenses/by/4.0/).

Abstract: In this paper, we propose a novel and facile methodology to chemically construct a thin and highly compliant metallic electrode onto a twisted and coiled nylon-6 fiber (TCN) with a three-dimensional structure via surface modification of the TCN eliciting gold-sulfur (Au-S) interaction for enabling durable electro-thermally-induced actuation performance of a TCN actuator (TCNA). The surface of the TCN exposed to UV/Ozone plasma was modified to (3-mercaptopropyl)trimethoxysilane (MPTMS) molecules with thiol groups through a hydrolysis-condensation reaction. Thanks to the surface modification inducing strong interaction between gold and sulfur as a formation of covalent bonds, the Au electrode on the MPTMS-TCN exhibited excellent mechanical robustness against adhesion test, simultaneously could allow overall surface of the TCN to be evenly heated without any significant physical damages during repetitive electro-thermal heating tests. Unlike the TC-NAs with physically coated metallic electrode, the TCNA with the Au electrode established on the MPTMS-TCN could produce a large and repeatable contractile strain over 12% as lifting a load of 100 g even during 2000 cyclic actuations. Demonstration of the durable electrode for the TCNA can lead to technical advances in artificial muscles for human-assistive devices as well as soft robots those requires long-term stability in operation.

Keywords: surface modification; electro-thermal; contractile strain; actuation durability

1. Introduction

Recently, as a new class of artificial muscle, twisted and coiled polymer actuators (TCPAs) have given much attention due to their excellent mechanical properties, high flexibility, cost-effectiveness, and large thermally-induced contractile deformation exceeding stroke of human skeletal muscles as simultaneously enabling periodic lifting a load over 20 MPa even under a light-weight structure [1]. Many researchers have introduced polymer artificial muscles possessing the structural resilience using various fibers, such as shape memory polymer [2], polymer bimorphs [3], polymer nanocomposites with carbon nanotube (CNT) or graphene oxide [4], natural rubbers [5], and hydrogels [6]. Technical advances in actuation mechanism, materials, and fabrication process enables the TCPAs to be utilized in prospective application fields of prosthetics [3], robotics [7,8], exoskeletons [9], energy harvesting [10], morphing skin [11], smart window [12], and smart fabrics [1,13,14]. Typically, the TCPAs are composed of a heating source and a coiled polymer, which is mainly formed by twist insertion of a monofilament fiber and then thermally annealing it to prevent from being untwisted. When uniaxially pre-stretched under load, heating induces the TCPAs to lift the load owing to their large contractile deformation in response to the anisotropic thermal expansion of the polymer fibers in the radial direction.

For studying thermo-mechanical deformation behavior of the TCPAs, researchers have adopted diverse heat sources, such as water, light, ambient air, and electric power [15–17]. Among these heating methodologies, particularly, electric power has been widely utilized due to its advantages over others in miniaturization and controllability. For the electric heating that occurs by electric current flowing through a conductor integrating onto the coiled polymer, diverse conductors have been employed in the form of a thin electrode, as well as a metallic wire (e.g., steel, copper) or a silver-plated polymer wire. The electrodes are formed on the surface of the coiled structure by not only painting silver paste [1,18] or conductive elastomer containing silver particles [19], electroless silver plating via chemical reduction of silver ions [20], and spray-coating silver nanowires (AgNWs) [21], but also consecutively winding a carbon nanotube (CNT) sheet drawn from a vertically-grown CNTs on a polymeric fiber [1]. In parallel, the conductive wire is formed by coiling with the polymeric fiber via twist insertion after winding it on the polymeric fiber [1,22–24]. These approaches have contributed to eliciting electro-thermally-induced contractile stroke from the TCPAs as providing benefits with respect to temperature controllability, stretchability and processability. However, the electric heating capability of the electrode for the TCPAs has been demonstrated by only implementing a few cyclic actuations [18,19]. In the case of the conductive wire, an achievable tensile stroke was limited to less than 10%, accompanied by slow heating response of the TCPA due to sparse coiling density of the wire [19,24]. Therefore, for practical use, there are still technical challenges in imparting mechanical robustness and electrical stability to the conductors during highly repetitive heating-cooling cycles and simultaneously reducing complexity in fabrication.

Here, we propose a novel and facile approach to chemically form a thin, mechanically-robust and electro-thermally-stable gold (Au) electrode on a twisted and coiled nylon-6 fiber (TCN) with a three-dimensional structure as exploiting gold-sulfur (Au-S) interaction achieved via surface modification chemically-attaching (3-mercaptopropyl)trimethoxysilane (MPTMS) onto the TCN. In this paper, we report studies on not only fabrication, chemical analysis, and mechanical and electro-thermal heating characteristic with morphological observation of the metallic electrodes, but also electro-thermally-induced actuation performance of the TCN actuators (TCNAs) during a large number of repetitive heating-cooling cycles.

2. Materials and Methods

2.1. Materials

Nylon-6 monofilament fishing line (Tournamenter SE No.3) with a diameter of 285 µm was purchased from Toray Industries, INC (Tokyo, Japan). Nylon-6 film with a thickness of 100 µm were purchased from Goodfellow Cambridge Ltd. (Huntingdon, United Kingdom), respectively. (3-mercaptopropyl)trimethoxysilane (MPTMS, 95%), ethyl alcohol (>99.5%), acetone (>99.5%), and isopropyl alcohol (IPA, 99.5%) were purchased from Sigma-Aldrich (St. Louis, MO, USA). Silver nanowires (AgNWs) solution (AgNWs were dispersed in IPA with a concentration of 0.5 wt %) was purchased from DS Hi-Metal (Ulsan, Korea). The AgNWs have average diameter of 40 nm and length of 20 µm. Deionized water (DI-water) with a resistivity of 18.2 MΩ·cm at 25 °C was achieved from a filtration system of Milli-Q MerckMillipore (Burlington, VT, USA).

2.2. Fabrication of a Twisted and Coiled Nylon-6 Structure

Twisted and coiled nylon-6 fibers (TCNs) were fabricated by twist insertion and thermal annealing processes sequentially as shown in Figure 1. By exploiting the fabrication process reported [1], the upper end of the nylon-6 fiber (length: 1200 mm) was clamped to the shaft of a motor and then a load of 19.4 MPa was applied to the bottom end of the fiber by attaching a weight of around 126 g. After inserting twist to the fiber as operating a rotational motor with a consistent speed of 350 rpm, we achieved the TCNs (length/diameter: around 270 mm/670 µm) with a consistent spring index of 2.35 (deviation: < 0.02). After uniaxially stretching to 10% and clamping with a metallic fixture at both ends, the TCN was thermally-annealed in a VDL23 vacuum oven of Binder (Tuttlingen, Germany) at 170 °C for 100 min.

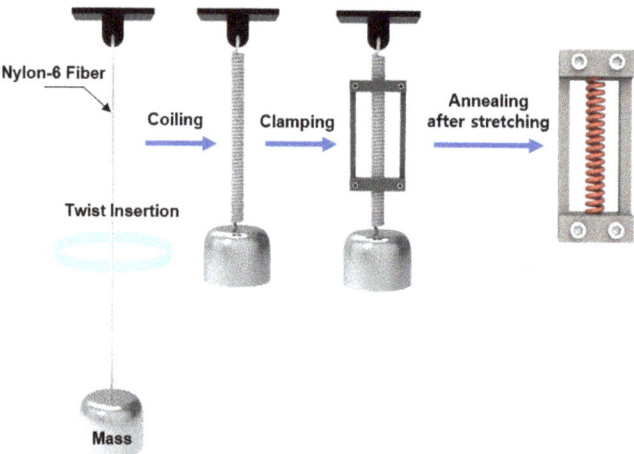

Figure 1. An illustrated fabrication process of the TCN structure.

2.3. Surface Modification of the TCN

The prepared TCN was precleaned by sequentially immersing it in acetone and IPA bath with sonication for 3 min and then drying at 60 °C in a vacuum oven for 1 h. The precleaned TCN was mechanically stretched to 40% and clamped to a couple of metallic fixtures because the stretching, which is much less than its elastic limit, could sufficiently provide space among the coiled structure stuck to each other for facilitating contact of UV/Ozone plasma onto its overall surface area. After placing the TCN at 10 cm distant from the UV lamp (wavelength: 254 nm) in a chamber of a UVC 300 UV/Ozone system of YUIL (Incheon, Korea), the surface of the TCN was exposed to UV light for 15 min at atmospheric pressure. The UV/Ozone treated TCN immediately immersed into 50 mM solution of MPTMS and deionized water, which has been reported as an effective concentration to moderately hydrolyze MPTMS molecules in water [25], and then it was kept in the solution under magnetic stirring at 40 °C for 30 min. Finally, the MPTMS treated TCN was sonicated in an ethanol bath to remove residual MPTMS and dried using nitrogen gas.

2.4. Construction of Metallic Electrode on the TCN

Gold nanoparticles were coated on the surface of both a MPTMS treated and a pristine TCN under the same pre-stretch condition by a Q300TD sputter coater of Quorum Technology (Lewes, United Kingdom) as setting the coating thickness to be 50 nm under a stage rotation with a constant speed. AgNWs were coated on the pristine TCN by a HP-TR1 spray gun of IWATA (Cincinnati, OH, USA). For the AgNWs coating, a frame holding the TCN was fixed on a motorized zig that can control rotational speed. 1 mL of the AgNWs solution was sprayed on the surface of the TCN as simultaneously rotating the zig with constant speed and then dried at room conditions.

2.5. Characterization

The surface characteristic of the nylon-6 material before after UV/Ozone plasma treatment was investigated by measuring water contact angle (WCA) via a DSA 25S drop shape analyzer of KRÜSS (Hamburg, Germany). For the measurement, in substitution for the TCN, we used a nylon-6 film with thickness: 350 μm. The morphological characteristic was measured by a Sirion 600 field emission scanning electron microscope (FESEM) of FEI (Hillsboro, OR, USA) and an Axico Sope A1 optical microscope of Carl ZEISS (Stuttgart,

Germany). The chemical analysis for the surface modification with MPTMS was performed by a K-Alpha X-ray photoelectron spectroscopy (XPS) of Thermo Fisher Scientific Inc. (Waltham, MA, USA) with 0.1 eV scanning step. Electrical resistance of the metallic electrode was measured by a 34450A digital multimeter of Keysight technologies (Santa Rosa, CA, USA).

2.6. Performance Test

Electro-thermal actuation performance of the TCPAs with three different electrodes were evaluated via a performance measurement system composed of a B2901A precision source/measure unit of Keysight technologies, a LK-HD500 laser displacement sensor of Keyence (Osaka, Japan), and a PI 640 thermal imaging camera of Optris (Berlin, Germany), which is shown in Figure 2. In the measurement system, the precision source/measure unit was used to provide an input voltage for electro-thermal heating as measuring an electric current. During the electro-thermal excitation, actuation response and temperature distribution for the TCPAs were simultaneously achieved from the laser displacement sensor and the thermal imaging camera.

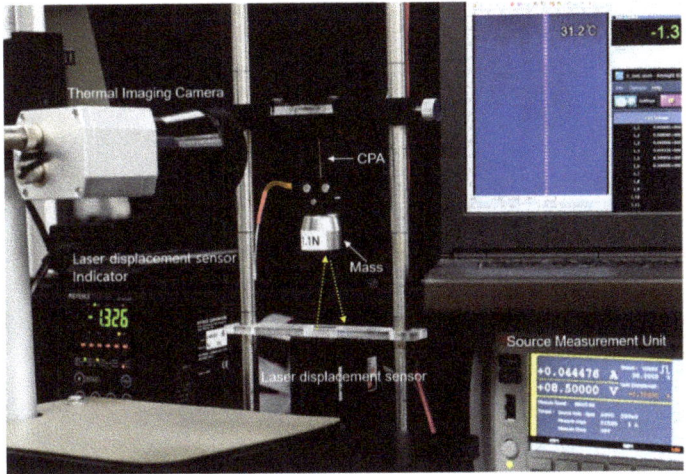

Figure 2. A photograph of performance evaluation system for the TCNA.

3. Results and Discussion

3.1. Surface Modification of TCN

The TCN was prepared by following the process described in Section 2.2. In order to establish a compliant electrode on the surface of the TCN, we implemented surface modification of the TCN with MPTMS that can lead to strong gold-sulfur (Au-S) covalent bonds between gold nanoparticles and thiol groups in MPTMS molecules. Figure 3 shows a scheme to chemically-attaching MPTMS molecules onto the surface of the TCN. At first, we investigated influence of the plasma treatment on surface characteristic of the nylon-6 material via measuring change in WCA responding to the plasma treatment. Here, we note that the test was performed by using a nylon-6 film instead of a TCN because the TCN with a small fiber diameter, which is much less than 1 mm, has a difficulty in exploiting sessile drop method. Prior to the test, the nylon-6 film was precleaned by the same process implemented before surface modification of the TCN. Figure 4 shows change in the WCA of the nylon-6 film depending on exposure time of UV/Ozone plasma. In the case of a pristine nylon-6 film, the water droplet on the surface formed a contact angle of 94.2°, indicating that the polymer intrinsically possesses hydrophobic nature. However, the surface turned into hydrophilic after exposure to UV/Ozone plasma. The hydrophilicity became enhanced as exposure time of UV/Ozone plasma increased. Particularly, implementation of the plasma treatment as long as 15 min changed the surface close to superhydrophilic, resulting in reduction of the WCA as low as 12.0°. It suggests that the UV/Ozone plasma treatment can be an effective methodology to modify nylon-6 surface with abundant hydroxyl and carboxyl moieties, which has been known as oxygen based polar groups enhancing hydrophilicity [26,27].

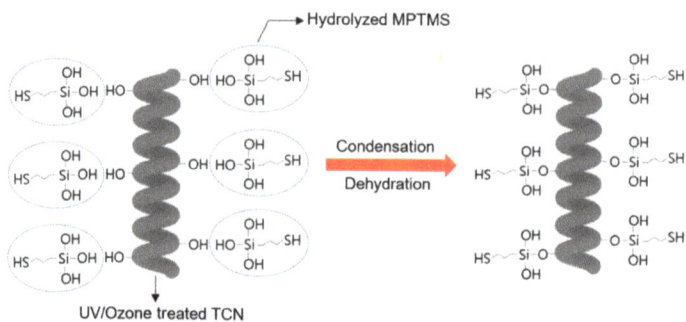

Figure 3. A schematic illustration of chemically-attaching MPTMS molecules onto the surface of a TCN.

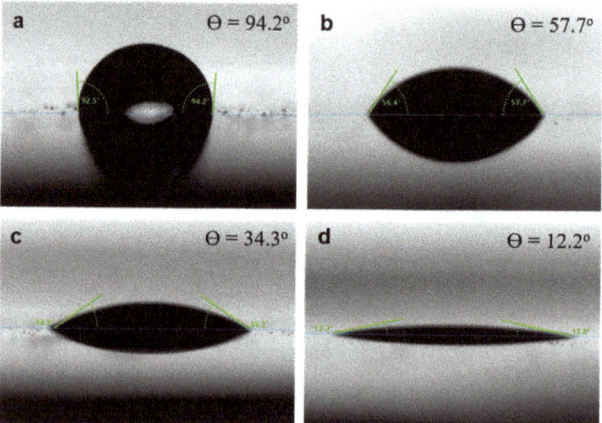

Figure 4. Change in the WCA of the nylon-6 films according to exposure time of UV/Ozone plasma: (**a**) without exposure, (**b**) 3 min, (**c**) 5 min, and (**d**) 15 min.

Secondly, in order to confirm the chemical surface modification of the nylon-6 film with MPTMS molecules, we investigated XPS spectra for the film before and after MPTMS treatment. In Figure 5a, the full scale XPS spectra revealed that both nylon-6 films have two common peaks at the binding energy of 533 and 285 eV, which are individually assigned to O 1s and C 1s. On the other hand, the MPTMS treated nylon-6 film exhibited distinguishable peaks for Si 2p and S 2p at each binding energy of 102.1 and 162.8 eV. As shown in Figure 5b, the binding energy peak for S 2p can be deconvoluted into two peaks for spin-orbit doublets of S $2p_{3/2}$ and S $2p_{1/2}$, corresponding to 162.5 and 163.8 eV, respectively. The binding energy peaks for S $2p_{3/2}$ and S $2p_{1/2}$ exhibited a 2:1 area ratio of the peaks with their deviation of 1.3 eV. In parallel, as shown in Figure 5c, a high resolution XPS spectra for Si 2p revealed a binding energy peak at 102.1 eV, which is assigned to Si-O-C bond, indicating that the surface of the nylon-6 film was modified to MPTMS molecules with thiol groups via condensation reaction of silanol groups with hydroxyl groups on the film surface [28,29]. In addition, based on the elemental concentration data taken from XPS survey spectra

(Table 1), we also confirmed that MPTMS treated nylon-6 film has remarkably higher Si and S contents than the pristine nylon-6 film.

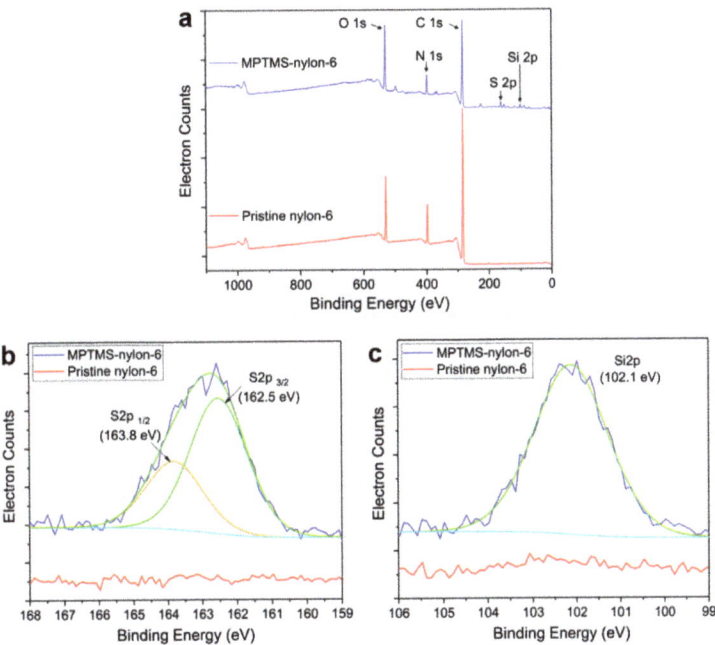

Figure 5. Comparison of XPS spectra of a pristine nylon-6 and a MPTMS-nylon-6 film: (**a**) full-scale spectra, (**b**,**c**) high-resolution spectra for S 2p and Si 2p. In the overlapped lines, the light blue and green line are repectively assigned to XPS spectra experimentally measured and after smoothing.

Table 1. Elemental composition of a pristine nylon-6 and a MPTMS-g-nylon-6 film achieved via XPS.

	C 1s	O 1s	N 1s	Si 2p	S 2p
Pristine nylon-6	75.87	11.63	12.05	0.26	0.18
MPTMS-nylon-6	65.70	20.17	8.54	3.22	2.38

3.2. Mechanical Robustness of Metallic Electrode on the TCN

By implementing chemical surface modification, we achieved the MPTMS chemically-attached TCN, which is termed as MPTMS-TCN. Since thiol groups on MPTMS molecules could form strong covalent bonds with gold nanoparticles, we investigated influence of the gold-sulfur (Au-S) interaction on securing mechanical robustness and stable electrical property of electrode on the TCN. For the study, we established each metallic electrode on the surface of the TCNs by not only sputtering gold nanoparticles on both a pristine TCN and a MPTMS-TCN, but also spray-coating AgNWs solution on a pristine TCN. Using a commercially-available 3M adhesive tape, we performed manual adhesion tests of the electrodes on the TCNs by following a stepwise process of fixing the electrode coated TCN on a glass substrate, attaching the adhesive tape, and peeling the tape off after pressurized rubbing it several times. In addition, as repeating the peel tests, we also measured change in electrical resistance with respect to initial resistance (R/R_0) for the electrodes. As shown in Figure 6, the SEM observation of each metallic electrode coated on the TCNs revealed that regardless of coating methodology and material, each electrode was uniformly constructed on the whole surface area of the TCN without any noticeable defect site. However, in spite of the uniform coating, the metallic electrodes have crucial difference in mechanical robustness against adhesion force (Figure 7). In the case of the physically coated Au and AgNWs electrodes (Figure 7a,b), the metallic particles were easily peeled off from the surface of the TCN even under soft finger touch. The mechanical loss in the electrode became significant after the adhesion test using the 3M adhesive tape. On the other hand, unlike physically coated electrodes, we rarely observe mechanical loss in the Au electrode on the MPTMS-TCN even after the same adhesion test (Figure 7c). Thanks to the excellent mechanical robustness, the Au electrode on the MPTMS-TCN only maintained stable electrical resistance ($R/R_0 < 2$) even after fifteen peel tests at the same area, while the physically coated electrodes easily lost their conductive characteristic in a few peel tests (Figure 7d). We believe that the electrical stability together with the excellent mechanical robustness against the adhesive force can be crucial evidence to support strong Au-S interaction enabled by the chemical surface modification.

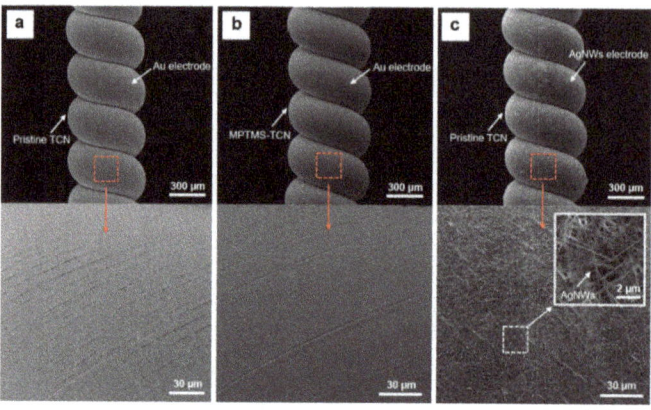

Figure 6. SEM surface images of metallic electrodes formed on the TCN: (**a**) Au electrode on a pristine TCN, (**b**) Au electrode on MPTMS-TCN, and (**c**) AgNWs electrode on pristine TCN.

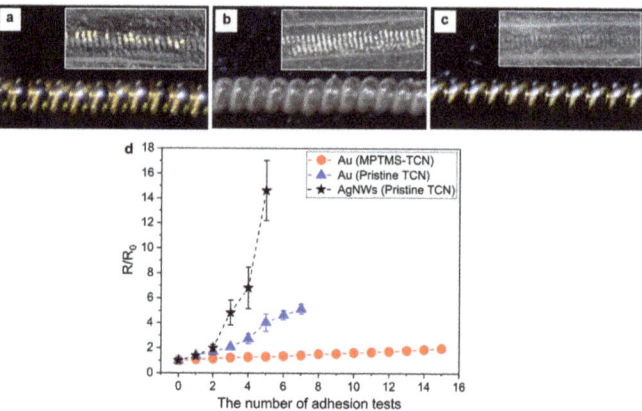

Figure 7. (**a**-**c**) Optical microscope images of metallic electrode on each TCN after adhesion test: (**a**,**b**) Au and AgNWs electrode on a pristine TCN and (**c**) Au electrode on MPTMS-TCN. (**d**) Comparison of change in electrical resistance with respect to initial resistance during repetitive peel tests. Each inset is photograph showing surface of adhesive tape peeled off from each electrode.

3.3. Electro-Thermally-Induced Deformation Behavior of the TCNAs

The TCNAs were prepared by constructing the metallic electrode on the surface of the TCN. According to materials for electrode and surface modification for the TCN, the TCNAs were classified as TCNA I, TCNA II, and TCNA III. The TCNA I was prepared by coating Au nanoparticles on the surface of the MPTMS-TCN. Unlike the TCNA I, the TCNA II and III were based on the pristine TCN. The sputtered Au nanoparticles and the spray-coated AgNWs were used as the electrode for TCNA II and III, respectively. Both the Au and AgNWs electrodes exhibited a fairly consistent initial electrical resistance (R_i) of 32 ± 1.6 Ω and 95 ± 3.8 Ω, respectively. There was no meaningful difference in the R_i of the Au electrodes whether the surface of the TCN was modified or not. Even after pressurized clamping of each TCNA with a couple of rigid frames and applying a load of 100 g, only small increase in their electrical resistance (R/R_i: < 5%) was observed, indicating that the metallic electrodes could secure stable electrical property before the actuation test. Under the same loading condition, we investigated electro-thermally-induced deformation behavior of the TCNAs by simultaneously measuring their contractile strain and surface temperature using the performance evaluation system, which is shown in Figure 2.

Figure 8a shows heating temperature dependent contractile strain of the TCNAs. Here, the contractile strain is defined as decrease in length of the TCNA divided by its length after loading, which is expressed as $\Delta L/L_{loading} \times 100$ (%). When the heating temperature increased as high as 150 °C, the contractile strain of all TCNAs increased quadratically, while their temperature-strain curves were clearly different with the electrode. Based on curve fitting of the experimental data, we confirmed that actuation behavior of the TCNAs could follow simple exponential functions (R-square ≈ 0.98) as representing a strong dependency on heating temperature. Meanwhile, the TCNAs with Au electrode (TCNA I and II) exhibits an analogous temperature-strain profile, producing the contractile strain as large as 16.8% with a sensitivity of 0.14%/°C and its high reproducibility (Pearson correlation coefficient: 0.983), regardless of the surface modification for chemically-attaching the MPTMS molecules onto the TCN, which can lead to enhancement of adhesion force between Au nanoparticles and the TCN. Unlike TCNA I and II, the TCNA with AgNWs electrode (TCNA III) could only produce a strain as high as 12%, which is 27% lower than that of the TCNA I and II, with a sensitivity of 0.1%/°C and relatively low reproducibility (Pearson correlation coefficient: 0.979), although the TCNAs consistently heated to 150 °C. It suggests that the use of AgNWs electrode cannot fully elicit an achievable stroke from the TCNA.

As shown in Figure 8b, the quadratic increasing tendency of the strain was consistently observed during repetitive actuation tests using five TCNAs each of the three forms (TCNA I, II, and III) at four different temperature conditions. The reachable magnitude of strain for each type of the TCNAs was also almost identical, revealing only a small deviation of strain, which is less than 0.4%. It indicates that our fabrication process can allow highly repetitive production of the TCNAs with a consistent performance. In parallel, to find the reason for electrode dependency in actuation performance, we monitored temperature distribution on the surface of the TCNAs via a thermal imaging camera during electro-thermal heating to 150 °C. As shown in Figure 8c, the Au electrode allows the TCNA to be heated with a fairly consistent temperature in whole surface area with a small deviation in temperature at the local surface areas, which is less than 7%. On the other hand, the AgNWs electrode exhibited area-dependent temperature distribution with a large deviation in temperature as high as 20%. The uneven temperature distribution can be strongly correlated with irregularity in coating density of AgNWs, which causes electrical resistance to be much different with area. As the result, presence of the underheated areas led to the reduction in stroke at the localized areas and it resulted in degradation of overall strain of the TCNA III. In addition, the temperature distribution affects recovery from the deformed state during a heating-cooling cycle. As compared to the TCNA I, the TCNA III exhibited relatively larger hysteresis loop during a heating-cooling cycle (Figure 8d), indicating that building a compliant electrode enabling uniform and sustainable electro-thermal heating is important to impart highly reversible actuation to the TCNA as suppressing hysteresis behavior.

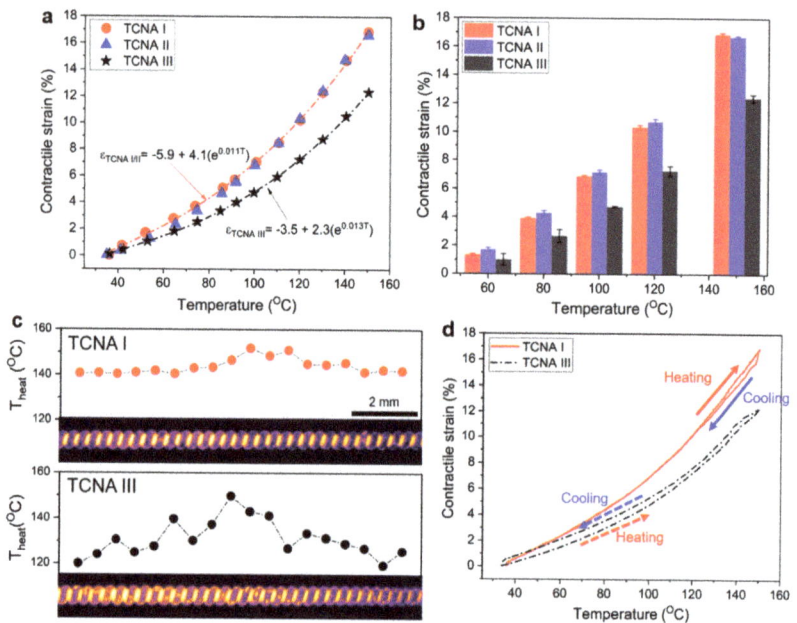

Figure 8. Electro-thermally induced actuation performance of the TCNAs with different electrode: (**a**) comparison of temperature-strain profiles of the TCNAs; (**b**) comparison of contractile strain with deviation measured at four different temperatures for five TCNAs each of TCNA I, II, and III; (**c**) thermal images with temperature distribution of the TCNAs heated to 150 °C; and (**d**) their hysteresis curves of the actuation performance during a heating-cooling cycle.

3.4. Durability Test of the TCNAs

Based on comparative study of electro-thermally-induced actuation performance of the TCNAs, we found that as compared to the AgNWs electrode, the Au electrode could contribute to eliciting larger tensile stroke with relatively small hysteresis during a heating-cooling cycle regardless of level in adhesive force between Au nanoparticle and TCN.

In order to investigate influence of the strong Au-S interaction on actuation durability, we monitored change in contractile strain of the TCNAs during repetitive actuations over 1000 heating-cooling cycles. For the test, we operated the TCNAs under the loading of 100 g by repeating an actuation cycle composed of electro-thermally heating to 150 °C, keeping for 20 s and then naturally cooling down for the same period of time. Figure 9 shows the result of durability test. During first 500 cycles, we consistently observed a decreasing tendency of contractile strain for all TCNAs. In the case of TCNA II and III, the contractile strain was rapidly reduced to around 40% as compared to the value at initial cycles. When the number of actuation cycles reached to over 1000, both TCNAs completely lost their actuation capability. On the other hand, thanks to surface modification of TCN enabling strong Au-S interaction, TCNA I exhibited stable contractile stroke without any significant performance degradation after 15% reduction in amplitude of contractile strain during first 500 actuation cycles. Moreover, the TCNA shows excellent actuation durability as maintaining a large and repeatable tensile stroke with amplitude of contractile strain over 12% even after 2000 cyclic actuations.

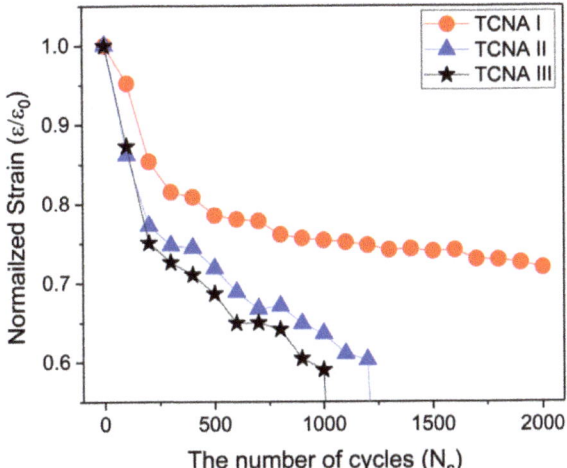

Figure 9. Comparison of electro-thermally induced contractile strain of the TCNAs during repetitive cyclic actuations.

Meanwhile, for clearly understanding the reason of difference in actuation performance of the TCNAs, we monitored change in temperature distribution on the TCNAs as increasing the number of actuation cycles to over 1000. As shown in Figure 10, when the number of actuation cycles reached 1000, the electrodes exhibited clear difference in electro-thermal heating performance with respect to heating capability and uniformity in temperature distribution. In the case of TCNA II and III, we observed that the local areas heating much below the set temperature gradually increased during repetitive cyclic actuations and their uneven temperature distribution became significant, accompanied by ΔT_h over 80 °C, as the number of cyclic actuations was closed to 1000. It caused degradation of overall tensile stroke. Here, ΔT_h is defined as the difference between the maximum and the minimum temperature in whole length of the TCNA. Particularly, at the specific area overheated to the temperature over 170 °C, the coiled structure began to irreversibly loosen and then fully unraveled, resulting in complete loss of actuation capability. Based on the SEM observation, we also found that the electrodes at the overheated area was significantly damaged, suffering from micro-cracks as well as partially peeled off from the TCN structure. It indicates that both physically coated electrodes hardly possess strong adhesion to TCN for resisting a large and repetitive electro-thermal deformation.

Unlike TCNA II and III, the TCNA I maintained not only consistent temperature distribution on its whole surface area with a remarkably small ΔT_h (<10 °C), but also smooth electrode surface similar to its initial state without any significant physical damage although 16% decrease in attainable temperature was unavoidable during first 500 heating-cooling cycles. The result suggests that the strong Au-S covalent bonds between gold nanoparticles and MPTMS-TCN can remarkably contribute to constructing a highly compliant electrode that secures high robustness against repetitive electro-thermally-induced deformation. Video S1 is a demonstration of the actuation performance of multiple TCNA I that is capable of producing contractile strain of 16.8% as lifting a load of around 1000 g in response to electro-thermal heating.

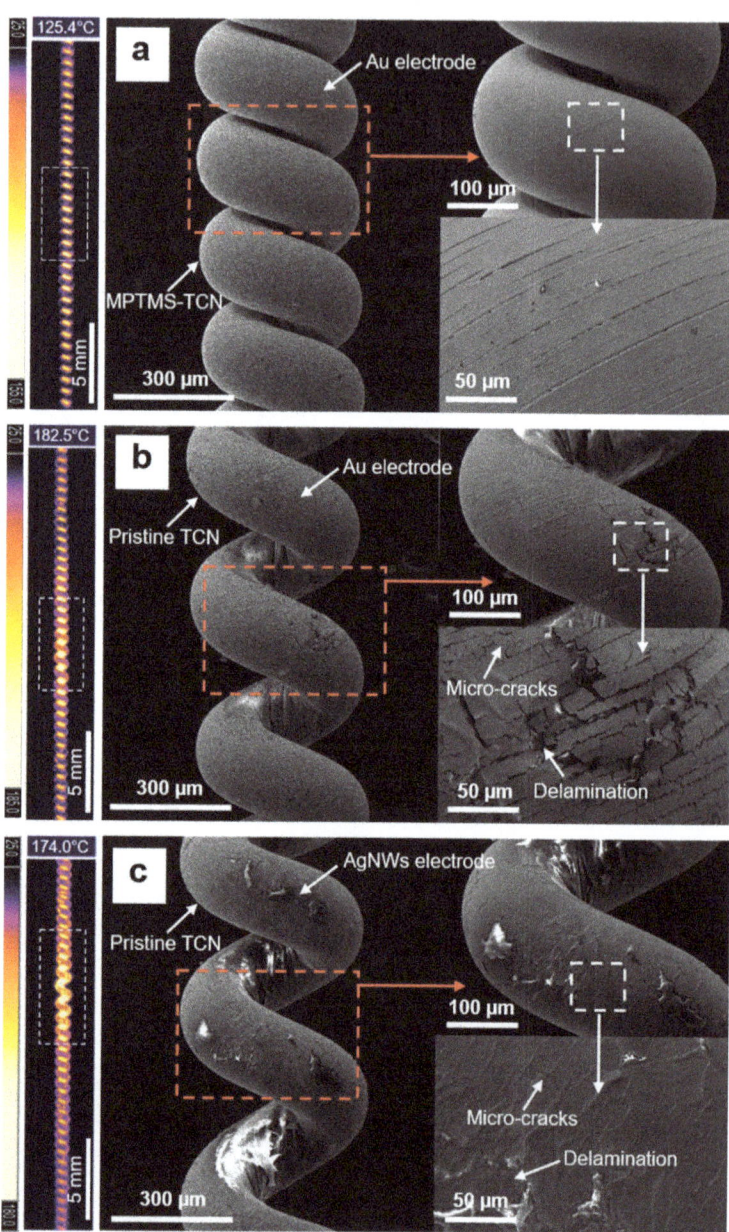

Figure 10. Thermal images with their maximum temperature and SEM surface images of the TCNAs after 1000 cyclic actuations: (**a**) TCNA I, (**b**) TCNA II, and (**c**) TCNA III.

4. Conclusions

In summary, we developed a novel and facile methodology to chemically construct a thin and highly compliant electrode for a TCNA with three-dimensional structure via Au-S interaction in a formation of covalent bonds between gold nanoparticles and thiol groups on MPTMS-TCN, which was achieved from surface modification chemically-attaching the MPTMS molecules on the surface of the TCN. Based on analysis of XPS spectra, we presented clear evidence of the surface modification of the nylon-6 with MPTMS molecules via condensation reaction of hydrolyzed MPTMS molecules with hydroxyl groups on UV/Ozone plasma treated nylon-6. By exploiting repetitive adhesion tests, we demonstrated that the Au electrode established on the MPTMS-TCN could secure excellent mechanical robustness as simultaneously maintaining stable electrical resistance ($R/R_0 < 2$) even after fifteen peel tests. Due to the benefit from the strong Au-S interaction, the compliant electrode not only enabled highly repetitive and uniform electro-thermal heating on the whole surface of MPTMS-TCN with only a small difference in temperature, which is less than 7%, but also could elicit a large and repeatable contractile strain over 12%, lifting a load of 100 g from the TCNA even during 2000 heating-cooling cycles.

Our future study will include finding ways to elicit contractile strain of the TCNA analogous to mammalian skeletal muscle with reversible tensile stroke over 20%. We also need to investigate not only TCNA with cost-effective metals instead of gold as the compliant electrode, but also integration of stretchable strain sensor onto the TCNA enabling real-time control in tensile stroke for practical use in promising applications such as soft robots and wearable human-assistive devices.

Supplementary Materials: The following is available online at https://www.mdpi.com/article/10.3390/polym14173601/s1, Video S1: Optical and thermal video in electro-thermally-induced actuation performance of multiple TCNA I.

Author Contributions: Conceptualization, S.Y.; methodology, analysis, S.M. and S.Y.; formal analysis, validation, data curation, S.M., S.Y., and S.K.P.; fabrication, M.C. and S.M.; resources, S.Y.; writing—review and editing, I.H. and S.Y.; visualization, I.H.; supervision, project administration and funding acquisition, S.Y. All authors have read and agreed to the published version of the manuscript.

Funding: This work was supported by internal grant of Electronics and Telecommunications Research Institute (ETRI) [22YS1200, Development of light driven three-dimensional morphing technology for tangible visuo-haptic interaction] and IITP grant (2017-0-00050), Development of Human Enhancement Technology for auditory and muscle support funded by the Korea government (MSIT).

Institutional Review Board Statement: Not applicable.

Informed Consent Statement: Not applicable.

Data Availability Statement: The data presented in this study are avaiable on request from the corresponding author.

Conflicts of Interest: The authors declare no conflict of interest. The funders had no role in the design of the study; in the collection, analyses, or interpretation of data; in the writing of the manuscript; or in the decision to publish the results.

References

1. Haines, C.S.; Lima, M.D.; Li, N.; Spinks, G.M.; Foroughi, J.; Madden, J.D.W.; Kim, S.H.; Fang, S.; Jung de Andrade, M.; Göktepe, F.; et al. Artificial Muscles from Fishing Line and Sewing Thread. *Science* **2014**, *343*, 868–872. [CrossRef] [PubMed]
2. Fan, J.; Li, G. High performance and tunable artificial muscle based on two–way shape memory polymer. *RSC Adv.* **2017**, *7*, 1127–1136. [CrossRef]
3. Kanik, M.; Orguc, S.; Varnavides, G.; Kim, J.; Gonzalez, T.; Akintilo, T.; Tasan, C.; Chandrakasan, A.; Fink, Y.; Anikeeva, P. Strain–programmable fiber–based artificial muscle. *Science* **2019**, *365*, 145–150. [CrossRef]
4. Yuan, J.; Neri, W.; Zakri, C.; Merzeau, P.; Kratz, K.; Lendlein, A.; Poulin, P. Shape memory nanocomposite fibers for untethered high–energy microengines. *Science* **2019**, *365*, 155–158. [CrossRef]
5. Wang, R.; Fang, S.; Xiao, Y.; Gao, E.; Jiang, N.; Li, Y.; Mou, L.; Shen, Y.; Zhao, W.; Li, S.; et al. Torsional refrigeration by twisted, coiled, and supercoiled fibers. *Science* **2019**, *366*, 216–221. [CrossRef] [PubMed]
6. Yoshida, K.; Nakajima, S.; Kawano, R.; Onoe, H. Spring–shaped stimuli–responsive hydrogel actuator with large deformation. *Sens. Actuators B Chem.* **2018**, *272*, 361–368. [CrossRef]
7. Wang, Y.; Qiao, J.; Wu, K.; Yang, W.; Ren, M.; Dong, L.; Zhou, Y.; Wu, Y.; Wang, X.; Yong, Z.; et al. High–twist–pervaded electrochemical yarn muscles with ultralarge and fast contractile actuations. *Mater. Horiz.* **2020**, *7*, 3043–3050. [CrossRef]

8. Roach, D.J.; Yuan, C.; Kuang, X.; Li, V.C.; Blake, P.; Romero, M.L.; Hammel, I.; Yu, K.; Qi, H.J. Long liquid crystal elastomer fibers with large reversible actuation strains for smart textiles and artificial muscles. *ACS Appl. Mater. Interfaces* **2019**, *11*, 19514–19521. [CrossRef]
9. Maziz, A.; Concas, A.; Khaldi, A.; Stålhand, J.; Persson, N.-K.; Jager, W.H. Knitting and weaving artificial muscle. *Sci. Adv.* **2017**, *3*, e1600327. [CrossRef]
10. Leng, X.; Hu, X.; Zhao, W.; An, B.; Zhou, X.; Liu, Z. Recent Advances in Twisted–Fiber Artificial Muscles. *Adv. Intell. Syst.* **2021**, *3*, 2000185. [CrossRef]
11. Lamuta, C.; He, H.; Zhang, K.; Rogalski, M.; Sottos, N.; Tawfick, S. Digital texture voxels for stretchable morphing skin applications. *Adv. Mater. Technol.* **2019**, *4*, 1900260. [CrossRef]
12. Li, Y.; Leng, X.; Sun, J.; Zhou, X.; Wu, W.; Chen, H.; Liu, Z. Moisture–sensitive torsional cotton artificial muscle and textile. *Chin. Phys. B* **2020**, *29*, 048103. [CrossRef]
13. Mirvakili, S.M.; Hunter, I.W. Artificial Muscles: Mechanisms, Applications, and Challenges. *Adv. Mater.* **2018**, *30*, 1704407. [CrossRef] [PubMed]
14. Jia, T.; Wang, Y.; Dou, Y.; Li, Y.; Andrade, M.J.; Wang, R.; Fang, S.; Li, J.; Yu, Z.; Qiao, R.; et al. Moisture sensitive smart yarns and textiles from self–balanced silk fiber muscle. *Adv. Funct. Mater.* **2019**, *29*, 1808241. [CrossRef]
15. Mendes, S.S.; Nunes, L.C.S. Experimental Approach to Investigate the Constrained Recovery Behavior of Coiled Monofilament Polymer Fibers. *Smart Mater. Struct.* **2017**, *26*, 115031. [CrossRef]
16. Cherubini, A.; Moretti, G.; Vertechy, R.; Fontana, M. Experimental Characterization of Thermally–Activated Artificial Muscles Based on Coiled Nylon Fishing Lines. *AIP Adv.* **2015**, *5*, 067158. [CrossRef]
17. Chen, J.; Pakdel, E.; Xie, W.; Sun, L.; Xu, M.; Liu, Q.; Wang, D. High–Performance Natural Melanin/Poly(Vinyl Alcohol–Co–Ethylene) Nanofibers/PA6 Fiber for Twisted and Coiled Fiber–Based Actuator. *Adv. Fiber Mater.* **2020**, *2*, 64–73. [CrossRef]
18. Mirvakili, S.M.; Ravandi, A.R.; Hunter, I.W.; Haines, C.S.; Li, N.; Foroughi, J.; Naficy, S.; Spinks, G.M.; Baughman, R.H.; Madden, J.D.W. Simple and Strong: Twisted Silver Painted Nylon Artificial Muscle Actuated by Joule Heating. In *Electroactive Polymer Actuators and Devices (EAPAD) 2014*; SPIE: Bellingham, WA, USA, 2014; Volume 9056. [CrossRef]
19. Hiraoka, M.; Nakamura, K.; Arase, H.; Asai, K.; Kaneko, Y.; John, S.W.; Tagashira, K.; Omote, A. Power–Efficient Low–Temperature Woven Coiled Fibre Actuator for Wearable Applications. *Sci. Rep.* **2016**, *6*, 36358. [CrossRef]
20. Park, J.; Yoo, J.W.; Seo, H.W.; Lee, Y.; Suhr, J.; Moon, H.; Koo, J.C.; Choi, H.R.; Hunt, R.; Kim, K.J.; et al. Electrically Controllable Twisted–Coiled Artificial Muscle Actuators Using Surface–Modified Polyester Fibers. *Smart Mater. Struct.* **2017**, *26*, 035048. [CrossRef]
21. Pyo, D.; Lim, J.-M.; Mun, S.; Yun, S. Silver–Nanowires Coated Pitch–Tuned Coiled Polymer Actuator for Large Contractile Strain under Light–Loading. *Int. J. Precis. Eng. Manuf.* **2018**, *19*, 1895–1900. [CrossRef]
22. van der Weijde, J.; Smit, B.; Fritschi, M.; van de Kamp, C.; Vallery, H. Self-Sensing of Deflection, Force, and Temperature for Joule–Heated Twisted and Coiled Polymer Muscles via Electrical Impedance. *IEEE/ASME Trans. Mechatron.* **2017**, *22*, 1268–1275. [CrossRef]
23. Padgett, M.E.; Mascaro, S.A. Investigation of Manufacturing Parameters for Copper–Wound Super–Coiled Polymer Actuators. In *Electroactive Polymer Actuators and Devices (EAPAD) XXI*; SPIE: Bellingham, WA, USA, 2019; Volume 10966. [CrossRef]
24. Zhou, D.; Zuo, W.; Tang, X.; Deng, J.; Liu, Y. A multi–motion bionic soft hexapod robot driven by self-sensing controlled twisted artificial muscle. *Bioinspir. Biomim.* **2021**, *16*, 045003. [CrossRef]
25. Kwon, Y.-T.; Kim, Y.-S.; Lee, Y.; Kwon, S.; Lim, M.; Song, Y.; Choa, Y.-H.; Yeo, W.-H. Ultrahigh conductivity and superior interfacial adhesion of a nanostructured, photonic–sintered copper membrane for printed flexible hybrid electronics. *ACS Appl. Mater. Interfaces* **2018**, *10*, 44071–44079. [CrossRef] [PubMed]
26. Lee, M.; Lee, M.S.; Wakida, T.; Tokuyama, T.; Inoue, G.; Ishida, S.; Itazu, T.; Miyaji, Y. Chemical Modification of Nylon 6 and Polyester Fabrics by Ozone–Gas Treatment. *J. Appl. Polym. Sci.* **2006**, *100*, 1344–1348. [CrossRef]
27. Drelich, J.; Chibowski, E.; Meng, D.D.; Terpilowski, K. Hydrophilic and Superhydrophilic Surfaces and Materials. *Soft Matter.* **2011**, *7*, 9804–9828. [CrossRef]
28. SelegArd, L.; Khranovskyy, V.; Soderlind, F.; Vahlberg, C.; Ahren, M.; Kall, P.O.; Yakimova, R.; Uvdal, K. Biotinylation of ZnO nanoparticles and thin films: A two–step surface functionalization study. *ACS Appl. Mater. Interfaces* **2010**, *2*, 21288–22135. [CrossRef]
29. Wang, Y.; Luo, S.; Ren, K.; Zhao, S.; Chen, Z.; Li, W.; Guan, J. Facile Preparation of Graphite Particles Fully Coated with Thin Ag Shell Layers for High Performance Conducting and Electromagnetic Shielding Composite Materials. *J. Mater. Chem. C* **2016**, *4*, 2566–2578. [CrossRef]

MDPI
St. Alban-Anlage 66
4052 Basel
Switzerland
www.mdpi.com

Polymers Editorial Office
E-mail: polymers@mdpi.com
www.mdpi.com/journal/polymers

Disclaimer/Publisher's Note: The statements, opinions and data contained in all publications are solely those of the individual author(s) and contributor(s) and not of MDPI and/or the editor(s). MDPI and/or the editor(s) disclaim responsibility for any injury to people or property resulting from any ideas, methods, instructions or products referred to in the content.

www.ingramcontent.com/pod-product-compliance
Lightning Source LLC
LaVergne TN
LVHW070658100526
838202LV00013B/997